Advances in Intelligent Systems and Computing

196

Editor-in-Chief

Prof. Janusz Kacprzyk
Systems Research Institute
Polish Academy of Sciences
ul. Newelska 6
01-447 Warsaw
Poland
E-mail: kacprzyk@ibspan.waw.pl

For further volumes:
http://www.springer.com/series/11156

Antonio Chella, Roberto Pirrone, Rosario Sorbello,
and Kamilla Rún Jóhannsdóttir (Eds.)

Biologically Inspired Cognitive Architectures 2012

Proceedings of the Third Annual Meeting of the BICA Society

Springer

Editors
Prof. Antonio Chella
Department of Chemical, Management,
Computer, Mechanical Engineering
Università di Palermo
Palermo
Italy

Prof. Roberto Pirrone
Department of Chemical, Management,
Computer, Mechanical Engineering
Università di Palermo
Palermo
Italy

Dr. Rosario Sorbello
Department of Chemical, Management,
Computer, Mechanical Engineering
Università di Palermo
Palermo
Italy

Dr. Kamilla Rún Jóhannsdóttir
Department of Psychology
Reykjavik University
Reykjavik
Iceland

ISSN 2194-5357
ISBN 978-3-642-34273-8
DOI 10.1007/978-3-642-34274-5
Springer Heidelberg New York Dordrecht London
e-ISSN 2194-5365
e-ISBN 978-3-642-34274-5

Library of Congress Control Number: 2012949570

Printed on acid-free paper

Springer is part of Springer Science+Business Media (www.springer.com)

Preface

This volume documents the proceedings of the Annual International Conference on Biologically Inspired Cognitive Architectures (BICA) 2012, which is the Third Annual Meeting of the BICA Society and the fifth annual BICA meeting. The series of BICA conferences started in 2008 under the umbrella of the Association for the Advancement of Artificial Intelligence (AAAI). In 2010, the BICA Society was eventually incorporated as a nonprofit organization - a scientific society with headquarters in the United States, with the mission of promoting and facilitating the transdisciplinary study of BICA. Aims of the BICA Society include creating a world-wide scientific infrastructure that supports multidisciplinary researches in addressing the challenge of creating a computational equivalent of the human mind (known as the BICA Challenge). The list of Founding Members of the BICA Society is too long to be included here, and includes many eminent scholars and scientists.

As outlined in the manifesto of the BICA Society published in the Proceedings of BICA 2010, we can reasonably predict that in the coming years robots and artifacts with general-purpose human-level intelligence will become available. It is therefore vital for scientists from different disciplines to join their efforts and share results in addressing this important goal, and to analyze the scientific and technological aspects of the challenge, including ethical and moral problems. The BICA Challenge calls for an integrated understanding of artificial and natural intelligent systems, including their biological functions, cognition and learning. The BICA paradigm is a new approach that integrates different disciplines, from neuroscience to cognitive science, artificial intelligence and robotics. This approach will allow us, on the one hand, to better understand the complex operation of the brain in order to suggest new horizons for research in neuroscience and psychology, and on the other hand, to use inspirations from these fields for modern robotics and intelligent agent design.

The BICA Challenge can be compared to greatest human challenges of the past (the Human Genome Program, the Apollo Moon Expedition, etc.). It is unlikely that a single laboratory, no matter how big and full of resources it may be, can succeed in this fantastic effort. This huge challenge can only be tackled by unified efforts of many laboratories, scientists, institutions, and research facilities around the world that continuously exchange their ideas, knowledge and results.

One of the aims of the BICA Society is to provide a common reference framework for researchers and scholars from different scientific communities, who may be talking different scientific languages and pursuing different research goals, yet implicitly or explicitly they contribute efforts to solving the BICA Challenge. The BICA Society promotes such interdisciplinary studies and allows researchers to freely share their ideas, tools and results in pursuit of the overarching goal.

All previous conferences of the BICA Society were held in the United States. For the first time, this year's BICA conference is held in Europe. The BICA Society chose Palermo as the Conference site, in recognition of support of the BICA Society by the *RoboticsLab* of the University of Palermo. The Conference is held from October 31st to November 3rd at the Grand Hotel Piazza Borsa in the center of Palermo.

Among participants of BICA 2012 are top-level scientists like Karlheinz Meier, Hiroshi Ishiguro, Giorgio Ascoli, Giulio Sandini, Igor Aleksander, Kristinn R. Thórisson, Ricardo Sanz, Pentti Haikonen, Christian Lebiere, Alexei Samsonovich, Michele Migliore, Soo-Young Lee, Raul Arrabales, Geraint A. Wiggins, and many others.

BICA 2012 received 80 submissions from more than 20 countries; the Program Committee accepted 44 papers and 7 extended abstracts that well represent the main two scientific areas of interest of the BICA community: (i) the creation and study of intelligent artifacts and (ii) the study and computational modeling of the human brain.

We would like to thank all members of the Organizing Committee and Program Committee for their precious help in reviewing submissions and in preparing an interesting and stimulating scientific program, and all members of the Local Organizing Committee from DICGIM and ICAR-CNR Palermo, for their generous work in organizing all the aspects of the conference.

We also want to thank our supporting scientific institutions: the BICA Society, the University of Palermo, the EU FP7 project "Humanobs", the "Istituto di Calcolo e Reti ad Alte Prestazioni" (ICAR-CNR), the "Centro Interdipartimentale di Tecnologie della Conoscenza" (CITC), the "Associazione Italiana per l'Intelligenza Artificiale" (AI*IA) and Informamuse. Moreover, let us thank the Municipality of Palermo, the Regional Provincia of Palermo, the Sicily Parliament Assembly and the Siciliy Governor Bureau.

Last but not the least, we would like to thank Alexei Samsonovich, President and Founder of the BICA Society, for continuous support and advices.

September 2012

Antonio Chella
Roberto Pirrone
Rosario Sorbello
Kamilla R. Jóhannsdóttir

Organization

BICA 2012 is organized by DICGIM (Department of Chemical, Management, Computer, Mechanical Engineering), University of Palermo in cooperation with the BICA (Biologically Inspired Cognitive Architectures) Society.

Chairs

Conference Chair

Antonio Chella	University of Palermo, Italy

Program Chairs

Roberto Pirrone	University of Palermo, Italy
Kamilla R. Jóhannsdóttir	University of Akureyri and Reykjavik University, Iceland

Local Organization Chair

Rosario Sorbello	University of Palermo, Italy

Organizing Committee

Igor Aleksander	Imperial College, London, UK
Ronald Arkin	Georgia Institute of Technology, Atlanta, USA
Raul Arrabales	Universidad Carlos III, Madrid, Spain
Stan Franklin	The University of Memphis, USA
Ashok K Goel	Georgia Institute of Technology, Atlanta, USA
Stephen Grossberg	Boston University, Boston, USA
Pentti Haikonen	University of Illinois at Springfield, USA
Owen Holland	University of Sussex, Brighton, UK

Christian Lebiere	Carnegie Mellon University, Pittsburgh, USA
Alexandr Letichevsky	Glushkov Institute of Cybernetics, Ukraine
Riccardo Manzotti	IULM University, Milan, Italy
Michele Migliore	National Research Council, Palermo, Italy
David Noelle	University of California, Merced, USA
Alexei Samsonovich	George Mason University, Fairfax, VA, USA
Ricardo Sanz	Universidad Politecnica de Madrid, Spain
Brandon Rohrer	Sandia National Laboratories, USA
Junichi Takeno	Meiji University, Japan
Rodrigo Ventura	Instituto Superior Tecnico, Lisbon, Portugal
Mary-Anne Williams	University of Technology, Sydney, Australia

Program Committee

Roberto Pirrone	University of Palermo (Program co-Chair)
Kamilla Johannsdottir	University of Akureyri and Reykjavik University (Program co-Chair)
Myriam Abramson	Naval Research Laboratory
Tsvi Achler	Los Alamos National Labs
Samuel Adams	IBM Research
Liliana Albertazzi	University of Trento
Yiannis Aloimonos	University of Maryland
Heather Ames	Boston University
Kenji Araki	Hokkaido University
Itamar Arel	University of Tennessee
Paul Baxter	University of Plymouth
Mark Bishop	Goldsmiths, University of London
Jonathan Bona	University at Buffalo, The State University of New York
Tibor Bosse	Vrije Universiteit Amsterdam
Bert Bredeweg	University of Amsterdam
Andrew Browning	Boston University
Andrea Carbone	ISIR/UPMC
Luigia Carlucci Aiello	University of Rome "La Sapienza"
Suhas Chelian	HRL Laboratories LLC
Robert Clowes	University of Sussex
Roberto Cordeschi	University of Rome "La Sapienza"
Massimo Cossentino	National Research Council, Italy
Antonella D'Amico	University of Palermo
Haris Dindo	University of Palermo
Simona Doboli	Hofstra University
Wlodek Duch	Nicolaus Copernicus University
Elena Fedorovskaya	Eastman Kodak Company

Marcello Frixione	University of Salerno
Karl Fua	High Performance Computing, Singapore
Salvatore Gaglio	University of Palermo
David Gamez	Imperial College, London
Ross Gayler	La Trobe University
Jaime Gomez	Universidad Politecnica Madrid
Claudius Gros	Intitute for Theoretical Physics
Steve Ho	University of Hertfordshire
Owen Holland	University of Sussex
Ian Horswill	Northwestern University
Mike Howard	HRL Laboratories, LLC
Eva Hudlicka	Psychometrix Associates
Christian Huyck	Middlesex University
David Israel	SRI International
Magnus Johnsson	Lund University
Benjamin Johnston	University of Sydney
Allan Kardec Barros	UFMA
Deepak Khosla	HRL Laboratories
Paul Kogut	Lockheed Martin
Jeff Krichmar	University of California, Irvine
Emmanuel Lesser	Nuance Communications
Simon Levy	Washington and Lee University
Giuseppe Lo Re	University of Palermo
Andràs Lörincz	Eotvos Lorand University
Thomas Lu	Jet Propulsion Lab
Evguenia Malaia	Indiana University
Maria Malfaz	UC3M
Martin Mcginnity	University of Ulster
Christophe Menant	
Emanuele Menegatti	University of Padua
Elena Messina	National Institute of Standards and Technology
Steve Morphet	SRC
Hector Munoz-Avila	Lehigh University
Waseem Naqvi	Raytheon
Guenter Neumann	DFKI
Rony Novianto	University of Technology, Sydney
Andrew Nuxoll	University of Portland
Andrea Omicini	University of Bologna
Marco Ortolani	University of Palermo
Enrico Pagello	University of Padua
Charles Peck	IBM T.J. Watson Research Center
Jacques Penders	Sheffield Hallam University
Alfredo Pereira	São Paulo State University
Giovanni Pezzulo	ILC-CNR
Robinson Pino	Air Force Research Laboratory

Roberto Poli	University of Trento
Michal Ptaszynski	Hokkai-Gakuen University
Paavo Pylkannen	University of Helsinki
Vladimir Redko	SRISA
Jim Reggia	University of Maryland
Scott Reilly	Charles River Analytics
Brandon Rohrer	Sandia National Laboratories
Pablo Roman	Universidad de Chile
Paul Rosenbloom	University of Southern California
Pier-Giuseppe Rossi	University of Macerata
Christopher Rouff	Lockheed Martin Advanced Technology Laboratories
Miguel Salichs	Univ. Carlos III, Madrid
Giulio Sandini	Italian Institute of Technology
Amy Santamaria	Alion Science and Technology
Ricardo Sanz	Universidad Politecnica de Madrid
Michael Schader	Yellow House Associates
Theresa Schilhab	University of Aarhus
Juergen Schmidhuber	IDSIA and TU Munich
Michael Schoelles	Rensselaer Polytechnic Institute
Valeria Seidita	University of Palermo
Mike Sellers	Online Alchemy, Inc.
Ignacio Serrano	CSIC
Antonio Sgorbissa	University of Genova
Timothy Smith	Raytheon
Javier Snaider	The University of Memphis
Donald Sofge	Naval Research Laboratory
Meehae Song	Simon Fraser University
Rosario Sorbello	University of Palermo
Terry Stewart	University of Waterloo
Andrea Stocco	University of Washington
Susan Stuart	University of Glasgow
Eugene Surowitz	New York University (visiting)
Junichi Takeno	Meiji University
Gheorghe Tecuci	George Mason University
Marco Temperini	University of Rome "La Sapienza"
Knud Thomsen	Paul Scherrer Institute
Guglielmo Trentin	National Research Council, Italy
Jan Treur	Vrije Universiteit Amsterdam
Alfonso Urso	National Research Council, Italy
Cesare Fabio Valenti	University of Palermo
Akshay Vashist	Telcordia Technologies
Rodrigo Ventura	IST, TU Lisbon
Craig Michael Vineyard	Sandia National Laboratories

Roseli Wedemann	Universidade do Estado do Rio de Janeiro
Andrew Weiss	
Neil Yorke-Smith Olayan	American University of Beirut
Changkun Zhao	Penn State University
Terry Zimmerman	SIFT Smart Information Flow Technologies

Local Organizing Committee

Rosario Sorbello	University of Palermo (Local Organization Chair)
Vincenzo Cannella	University of Palermo
Massimo Cossentino	ICAR-CNR, Palermo
Antonella D'Amico	University of Palermo
Haris Dindo	University of Palermo
Ignazio Infantino	ICAR-CNR, Palermo
Giuseppe Lo Re	University of Palermo
Marco Ortolani	University of Palermo
Arianna Pipitone	University of Palermo
Riccardo Rizzo	ICAR-CNR, Palermo
Valeria Seidita	University of Palermo

Sponsoring Institutions

BICA Society
University of Palermo, Italy
EU FP7 Project Humanobs
Istituto di Calcolo e Reti ad Alte Prestazioni (ICAR), CNR, Italy
Centro Interdipartimentale di Tecnologie della Conoscenza (CITC), Italy
Informamuse s.r.l., Italy

Supporting Institutions

Associazione Italiana per l'Intelligenza Artificiale (AI*IA)
Comune di Palermo
Provincia Regionale di Palermo
Assemblea Regionale Siciliana
Presidenza della Regione Siciliana

Contents

Invited Papers

Conference Papers

Extended Abstracts

Back to Basics and Forward to Novelty in Machine Consciousness

Igor Aleksander and Helen Morton

Imperial College, London
{igor.aleksander,helen.morton}@imperial.ac.uk

Abstract. Machine consciousness has emerged from the confusion of an oxymoron into an evolving set of principles which, by leaning on information integration theories, define and distinguish what is meant by 'a conscious machine'. This paper reviews this process of emergence by indicating how it is possible to break away from the Chalmers 'hardness' of a computational consciousness by a general concept of A becoming conscious of B where both are formally described. We highlight how this differs from classical AI approaches, by following through a simple example using the specific methodology of weightless neural nets as an instance of a system that owes its competence to something that can be naturally described as 'being conscious' rather depending on the use AI algorithms structured by a programmer.

Keywords: Machine consciousness, consciousness theory, weightless neural networks.

1 Introduction: Some Basics

If a scientist suggests that some artificial device could become conscious he is likely to be accused of philosophical naiveté and charlatanism. And yet, this may be precisely the route that will remove ancient obstacles that stand in the way of answering the simple question, 'what makes me conscious?' The question drags in its wake Chalmers' notorious hard problem [1] which requires a theory that links the first and third person perspectives. "Machine Consciousness" practitioners do this by re-castsing the first person problem entirely into the third person without losing the thrust of the question. That is, asking "how does A become conscious of B?" instead of "how do I become conscious of anything?" is amenable to theory and design. To be more precise, A is imagined as some form of constructed artifact, a robot maybe, and B is a world in which A is situated. How can conscious behavior of A be discerned and analyzed?

2 AI and Consciousness

Unfortunately, the history of Artificial Intelligence is strewn with such scenarios which are not convincing in the consciousness discussion. Take a typical example: a planetary

A. Chella et al. (Eds.): Biologically Inspired Cognitive Architectures 2012, AISC 196, pp. 1–6.
springerlink.com

exploration vehicle. The robotic explorer needs to data on how to deal with contingencies that its creator/programmer may have thought to be important. A build-up of relevant experience is the drive behind paradigms of ‘machine learning’. Have machine learning algorithms impacted on a study of consciousness? Tom Mitchell’s original definition [2] of machine learning is “A computer program is said to learn from experience E with respect to some class of tasks T and performance measure P, if its performance at tasks in T, as measured by P, improves with experience E”. The task for the robot may be to avoid rocks in its path, an ‘experience’ it might have encountered in the Arizona desert earlier. Performance would be stored much like the pictures in a computer taken with a digital camera and labeled as (say) ‘rock to the left of centre, move 30 degrees right to avoid it’ acquired through many successes and failures. This stored element of experience could be decoded to send signals to the appropriate parts of the steering and motion mechanisms to avoid the rock. Actually doing this in a test may lead an observer to say ‘gee – the robot is conscious of the rock to the left in its field of vision’. Proof of successful learning ‘to become conscious of a rock’ is claimed to be in the performance of the action of avoiding the rock.

Doubting that consciousness is involved in this process comes from realizing that one’s own (1st person) sensations involve mental states that feel active and engaged with the world. Mental states feel complex and comprehensive involving most of the senses. Aristotle put this down to the needs of living organisms to move autonomously and are brought to life through the biological process of procreational living which develop a sense of needing to be conscious of their worlds. This because the procreational device has a need for nutrition and hence a need for a ‘soul’ (that which distinguishes a corpse from a functioning body) which has the property of driving the body into a search for food which then sustains the body so that it can sustain its soul and stay alive. While we may not exactly put it in this body/soul way, it may be precisely this non-biological character of a robot that makes us shrink from believing that a robot that works on a recall of stored sensory images might have a ‘soul’ or what may now be called a conscious mind. We would call it a dynamic state structure that accurately represents the world and the organism in it. This is not often found in AI-based robots leading to the second and possibly principal objection to the AI robot. Whatever intelligent behavior it might display has to be the result of a simulation of what a programmer thinks the mechanics of appropriate behavior are. Now, one of the fascinating features of consciousness is that it seems to be a product (possibly emergent) of a complex system of neural machinery which is activated as the organism advances its engagement with the world in which it is situated. The complexity and richness of the first person mental state must be supported by an underlying complexity and richness of the supporting mechanisms.

So the third person question that might reveal what it is for ‘me’ to become conscious is “what kind of mechanisms must any A possess to become conscious of B”. It seems to us that the only way that one can evaluate an answer to the ‘what kind of mechanism’ question is a first person event. It might even satisfy different individuals in different ways. We are not satisfied by AI designs because we *are conscious* of the fact that what we consciously feel that consciousness doesn’t work like that. So it’s being conscious of how things work that may be triggering the evaluation of whether something is conscious or not. We now examine what the implications of this are for conscious machine designers.

3 The Task for the Conscious Machine Designer

The bar appears to be set as high as it could be. The first step is to examine the "No life no consciousness" dictum. All this tells us is that what consciousness may be *for* is to allow us to manage those things that are important to *sustain* life. But we are also conscious of the phases of the moon, the joy of hearing a passage of a Shostakovich symphony, or the nuisance of a computer malfunction. These are all examples of conscious mental states which may not be even weakly linked to life-sustenance. So 'life' is a way of saying that evolution has created machinery that sustains functional, mental states which are necessary for survival. This requirement gives such mental states very wide representational powers which bring in the phases of the moon and the like. But what makes such states special and different from the stored recordings of AI? Firstly, they are the *results* of active neural processes, the states of a dynamic system which differ from stored recordings.

Secondly they are not the arbitrarily chosen symbolic encodings selected from random firing signals produced by neurons. They have 'meaning' which arbitrary random states do not. This too is pursued further below. This boils down to two central issues about the character of the mental state which need to be understood. First, how does the world get reflected into an internal structure of states which, by being representative of the world, can be said to be the elements of intentional thought *about* that world, and, second, what is it that sustains the intentional mental states and makes them *feel* representative. The next two sections enter this discussion.

3.1 Iconic Learning

For many years (since 1997 at least, [3]) we have intuited that for the states of a neural state machine to have the reality-representational powers, assumed to be essential for conscious mental states, they must be created through a process of 'iconic learning'. Here the input to a dynamic neural network determines not only the input but is also reproduced as the state of the network (and made re-entrant in some versions of the method). Consider a sensory pattern *W* at the input terminals of a neural state machine (*n* neurons, one neuron per state variable *j* sensory inputs per neuron, *k* feedback inputs per neuron from the output of other neurons). For iconic learning there is one additional input per neuron which samples *W* and fixes the state of the neuron to be the value of that sample. This creates a state S_w which we call an iconic representation of *W*. The system can reflect trajectories of inputs *W* or re-entrant states for individual patterns in *W*. A contemporary view of the concept is available [4].

However, it is clear that iconic learning is a long way from being sufficient to suggest that it captures consciousness in the structure of its states. The major difficulty is that depending on circuit parameters, the neural state machine can fail to generate properly reflective states. Figure 1 shows the behavior of a 'square net' of 10,000 binary neurons (j=k=30) iconically trained to form attractors for the pictures of 6 politicians observing (left square) a very noisy version of one such politician and entering the appropriate undistorted attractor. It could, in a highly overstated way, be said that the net enters a conscious mental state corresponding to the visual experience of the picture shown. In this situation the net remains in the attractor even if the input is replaced by noise, shifting its 'thought' only if another known politician appears at the input. For the second line we have (j=k=10). The discovery of the appropriate attractor fails despite the iconic training, solely due to a reduced connectivity.

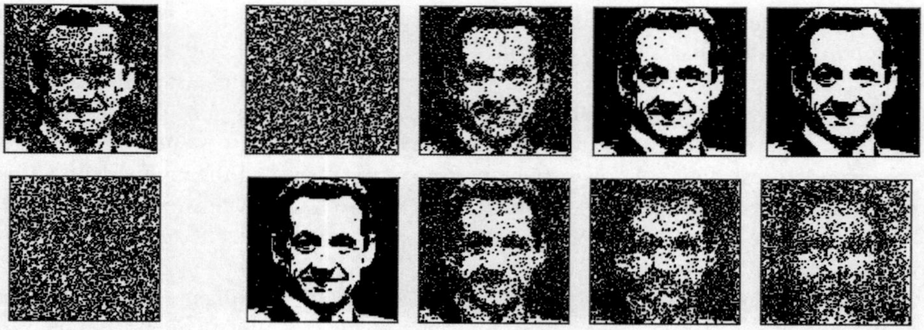

Fig. 1. Top: proper discovery of an attractor, bottom: failure of an iconically trained network

A theory is therefore required which defines a conscious mental state not only as an attractor in a neural state machine, but also in terms of the necessary properties of the network which make the attractor sustainable and uniquely representative.

3.2 Information Integration

Information Integration theory due to Giulio Tononi and his colleagues [5] was borne in Tononi's suggestion that there was something technically special about conscious experiences. First, every sensory state must be distinguished from the rest of experience. That is the state generates information. The often quoted example is that when a TV screen switches from black to white 'observed' by a light-sensitive diode, the diode generates only one bit of information. But a human (or similar) observer, measures this against a vast set of other states that have been experienced on the screen, hence the human apparatus resolves uncertainty among a much larger set of screen states and, therefore generates information. But a second condition is that the mental state is indivisible. That is, the experince of fish and chips is not composed of the thought of fish and the thought of chips separately and similarly for strawberries and cream. Otherwise having experienced the two, would imply that strawberries and chips as well as fish and cream have also been experienced. We intuit that this not true. Therefore a mental state needs not only to be informational but also integrated. The focus of Information Integration theory is to trace the level of informational integration to the connections among the state variables (say, neurons) of the network. This is perfectly possible as summarized in [5] even if it is computationally tortuous [6].

4 World Relatedness

4.1 Qualia?

Along the methods for measuring information integration, Balduzzi and Tononi have developed a theory of qualia [7]. However, being based on a geometrical argument of the way that an integrated state is built up from information transactions of progressively greater groups of neurons within active complexes, we have argued [8] that there is no attempt here to link this informational process to the world-relatedness that a neural support for qualia is thought to require. Therefore, we argue that the iconic mechanism in 3.1 above is a practical route to creating a world-related state structure in a neural system which adequately integrates information. But how does one ensure that the states crated by iconic learning allow integration? An example might help.

4.2 An Example: How Does a General Entity Become Conscious of a World in Which It Is Situated?

Here we give an example of how asking the above question leads to a definition of one example of machine consciousness. Let there be a world W and an entity E which at a moment in time t receives sensory input from a window w_t positioned at p on W. The output of E is an action a_t taken from a finite set of such actions which re-positions the window to provide a new sensory input w_{t+1} associated with the new position p', of the window. We express this in discrete automata theoretic terms as follows

$W = \{w_1, w_2, \ldots w_z\}$, where z is a finite number of positions of window w in the complete set of postitions W that define the world.

E is an automaton conventionally defined by

$$(Q \times W) \xrightarrow{\delta} Q', Q \text{ is a set of states}, Q' \text{ is the 'next' set of states}, \\ \times \text{ is a cartesian product and } \delta \text{ is a mapping function of the net.} \quad (1)$$

Also $Q \xrightarrow{\gamma} A$ where $A = \{a_1, a_2 \ldots a_v\}$ is the set of v possible actions with γ as a mapping function.

During an *exploratory* phase we assume that at any point in time t , the action is chosen by some arbitrary selection from A and an element of δ is fixed in a distributed fashion in the network by Iconic learning. In this way a state structure is built up in the net which represents the exploration of E in W. But what is needed for E to qualify for being *conscious* of W? To answer this, we assert (i) that the states structures of Q should be world-related and (ii) that they should be information-integrated. While (i) is assured by the Iconic learning, there is no guarantee that the states of Q are integrated. So here is a simple **assertion**:

> *In the above scenario, the states of Q are integrated in the sense of being unique and indivisible* ***if time t is included in the iconic coding of*** *Q and if the circuit level of integration is made sufficiently high.*

This becomes evident as no state can be repeated in the iconic training process an no two values of t can be the same. As a corollary, it is also the case that for a past history of Λ training steps each state generates information by resolving an uncertainty order Λ. Finally we simply assert that the state structure of Q becomes that of a probabilistic automaton if the time and input signals in E are replaced by noise (see the appendix in Author Summary. In prior work, we suggested that consciousness has to do with integrated information, which was defined as the amount of information generated by a system in a given state, above and beyond the information generated independently by its parts. In the present paper, we move from computing the quantity of integrated information to describing the structure or quality of the integrated information unfolded by interactions in the system. We take a geometric approach, introducing the notion of a quale as a shape that embodies the entire set of informational relationships generated by interactions in the system. The paper investigates how features of the quale relate to properties of the underlying system and also to basic features of experience, providing the beginnings of a mathematical dictionary relating neurophysiology to the geometry of the quale and the geometry to phenomenology.

5 Conclusion: A Challenge

We have argued that one can move forward from the basic but unsatisfactory approaches of machine learning for building conscious experience in an artifact. We use an entity E to explore a world W. Then through being a neural automaton capable of time-indexed iconic learning and a high level of information integration, E can build both a perceptual model that informs internally of its presence in W which as well as a contemplative one which enables planning to act in W. The challenge is to express in what way is this ***not*** described by the statement: ***E* becomes conscious of *W*?**

References

1. Chalmers, D.J.: The Conscious Mind: In Search of a Fundamental Theory. Oxford University Press (1996)
2. Mitchell, T.: Machine Learning. McGraw Hill, New York (1997)
3. Aleksander, I.: Iconic Learning in Networks of Logical Neurons. In: Higuchi, T., Iwata, M., Weixin, L. (eds.) ICES 1996. LNCS, vol. 1259, pp. 1–16. Springer, Heidelberg (1997)
4. Aleksander, I., Morton, H.B.: Aristotle's Laptop: The Discovery of our Informational Minds. World Scientific Press, Singapore (2012)
5. Tononi, G.: Consciousness as Integrated Information: a Provisional Manifesto. Biological Bulletin 215, 216–242 (2008)
6. Gamez, D., Aleksander, I.: Accuracy and performance of the state-based phi and liveliness measures of information integration. Consciousness and Cognition 20(4), 1403–1424 (2011)
7. Balduzzi, D., Tononi, G.: Qualia: The Geometry of Integrated Information. PLoS Computational Biology 5(8) (2009)
8. Beaton, M., Aleksander, I.: World-Related Integrated Information: Enactivist and Phenomenal Perspectives. International Journal of Machine Consciousness (in Press, 2012)

Characterizing and Assessing Human-Like Behavior in Cognitive Architectures

Raúl Arrabales, Agapito Ledezma, and Araceli Sanchis

Carlos III University of Madrid

Abstract. The Turing Test is usually seen as the ultimate goal of Strong Artificial Intelligence (Strong AI). Mainly because of two reasons: first, it is assumed that if we can build a machine that is indistinguishable from a human is because we have completely discovered how a human mind is created; second, such an intelligent machine could replace or collaborate with humans in any imaginable complex task. Furthermore, if such a machine existed it would probably surpass humans in many complex tasks (both physically and cognitively). But do we really need such a machine? Is it possible to build such a system in the short-term? Do we have to settle for the now classical narrow AI approaches? Isn't there a more reasonable medium term challenge that AI community should aim at? In this paper, we use the paradigmatic Turing test to discuss the implications of aiming too high in the AI research arena; we analyze key factors involved in the design and implementation of variants of the Turing test and we also propose a medium term plausible agenda towards the effective development of Artificial General Intelligence (AGI) from the point of view of artificial cognitive architectures.

1 Introduction

Alan Turing approached the problem of the definition of intelligence in artificial agents using the potential analogy between humans and machines [24]. The Turing test was originally conceived as a pragmatic way to decide if a machine was intelligent or not. In the original test conceived by Turing a human judge is given the task to tell apart a human and a machine by having a conversation with both by means of a text-based chatting system that preserves the anonymity of the subjects. In the case the human judge is not able to tell them apart it is said that the machine has passed the Turing test, and this is something that has never happened (we disregard here restricted Turing tests like PARRY [9]).

Given the great complexity of the original Turing test – parallel to that of the human-like behavior itself, a good number of variant tests have been proposed, most of them being restricted Turing tests, but also extended Turing tests have been proposed [13, 14, 19]. In this work, we will focus exclusively in restricted Turing tests, given that extended variants are even less likely to be practicable in the short and mid terms.

Reverse Turing tests, where the subject under test is a human pretending to be a machine [8], or a machine performing a Turing test on a human (as in Web pages Captchas [2]), are not of interest for this study and therefore not analyzed.

A. Chella et al. (Eds.): Biologically Inspired Cognitive Architectures 2012, AISC 196, pp. 7–15.
springerlink.com

In the following sections we describe some common variants of the Turing test, analyzing their strategy to make the test simpler and more realistic in the short term. Specifically, we illustrate this approach with the specific example of the BotPrize competition [15], an adaptation of the Turing test aimed at assessing intelligence in video game characters. Then, we discuss about the role of the human's Theory of Mind (ToM) capability [25] in the assessment process, and the great bias that it induces in the judges. In light of this problem, we present what we think is a critical distinction between believability and indistinguishability, defining the latter as a much stronger claim in this context. Using these two concepts we propose a set of key factors that must be carefully taken into account in the design of a restricted and practical human-like behavior tests, explaining the preferred strategy in each case in order to prevent or diminish the strong effect of judge's ToM bias. Finally, we confront this list of key design factors to the roadmap to Artificial General Intelligence (AGI) implicit in ConsScale [5], proposing specific steps towards the design of new pragmatic variants of the Turing test.

2 Human-Like Behavior Tests

The Turing test can be modified, adapted, and applied to a variety of different problem domains. We believe any derived test can be still referred as to a variant of the Turing test as long as it maintains the challenge of telling apart humans and machines. In other words, we think a test can be regarded as a variant of the Turing test if it checks for indistinguishability between humans and artificial agents, no matter in what environment the test is performed or what particular conditions are set up. Of course, these parameters would characterize the complexity, scope, and meaningfulness of the particular test, but it would always be essentially a Turing test if it is based on the "indistinguishability check".

For instance, the Turing test has been applied to Non-Player Characters (NPC) in computer video games. Specifically, the 2K BotPrize competition is an adaptation of the Turing test for first person shooter (FPS) video games [15]. The objective of a FPS video game is to kill as many opponents as possible without being killed by the other players (this is the so-called deathmatch game mode). The objective of the competition is to develop NPCs able to behave the same way human players do, and therefore become indistinguishable from humans. Both NPCs and human players connect to the game server anonymously by means of a computer network (see Fig.1).

All variants of the Turing test are based on an assessment process usually carried out by a number of human observers/judges. Bearing this in mind, it is important to pay attention to the Theory of Mind (ToM) mechanisms that are present in human observers [1]. In the following section, the role of ToM is analyzed from the point of view of Turing test judges.

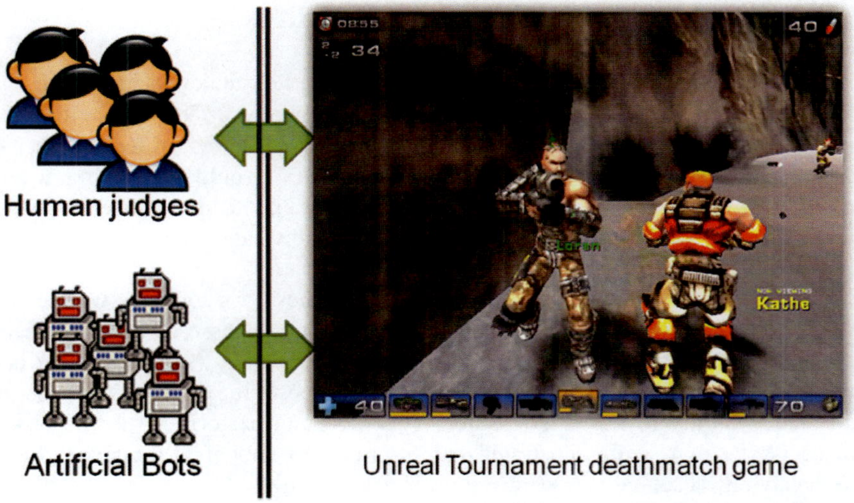

Fig. 1. In the BotPrize competition testing environment an equivalent number of NPCs and human players enter a deathmatch game. Human players (or judges) are assigned the task to both play the game and at the same time try to identify other characters either as humans or bots. The artificial bots are expected to be programmed to play the game as humans typically do, so they cannot be distinguished from regular human players. The level of "humanness" of the bots is calculated based on a statistically significant number of votes issued by the various judges over several game sessions in different virtual environments.

3 The Role of Theory of Mind in the Assessment Process

Usually the assessment carried out in any variant of the Turing test is based on the (active) observation performed by human judges. The bias induced by the very ToM mechanism present in humans cannot be neglected. In fact, it is a very important factor as it plays a central role in the judgment process.

ToM is the capability of humans of attributing mental states both to one's own self and to other selves, i.e. perceiving others in a similar way the subject is self-perceived [1, 21]. In terms of the Theory of Mind mechanisms, artificial agents failing to pass the Turing test in a specific domain could be regarded as autistic agents.

Human brain is designed to grasp the intention of others [20]. In simple tasks or observation during reduced periods of time, human observers can easily attribute intentional states to sequences of pseudorandom actions [10]. This effect has to be seriously considered when designing any human-like behavior testing protocol.

Shared intentionality is also an important factor during the assessment process: an agent able to develop a ToM capability is in turn more likely to be perceived as human. In other words, human observers expect other human-like agents to demonstrate the same ToM mechanisms that they themselves have. Shared intentionality and social perception are always present in the assessment process and need to be carefully considered when designing the testing protocol [7, 23].

4 Believability versus Indistinguishability

In the domain of video game NPCs believability is a key factor for improving game play user experience and engagement, as well as replay value (or replayability). Usually, virtual characters are disappointing because their behaviour is either very basic or outperforming [18]. Ideally, a believable NPC would behave like a human player, however this very aspect of behavior is difficult to define specifically. We need to answer the following question: are believability and indistinguishability the same thing?

In this context we need to differentiate between these two terms as we see an important aspect that should be taken into account during the assessment: sustainability and coherence over time of the behavior profile of the agent (something that is also pointed out in the Total Turing Test [13]). In this sense, the following distinction is made: believability refers to the ability of the agent to perform a specific task in a human-like fashion, while indistinguishability refers to a coherent sequence of believable behaviour sustained across different tasks and environments. In other words, a given agent is said to demonstrate human-like behavior when observed over time shows a characteristic, intelligent, commonsensical, and usually unique, behavior profile. This sustained believable behavior across different tasks, dynamic environments and over a significant period of time is shaped by the agent-specific personality traits.

Taking the former discussion into account, assessing the indistinguishability of an agent in terms of specific simple behaviors could be misleading, as a complete behavior profile can only be outlined when the agent is confronted with a complex enough sequence of tasks and environments.

As illustrated in the BotPrize competition current state of the art NPCs can show intelligence in some sense, but they cannot match human-like behavior [15]. In general, playing with other humans is still more engaging than playing with NPCs [18]. In other words, while current NPCs demonstrate believability for some particular situations, they lack indistinguishability.

5 Key Factors in the Design of Human-Like Behavior Tests

Having identified the human ToM bias and the problem of indistinguishability we may now proceed onto analyzing which could be the best practices in the design of a restricted Turing test oriented towards the detection of human-like behavior. We have also identified a set of key factors that have to be carefully evaluated.

- Complexity

The problem of determining the level of complexity required in order to effectively assess the indistinguishability of an agent is dependent on the problem domain. Actually, the differentiation between believability and indistinguishability illustrates the need for a minimum richness in the testing environment and the potential influence of the agent in such an environment.

- Anonymity

Of course anonymity has to be preserved. This aspect seems to be straightforward, but there can be many domain-specific traits and hallmarks that could help judges to identify artificial agents without the need to go through an exhaustive analysis and observation of a full behavioral profile. For instance, in the case of video game maps where NPCs use pre-defined directed graphs and path nodes to apply classical minimal path calculation algorithms like A*, the observer can easily discover artificial agents that apply this technique, as they all usually follow the same path, i.e. they walk or run along the invisible navigation graph.

- The role of the judge

The assessment process (or judging protocol) of the human judge involved in a Turing-like test can be in general described as a mind-reading task. This is a consequence of the very design of the human brain [11]. The sequences of actions perceived in the agents under analysis are rarely objectively considered as unrelated events, i.e. the brain of the human observer is incessantly looking for a model of mind (intentionality, personality, etc.) that explains observed behavior. For this reason, agents need to be exposed to the observation of judges during sufficient time and across multiple and different tasks, so the observer can double check for consistency and coherency in the observed behavior profile.

Another common problem in the assessment process is variability in observer criteria. Different human observers can provide very dissimilar assessments. Therefore, it is required to have a significant number of judges and ideally their individual results should be statistically consistent.

- Problem domain

The problem domain and the nature of the tasks that agents are expected to perform can have a deep impact on the results of the test. For instance, when several tasks are available simultaneously, agents are probably expected to demonstrate set shifting capabilities. However, this ability to interleave between different tasks can likely contribute to a more complex assessment as the observer would need much more time and cognitive resources (attention, memory) to properly evaluate the behavior of a given agent.

- Social dynamics

The number of agents involved in the test is another important issue. Whenever social interaction is part of the expected behavior of the agent, a proper number of agents have to be available in the environment. Furthermore, a good proportion of them have to be anonymous real human agents.

- Environment dynamics

If human-like learning and adaptation mechanisms are expected to be tested in the agents, the testing arena has to be designed in such a way that provides the necessary conditions and triggers for adaptation to occur and be regarded as a benefit for the agent. In other words, the goals sought by the agents should not always be achievable the same way.

- Proposed approach for human-like behavior tests

Taking into consideration all the aspects covered in the former section it is clear that the design of new tests for human-like behavior is not straightforward. We propose to address this challenge taking the cognitive dependencies defined in ConsScale as a design guideline. ConsScale is a biologically inspired scale designed to assess the level of cognitive development in artificial creatures (though applicable to biological organisms as well) [5]. The scale evaluates the presence of cognitive functions associated with consciousness. In this context, consciousness is considered as a super-function that synergistically aggregates a number of cognitive skills.

As a framework for characterizing the cognitive power of a creature, ConsScale includes the definition of an ordered list of cognitive levels arranged across a developmental path. This arrangement is inspired on the ontogeny and phylogeny of consciousness in biological organisms. The main levels defined in ConsScale are: 2 (reactive), 3 (adaptive), 4 (attentional), 5 (executive), 6 (emotional), 7 (self-conscious), 8 (empathic), 9 (social), 10 (human-like), and 11 (super-conscious).

The basic assumption is that there exists different instantiations of minds that could be characterized and classified by their level of cognitive development. Additionally, a correlation is assumed between the level of cognitive development and the associated cognitive profile. i.e., the more developed an agent is the more complex (or closer to human-like) is its overall behavior. The specific methodology to apply the scale and rate particular agents is explained elsewhere [3, 5].

ConsScale establishes a hierarchy of cognitive development levels in virtue of the definition of relations of dependency between specific cognitive skills. For instance, CS7,4 (the self-recognition capability) depends on CS6,4 (the ability to hold a precise and updated map of body schema). There are more than 120 cognitive dependency relations defined in current version of ConsScale[1]. These dependency relations can be used to design well structured tests that prevent cheating. Using again the example of CS7,4, the corresponding test for this cognitive skill would be the mirror test as typically applied to humans, other animals, and even machines [22]. However, in order to make sure that a robust and reliable behavioral test is performed, one that focuses more on indistinguishability than believability, ConsScale cognitive dependencies can be taken into account. Thus, in the case of the self-recognition mirror test, misleading results (see [12] for a detailed discussion about this issue) could be prevented focusing on CS6,4. In other words, a mirror test for machines inspired in ConsScale cognitive dependencies would test agents for self-recognition iteratively after subsequent changes in agent's physical body to make sure there are no pre-programmed responses and the agent really adapts to self-recognize effectively in different situations (including changes in own body).

[1] `http://www.consscale.com`

Taking again the BotPrize competition as an illustrative example of a limited Turing test, we have analyzed some remarkable state of the art entries from BotPrize 2010 competitions using ConsScale FPS [6] (an adaptation of ConsScale to the domain of FPS video games). The agents evaluated using ConsScale FPS are: ICE-2010 [17], UT^2 [16], CC-Bot2 [4], and Discordia [Rosenthal and Bates, personal communication]. Figure 2 shows the ConsScale FPS cognitive profile of these four FPS bots. All the bots, except Discordia, are classified as level 2 (reactive agents) because they don't fulfill all cognitive skills in level 3.

Apart from the qualitative measure provided by ConsScale levels of cognitive development (level 2 or reactive to level 11 or super-conscious), a related quantitative score can be also calculated, the so-called CQS (or ConsScale Quantitative Score). CQS possible values range from 0 to 1000 (see [5] for more details about CQS calculation). The CQS for the four analyzed bots is very low, slightly over 0.18, which is the CQS value of a canonical level 3 (adaptive) agent. ICE-2010 has the higher CQS (0.26) because it has the highest degree of accomplishment in level 3. However, it does worse in terms of CQS for the immediately higher levels. The reason why ICE-2010's CQS is higher than those of the other bots is because the ConsScale quantitative measure rewards those agents that follow the roadmap proposed by the scale.

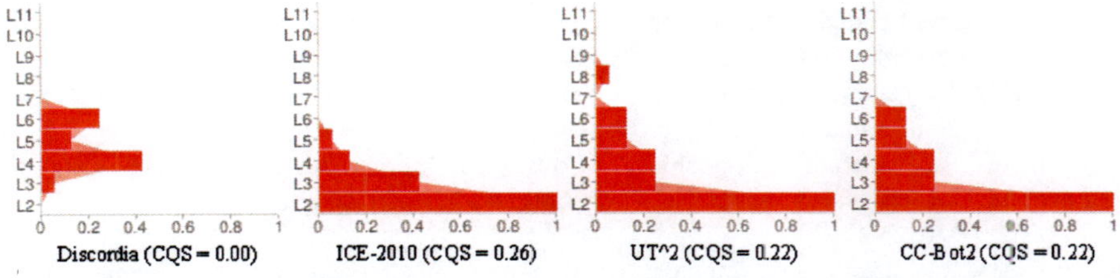

Fig. 2. These graphs depict the ConsScale FPS cognitive profiles of four different agents designed with the aim to pass the limited Turing test defined in the BotPrize competition. The arrangement of the graphs from left to right corresponds to the position achieved in the BotPrize 2010 competition. CC-Bot2 achieved the first place with a humanness ratio of 31.8%, UT^2 was classified second place with a humanness ratio of 27.2%, ICE-2010 was the third best bot rated as 23.3% human, and finally Discordia reached a 17.7% humanness ratio.

In general, the three more advanced cognitive profiles differ very little both in terms of CQS and the qualitative level of cognitive development. Actually, cognitive profiles are almost identical in the case of UT^2 and CC-Bot2. This analysis demonstrates that there is a clear correlation between ConsScale assessment and the evaluation carried out by the judges during the BotPrize competition. This fact supports the idea of using ConsScale as a design guideline to design more human-like agents. To our best knowledge, no agent has been ever able to reach humanness ratios over 50% applying the BotPrize testing protocol. Observing current correlations, we could speculate that a bot able to perform at such level would require a minimum ConsScale level above 2 or 3. Higher ConsScale levels wouldn't be actually required in a setting like the current BotPrize competition, where several key typical human-like cognitive skills are not really considered (such as accurate verbal report and complex linguistics, social interaction, team collaboration, etc.).

In terms of indistinguishability, it can be said that the more an agent fulfills the developmental path suggested in ConsScale, the more likely it can obtain better evaluations from a Turing test judge. Of course, the Turing test design factors considered in the former section are also critical and can greatly affect the testing results. However, exploiting the cognitive dependencies and their associated synergy seems to be a good approach to increase the probability of sustaining human-like believable behaviors over time.

In conclusion, well-designed human-like behavior tests (or variants of the Turing test) require paying attention both to the key factors discussed in this paper and also to the multiple cognitive function dependencies, like those described in ConsScale.

References

1. Adolphs, R.: How do we know the minds of others? Domain-specificity, simulation, and enactive social cognition. Brain Research 1079(1), 25–35 (2006)
2. Ahn, L.V., Blum, M., Hopper, N.J., Langford, J.: CAPTCHA: Using Hard AI Problems for Security. In: Biham, E. (ed.) EUROCRYPT 2003. LNCS, vol. 2656, pp. 294–311. Springer, Heidelberg (2003)
3. Arrabales, R., Ledezma, A., Sanchis, A.: ConsScale FPS: Cognitive Integration for Improved Believability in Computer Game Bots. In: Believable Bots. Springer (2011a)
4. Arrabales, R., Muñoz, J., Ledezma, A., Sanchis, A.: A Machine Consciousness Approach to the Design of human-like bots. In: Believable Bots Springer (2011b)
5. Arrabales, R., Ledezma, A., Sanchis, A.: ConsScale: A Pragmatic Scale for Measuring the Level of Consciousness in Artificial Agents. Journal of Consciousness Studies 17(3-4), 131–164 (2010)
6. Arrabales, R., Ledezma, A., Sanchis, A.: Assessing and Characterizing the Cognitive Power of Machine Consciousness Implementations. In: Biologically Inspired Cognitive Architectures II. AAAI Fall Symposium Series, p. 16 (2009)
7. Becchio, C., Bertone, C.: Beyond Cartesian Subjectivism: Neural Correlates of Shared Intentionality. Journal of Consciousness Studies 12(7), 20–30 (2005)
8. Boden, M.A.: Mind as Machine: A History of Cognitive Science. Oxford University Press, NY (2006)
9. Colby, K.M.: Modeling a paranoid mind. Behavioral and Brain Sciences 4(04), 515 (1981)
10. Freeman, W.J.: The Neurodynamics of Intentionality in Animal Brains Provide a Basis for Constructing Devices that are Capable of Intelligent Behavior. In: NIST Workshop on Metrics for Intelligence: Development of Criteria for Machine Intelligence. National Institute of Standards and Technology (2000)
11. Gallese, V., Goldman, A.: Mirror neurons and the simulation theory of mind-reading. Trends in Cognitive Sciences 2(12), 493–501 (1998)
12. Haikonen, P.O.A.: Reflections of Consciousness: The Mirror Test. In: Proceedings of the 2007 AAAI Fall Symposium on Consciousness and Artificial Intelligence, pp. 67–71 (2007)
13. Harnad, S.: The Turing Test is not a trick: Turing indistinguishability is a scientific criterion. SIGART Bull. 3(4), 9–10 (1992)
14. Harnad, S., Scherzer, P.: First, scale up to the robotic Turing test, then worry about feeling. Artificial Intelligence in Medicine 44(2), 83 (2008)

15. Hingston, P.: A Turing Test for Computer Game Bots. IEEE Transactions on Computational Intelligence and AI in Games, 169–186 (September 2009)
16. Karpov, I., Schrum, J., Miikulainen, R.: Believable Bots via Human Trace Data. In: Hingston, P. (ed.) Believable Bots. Elsevier (2011)
17. Kojima, A., Hirono, D., Sato, T., Murakami, S., Thawonmas, R.: Believable Bots via Multiobjective Neuroevolution. In: Hingston, P. (ed.) Believable Bots. Elsevier (2011)
18. Nareyek, A.: AI in Computer Games. Queue 1(10), 58–65 (2004)
19. Oppy, G., Dowe, D.: The Turing Test. In: Zalta, E.N. (ed.) The Stanford Encyclopedia of Philosophy (Spring 2011)
20. Pelphrey, K.A., Morris, J.P., McCarthy, G.: Grasping the Intentions of Others: The Perceived Intentionality of an Action Influences Activity in the Superior Temporal Sulcus during Social Perception. Journal of Cognitive Neuroscience 16(10), 1706–1716 (2004)
21. Perner, J., Lang, B.: Development of theory of mind and executive control. Trends in Cognitive Sciences 3(9), 337–344 (1999)
22. Takeno, J., Inaba, K., Suzuki, T.: Experiments and examination of mirror image cognition using a small robot. In: Proceedings of the 2005 IEEE International Symposium on Computational Intelligence in Robotics and Automation, pp. 493–498 (2005)
23. Tomasello, M., Carpenter, M.: Shared intentionality. Developmental Science 10(1), 121–125 (2007)
24. Turing, A.: Computing Machinery and Intelligence. Mind (49), 433–460 (1950)
25. Vygotsky, L.S.: Mind in Society: The Development of Higher Psychological Processes. Harvard University Press (1980)

Architects or Botanists? The Relevance of (Neuronal) Trees to Model Cognition

Giorgio Ascoli

Director, Center for Neural Informatics, Structure, and Plasticity
Molecular Neuroscience Dept., Krasnow Institute for Advanced Study
George Mason University, Fairfax, VA

Abstract. The only known cognitive architecture capable of human-level (or rat-level) performance is the human (rat) brain. Why have cognitive architectures equivalent to those instantiated by mammalian brains not been implemented in computers already? Is it just because we have not yet found the right 'boxes' to model in our data flow diagrams? Or is there a mysterious reason requiring the underlying hardware to be biologically-based? We surmise that the answer might lie in between: certain essential biological aspects of nervous systems, not (yet) routinely implemented in computational models of cognition, might prove to be necessary functional components. One such aspect is the tree-like structure of neuronal dendrites and axons, the branching inputs and outputs of nerve cells. The basic roles of dendritic and axonal arbors are respectively to integrate synaptic signals from, and to propagate firing patterns to, thousands of other neurons. Synaptic integration and plasticity in dendrites is mediated by their morphology and biophysics. Axons provide the vast majority of cable length, constituting the main determinant of network connectivity. Together, neuronal computation and circuitry provide the substrates for the activity dynamics instantiating cognitive function.

An additional consequence of the tree-shape of neurons is far less recognized. Axons and dendrites determine not only which neurons are connected to which, and thus the possible activation patterns (mental states), but also which new connections might form given the right experience. Specifically, new synapses can only form between axons and dendrites that are in close proximity. This fundamental constraint may constitute the neural correlate of the common observation that knowledge of appropriate background information is necessary for memory formation. Specifically, neurons that are near each other (in connectivity space) encode related information. This biologically inspired design feature also enables cognitive architectures (e.g. neural networks) to learn real associations better than spurious co-occurrences, resulting in greater performance and enhanced robustness to noise.

A. Chella et al. (Eds.): Biologically Inspired Cognitive Architectures 2012, AISC 196, p. 17.
springerlink.com

Consciousness and the Quest for Sentient Robots

Pentti O.A. Haikonen

Department of Philosophy
University of Illinois at Springfield
One University Plaza
Springfield, IL, 62703, USA
pentti.haikonen@pp.inet.fi

Abstract. Existing technology allows us to build robots that mimic human cognition quite successfully, but would this make these robots conscious? Would these robots really feel something and experience their existence in the world in the style of the human conscious experience? Most probably not. In order to create true conscious and sentient robots we must first consider carefully what consciousness really is; what exactly would constitute the phenomenal conscious experience. This leads to the investigation of the explanatory gap and the hard problem of consciousness and also the problem of qualia. This investigation leads to the essential requirements for conscious artifacts and these are: 1.) The realization of a perception process with qualia and percept location externalization, 2.) The realization of the introspection of the mental content, 3.) The reporting allowed by seamless integration of the various modules and 4.) A grounded self-concept with the equivalent of a somatosensory system. Cognitive architectures that are based on perception/response feedback loops and associative sub-symbolic/symbolic neural processing would seem to satisfy these requirements.

1 Introduction

Are we any nearer to sentient robots? Can we create conscious robots by designing systems that emulate cognitive functions and integrate information, maybe within the frameworks of different cognitive architectures? Is this only a matter of system complexity? We may think so and the large number of proposed cognitive architectures [23] would also indicate a trend towards that proposition.

Indeed, the recent surge of research has brought forward new insights and ideas that should not be overlooked. The philosophical studies of Baars (e.g. [4]), Boltuc (e.g. [6]), Block (e.g. [5]), Dennett (e.g. [9]), Harnad (e.g. [16]), Sloman (e.g. [26]) and others have influenced many practical approaches and the important works of Aleksander & Morton (e.g. [2]), Hesslow (e.g. [17]), Kinouchi (e.g. [20]), Manzotti (e.g. [22]), Shanahan (e.g. [25]), to name a few, are well-known. From the engineering point of view the empirical study of consciousness is an exercise in embodied cognitive robotics. Here Chella (e.g. [8]), Holland (e.g. [18]), Kawamura (e.g. [19]), Sanz (e.g. [24]), Takeno (e.g. [27]) and others have done important research.

A. Chella et al. (Eds.): Biologically Inspired Cognitive Architectures 2012, AISC 196, pp. 19–27.
springerlink.com

Therefore, we should be confident that progress is being made and conscious cognitive robots should be just around the corner. However, there may be a catch. Things may not be that easy. We may build robots that are able to execute a large number of cognitive functions and may in this way be able to mimic human behaviour successfully, but is something still missing? Do our robots really have "somebody inside" or are they only cleverly tuned automata? Do these robots really feel something and experience their existence in the world in the style of human conscious experience? And, what exactly would constitute this phenomenal conscious experience, what is consciousness actually? This is a fundamental question and a difficult one that is easily pushed aside for that very reason. However, there will be no true conscious machines unless this issue is satisfactorily solved.

2 The Problem of Consciousness

It has been said that we all know, what consciousness is; it is the state in which we are, when we are not unconscious. Apart from that consciousness is easily assumed to be a phenomenon that nobody can really explain.

On the other hand, it has also been proposed that consciousness were computational [10] and were related to "information integration" [28, 25]. This is not necessarily so. Rather than seeing the problem of consciousness as a problem of cognitive functions, skills and their integration, it should be seen as a problem of phenomenal experience.

It is a fact that the brain is a neural network and our thoughts and feelings are based on the activity of neurons, synapses and glia. On the other hand, it is also fact that we do not perceive the activity of the brain in this way, as the activity patterns of neurons. Instead, these activity patterns appear to us as our sensory percepts, thoughts, sensations and feelings directly. However, most of the activity of the brain does not have any internal appearance at all.

A typical example of this process is vision. The lens of the eye projects an image of the outside world on the retina. Light-sensitive receptor cells at each point of the retina transduce the light intensity into corresponding neural signal patterns. However, we do not "see" these neural signal patterns as such, instead our direct and immediate mental impression is that of an external world out there and its visual qualities. This constitutes our conscious awareness of the visual environment.

More generally, a perception process involves the transduction of the sensed physical phenomenon into neural signal patterns. These patterns have the internal appearance of apparent physical qualities as such, externalized to the outside world or to a body part. Visually seen objects are out there, sound sources are out there and if we hurt our finger or some other part of the body, the pain appears to be in the hurt body part, not inside the brain. The related neural activity is in the brain, but the actual appearance that is related to the sensed physical phenomenon, appears to be out there. We experience the world with its apparent qualities to be around us, not inside our brain. This appearance is most useful as it allows direct and easy interaction with the world.

How does the neural activity of the brain give rise to this kind of subjective experience and appearance? This is the hard problem of consciousness [7]. We can inspect the neural processes of the brain with various instruments and we can study and understand the physics of these processes to a great detail. On the other hand, we can study cognition and understand its functions, also to a great detail. We may also find the correspondence between the neural functions and cognitive functions; the neural correlates of cognitive functions.

3 Qualia and the Internal Experience

But then, there is the question of the internal appearance of the neural activity and the subjective experience. Why do our percepts appear as they are? Why does blue appear as blue, why does sweet taste like sweet, why does pain hurt? The qualitive appearances of our percepts are called qualia and no percepts seem to be without qualia. Again, we can find neural correlates of qualia, but we have not been able to explain, why these neural processes would give rise or be perceived as some subjective experience or qualia. This problem is called the explanatory gap [21]. This explanatory gap appears in the philosophy of the mind as the hard problem [7] and the mind-body problem, which has resulted in the nowadays not so popular dualistic theories of mind.

Could it be possible that the subjective experience would be an inherent property of biological neural activity patterns? That would appear to be an easy answer. In that case we might study cell biology and find the details of the process there, maybe. But then, this situation would exclude the artificial production of consciousness in non-biological systems.

However, if indeed, the subjective experience were an inherent property of biological neural signal patterns, then why does most of the activity of the brain remain subconscious, without any subjective experience at all? This observation would seem to show that the biological basis alone were not a sufficient or maybe even a necessary condition for the subjective experience and qualia. Instead, there must be something special in those neural signal patterns that appear as subjective experience. And there is. Introspection shows that the content of our conscious subjective experience consists of percepts with qualia. Consequently, the neural activity patterns perceived in this way are products of sensory perception. The problem of conscious perception would thus be reduced into the solving of the conditions for neural representations that could carry the qualities of the sensed entities in the form of qualia.

Qualia are direct, they need no interpretation or any additional information in order to be understood. Red is red and pain is pain directly, there is no need to evoke some learned meaning for those. The directness of qualia gives also an indication of the nature of the neural transmission that carries qualia. The transmission must be direct and transparent so that only the carried information conveys the effect, while the carrying neural machinery remains hidden. The directness of qualia also excludes the symbolic representation of sensory information. A symbolic representation would be a description, while qualia are the experience, the experienced qualities of the sensed entities.

In the brain there are no sensors that could perceive neurons and their firings as such. Therefore, only the effects of the neural signals are transmitted. The neural activity cannot appear or be perceived as neural firings, yet it has effects. In sensory perception circuits these effects are related to the qualities of the perceived entities. Therefore, it would seem that on the system level, "in the mind", the only possibility is the appearance of sensory signals as externally grounded qualia. All inner neural activity does not have this direct relationship to perception and without this it remains "sub-conscious".

4 Introspection

However, not all of our conscious mental content is related to direct sensory perception. Our inner speech is inside, our imagery is inside and our feelings are inside. Yet, there are no sensors inside the brain that could sense the internal neural activity patterns. If consciousness were related only to sensory perception, then how could we become aware of our own thoughts? This is the problem of introspection.

Therefore, is the hypothesis about consciousness as a sensory perception–related phenomenon wrong? Not necessarily. The problem of introspection can be solved by associative feedback loops that return the products of the internal processes into virtual percepts so that the mental content can be observed in terms of sensory percepts. In this way the perception process facilitates also the awareness of mental content. Introspection shows that we perceive our mental content in the form of virtual sensory qualities; we perceive and hear our verbal thoughts in the form of silent inner speech and perceive our imaginations as vague imagery. Without hearing the inner speech we would not know what we think. These kinds of feedback loops have been proposed by the author [11-15] and others, e.g. Chella [8] and Hesslow [17], who has proposed that thinking is simulated perception.

5 Reportability and Information Integration

Conscious states are reportable states that can be remembered for a while. A report is a "message" or a "broadcast" that is sent to various parts of the brain or an artificial cognitive apparatus. The message must be received and have effects on the receiving part, such as the forming of associative memories. These effects may also manifest themselves as motor actions including motions. The motions may include actions like reaching out or the turning of the head. Voiced reactions and verbal comments are possible forms of reports, like "Ouch" and "I see a black cat". The report may be about perceived external entities and conditions or internal ones, like "I feel pain" or "I feel like drinking soda".

The verbal reports may also remain silent, as a part of the inner imagery or silent speech. A verbal or non-verbal internal report may affect the behaviour of the subject and may, for instance, lead to the planning of new actions. Reporting involves the activation of cross-connections between the various parts of the brain. This kind of cross-coupling is sometimes called information integration. Tononi [28] has proposed that the degree of information integration could be used as a measure of consciousness. Information aspects of consciousness have been treated also by Aleksander and Morton [2].

6 The Concept of Self

We perceive the world with its apparent qualities to be around us, not inside the brain, but what is inside? Inside is the mind and the *impression of I*, that appears as the conscious, perceiving, feeling and thinking self. We are aware of ourselves, we are self-conscious. We are the system that is aware of the perceived qualia.

Which comes first, the perceiving self or the qualia? They come together. It is proposed that the qualia are a property of certain kinds of embodied perceptual systems and the impression of self arises from the qualia of self-percepts and also from the introspected mental contents. The "I" is associated with the body and its sensations. The somatosensory system operates as fundamental grounding to the self-concept, because it provides continuous information about the state of the body. The body and its sensations are with us wherever we go, while the environment changes. Thus the body is a fixed point of reference that can be associated with all things that make us an individual; our personal memories, needs, desires, etc. But more directly; the body is the vantage point for our percepts. The somatosensory percepts about the body relate directly to the system-self. For instance, when we feel pain, it is us who feel the pain, because the pain itself creates that impression. Thus, the conscious, perceiving, feeling and thinking self is a rather simple product of an embodied perceptual system.

What is Consciousness?

According to the previous chapters, consciousness is seen as the condition of having reportable qualia-based (phenomenal) perception. As such, it is only an internal appearance, not an executing agent. The difference between a conscious and non-conscious subject is the presence of the internal qualia-based appearance of the neural activity, which manifests itself as the direct perception of the external world with its qualities, the perception of the body with its sensations and the introspective perception of mental content such as imagery, inner speech and feelings. A non-conscious agent may process similar information, but without the experience of the direct internal appearance. It does not perceive its internal states in terms of world qualities, it does not perceive them at all.

Consciousness is not directly related to cognitive abilities or intelligence. Most probably even the simplest animals experience their consciousness in the same way as humans, as the direct qualia-based appearance, but the scope, content and fine structure of their conscious experience may be extremely narrow. However, a system that supports consciousness, the phenomenal internal appearance of its internal states, may be better suited for the realization of genuine general intelligence.

Consciousness and qualia go together. Without experienced qualia there is no consciousness. Qualia are sub-symbolic and therefore cannot be created by symbolic means. Therefore there are no algorithms that could create consciousness. Programmed artifacts cannot be conscious. This statement does not exclude conscious awareness of symbolic representations in otherwise conscious agents.

7 Sub-symbolic and Symbolic Processing

The creation of artificial conscious robots involves some system-technical issues beyond the fundamental issues of perception with internal appearances and qualia.

Human cognition utilizes sub-symbolic qualia, but is also based on symbolic information processing. There could not be any language or mathematics without the ability to use symbols. Yet, qualia are sub-symbolic and the very preconditions for qualia would seem to exclude symbolic processing. This problem has been evident in the traditional artificial neural networks. These networks operate in sub-symbolic ways and are unable to run programs. On the other hand, digital computers run programs, but are not able to process sub-symbolic representations directly. Thus, a gap between sub-symbolic and symbolic systems would seem to exist. The brain is able to bridge this gap, but how should this bridging be executed in artificial systems?

There have been some hybrid approaches, combinations of neural networks and program-based digital computers. Hybrid solutions are easily clumsy and too often combine the shortcomings of the component systems. The brain is definitely not a hybrid system.

An elegant non-hybrid solution to the sub-symbolic/symbolic problem is offered by the use of associative neural networks with distributed representations [13, 15]. These operate with sub-symbolic neural signals and signal patterns, which can be associated with each other. Via this association sub-symbolic signal patterns may become to represent entities that they do not directly depict; they become symbols for these entities. In this way, for instance, neural signals that are caused by heard sound patterns can be used as words that have associated meanings that are not naturally related to the sounds themselves.

8 Criteria for Consciousness

According to the aforesaid, the first and necessary criterion for consciousness is the presence of the phenomenal, qualia-based internal appearance of the neural activity. Unfortunately this internal appearance is subjective and cannot be directly measured unless some ingenious methods are invented. Therefore some indirect criteria for consciousness must be used. Such criteria have been developed by various researchers and include, for instance, the Aleksander's axioms [1] that try to define prerequisites for conscious systems and the ConsScale of Arrabales [3] that tries to determine the scope of consciousness by evaluating the presence of a number of behavioural and cognitive functions.

9 Requirements for Cognitive Architectures

The afore presented issues lead to the outlines and requirements for a cognitive architecture for artificial brains. The essential requirements relate to 1.) The realization of a perception process with qualia and percept location externalization, 2.) The realization of the introspection of the mental content, 3.) The reporting allowed by seamless integration of the various modules and 4.) A grounded self-concept with the equivalent of a somatosensory system. In addition, the architecture should facilitate the transition from sub-symbolic to symbolic processing.

It is proposed, and there is also some experimental proof that an architecture based on perception/response feedback loops and associative sub-symbolic/symbolic neural processing would satisfy these requirements. The author's "Haikonen Cognitive Architecture" (HCA) is an architecture realized in this way [13, 15].

10 An Example: The Robot XCR-1

It was argued afore that qualia cannot be simulated and consciousness cannot be created by programs. Therefore hardware experiments with embodied robots are necessary.

The author's Experimental Cognitive Robot XCR-1 is designed as a simple test bed for machine cognition experiments that involve direct perception and sensorimotor integration with motor action generation and control using the HCA architecture and associative neural processing [14, 15].

The robot XCR-1 is an autonomous, small three-wheel robot with gripper arms and hands, and simple visual, auditory, touch, shock and petting sensors. The XCR-1 has a limited natural language self-talk that reflects its instantaneous percepts and actions. It recognizes some spoken words and has a limited vocabulary self-talk and this allows limited verbal learning.

The sub-symbolic/symbolic processing properties of the HCA cannot be fully produced by computer programs. The HCA is also a parallel processing system and this is another benefit that would be lost if microprocessors were used. Therefore the robot XCR-1 utilizes hardwired neural circuits instead of the more common program-driven microprocessors and possible links to a master computer, which would execute the more demanding computations. Thus, the XCR-1 is not program-driven and represents a completely different approach to autonomous self-controlled robots.

The XCR-1 has a set of motor action routines and hard-wired reactions to stimuli. These routines and the hard-wired reactions can combine in various ways depending on the situation. Cognitive control may override reactions and modify the robot's behavior.

The XCR-1 utilizes the author's System Reactions Theory of Emotions (SRTE) [12, 13] for emotional learning and motivation. Corporal reward and punishment can be used to evoke pleasure and pain equivalent states. These can be associated with ongoing situations and in this way the robot will learn to seek or avoid the associated situations.

It is not claimed that the robot XCR-1 were conscious. The main purposes of this project have been the verification of the feasibility of associative information processing and the main principles of the HCA. However, XCR-1 would seem to satisfy certain requirements for consciousness, at least to a minimal degree, including:

- Direct sub-symbolic perception process
- The externalization of non-contact percepts
- Sensorimotor integration and information integration
- Attention
- Introspection of mental content in terms of sensory features via feedback loops
- Responses and reports including self-talk

- Emotional learning, control and motivation
- Somatosensory system for the grounding of the self-concept
- The transition from sub-symbolic to symbolic processing

11 Conclusions

It is argued here that consciousness is qualia-based perception. Qualia are the way in which the perception-related neural activity appears to us; not as the firing patterns of neurons, but as the apparent, externalized qualities of the world and the body. Therefore we experience us as being a vantage point in the world. All the other proposed aspects of consciousness build on the qualia-based perception. Feedback loops allow introspection, the awareness of mental content. Associative cross-connections, also known as information integration, allow internal and external reporting, memory making and recall as well as symbolic processing. Artificial realization of these call for direct, embodied perception and associative neural processing with a special cognitive architecture. All these requirements are implementable with available technology. Consequently, true conscious robots could be built in the foreseeable future, provided that the real hard problem of machine consciousness can be solved, namely money; the financing of the conscious robot design and development projects.

References

1. Aleksander, I., Dunmall, B.: Axioms and Tests for the Presence of Minimal Consciousness in Agents. In: Holland, O. (ed.) Machine Consciousness. Imprint Academic, UK (2003)
2. Aleksander, I., Morton, H.: Aristotle's Laptop. World Scientific, Singapore (2012)
3. Arrabales, R., Ledezma, A., Sanchis, A.: The Cognitive Development of Machine Consciousness Implementations. IJMC 2(2), 213–225 (2010)
4. Baars, B.J.: In the Theater of Consciousness. Oxford University Press, Oxford (1997)
5. Block, N.: On a Confusion about a Function of Consciousness. BBS 18(2), 227–287 (1995)
6. Boltuc, P.: The Philosophical Issue in Machine Consciousness. IJMC 1(1), 155–176 (2009)
7. Chalmers, D.J.: Facing Up to the Problem of Consciousness. JCS 2(3), 200–219 (1995)
8. Chella, A.: Perception Loop and Machine Consciousness. APA Newsletter on Philosophy and Computers 8(1), 7–9 (2008)
9. Dennett, D.: Are we explaining consciousness yet? Cognition 79, 221–237 (2001)
10. Edelman, S.: Computing the mind; how the mind really works. Oxford University Press, Oxford (2008)
11. Haikonen, P.O.: An Artificial Cognitive Neural System Based on a Novel Neuron Structure and a Reentrant Modular Architecture with implications to Machine Consciousness. Dissertation, Helsinki University of Technology (1999)
12. Haikonen, P.O.: The Cognitive Approach to Conscious Machines. Imprint Academic, UK (2003)
13. Haikonen, P.O.: Robot Brains. John Wiley & Sons, UK (2007)
14. Haikonen, P.O.: XCR-1: An Experimental Cognitive Robot Based on an Associative Neural Architecture. Cognitive Computation 3(2), 360–366 (2011)

15. Haikonen, P.O.: Consciousness and Robot Sentience. World Scientific, Singapore (2012)
16. Harnad, S.: The Turing Test Is Not a Trick: Turing Indistinguishability Is a Scientific Criterion. SIGART Bulletin 3(4), 9–10 (1992)
17. Hesslow, G.: Conscious thought as simulation of behaviour and perception. Trends in Cognitive Sciences 6(6), 242–247 (2002)
18. Holland, O., Knight, R., Newcombe, R.: A Robot-Based Approach to Machine Consciousness. In: Chella, A., Manzotti, R. (eds.) Artificial Consciousness. Imprint Academic, UK (2007)
19. Kawamura, K., Gordon, S.: From intelligent control to cognitive control. In: Proc. 11th International Symposium on Robotics and Applications, ISORA (2006)
20. Kinouchi, Y.: A Logical Model of Consciousness on an Autonomously Adaptive System. IJMC 1(2), 235–242 (2009)
21. Levine, J.: Materialism and qualia: the explanatory gap. Pacific Philosophical Quarterly 64, 354–361 (1983)
22. Manzotti, R., Tagliasco, V.: An Externalist Process-Oriented Framework for Artificial Consciousness. In: Proc. AI and Consciousness: Theoretical Foundations and Current Approaches, AAAI Fall Symposium 2007. AAAI Press, Menlo Park (2007)
23. Samsonovich, A.V.: Toward a Unified Catalog of Implemented Cognitive Architectures. In: Samsonovich, A.V., Johannsdottir, K.R., Chella, A., Goertzel, B. (eds.) Biologically Inspired Cognitive Architectures 2010. IOS Press, Amsterdam (2010)
24. Sanz, R., López, I., Bermejo-Alonso, J.: A Rationale and Vision for Machine Consciousness. In: Chella, A., Manzotti, R. (eds.) Artificial Consciousness. Imprint Academic, UK (2007)
25. Shanahan, M.: Embodiment and the Inner Life. Oxford University Press, Oxford (2010)
26. Sloman, A.: An Alternative to Working on Machine Consciousness. IJMC 2(1), 1–18 (2010)
27. Takeno, J., Inaba, K., Suzuki, T.: Experiments and examination of mirror image cognition using a small robot. In: Proc. 6th IEEE International Symposium on Computational Intelligence in Robotics and Automation (CIRA 2005), pp. 493–498 (2005)
28. Tononi, G.: Consciousness as Integrated Information: A provisional Manifesto. Biological Bulletin 215(3), 216–242 (2008)

Biological Fluctuation "Yuragi" as the Principle of Bio-inspired Robots

Hiroshi Ishiguro

Professor of Osaka University
Group Leader of ATR Hiroshi Ishiguro Laboratory

Abstract. The current robotics and artificial intelligence based on computer technology is very different from biological systems in principle. They suppress the noise by spending much energy and obtain clear states such as on/off and 1/0. One the other hand, the biological system utilizes noise naturally existing in the system and in the environment. The noise is called biological fluctuation "Yuragi (in Japanese)." We are studying applications of Yuragi to robotic systems as collaborative works between robotics and biology in Osaka University.

A. Chella et al. (Eds.): Biologically Inspired Cognitive Architectures 2012, AISC 196, p. 29.
springerlink.com

Active Learning by Selecting New Training Samples from Unlabelled Data

Ho Gyeong Kim, Cheong-An Lee, and Soo-Young Lee

Department of Electrical Engineering and Brain Science Research Center
Korea Advanced Institute of Science and Technology
373-1 Guseong-dong, Yuseong-gu, Daejeon 305-701, Korea

Abstract. Human utilizes active learning to develop their knowledge efficiently, and a new active learning model is presented and tested for automatic speech recognition systems. Human can self-evaluate their knowledge systems to identify the weak or uncertain topics, and seek for the answers by asking proper questions to experts (or teachers) or searching books (or webs). Then, the new knowledge is incorporated into the existing knowledge system.

Recently this active learning becomes very important to many practical applications such as speech recognition and text classification. On the internet abundant unlabelled data are available, but it is still difficult and time consuming to get accurate labels for the data. With the active learning paradigm, based on uncertainty analysis, a few data will be identified to be included in the training database and the corresponding labels will be asked to users. The answers will be incorporated into the current knowledge base by an incremental learning algorithm. This process will be repeated to result in a high- accuracy classification system with minimum number of labelled training data.

The active learning algorithm had been applied to both a simple toy problem and a real-world speech recognition task. We introduced a uncertainty measure for each unlabelled data, which is calculated from the current classifier. The developed algorithm shows better recognition performance with less number of labelled data for the classifier training. In the future we will also incorporate a smooth transition on the selection strategy based on the exploitation-exploration trade-off. At the early stage of learning human utilizes exploitation while exploration is applied at the later stage.

A. Chella et al. (Eds.): Biologically Inspired Cognitive Architectures 2012, AISC 196, p. 31.
springerlink.com

Biologically Inspired beyond Neural: Benefits of Multiple Modeling Levels

Christian Lebiere

Human-Computer Interaction Institute
School of Computer Science
Carnegie Mellon University
Pittsburgh, PA

Abstract. Biologically inspired cognitive architectures can adopt distinct levels of abstraction, from symbolic theories to neural implementations. Despite or perhaps because of those widely different approaches, they can constrain and benefit from each other in multiple ways. The first type of synergy occurs when a higher-level theory is implemented in terms of lower-level mechanisms, bringing implementational constraints to bear on functional abstractions. For instance, the ACT-RN neural network implementation constrained future developments of the ACT-R production system cognitive architecture in biologically plausible directions. The second type of synergy is when cognitive architectures at distinct levels are combined, leading to capabilities that woudn't be readily available in either modeling paradigm in isolation. The SAL hybrid architecture, a Synthesis of the ACT-R cognitive architecture and the Leabra neural architecture, provides an illustration through its combination of high-level control and low-level perception. The third type of synergy results when the same task or phenomena are modeled at different levels, bringing insights and constraints across levels. Models of the sensemaking processes developed in both ACT-R and Leabra illustrate the deep correspondence between mechanisms at the symbolic, subsymbolic and neural levels.

A. Chella et al. (Eds.): Biologically Inspired Cognitive Architectures 2012, AISC 196, p. 33.
springerlink.com

Turing and de Finetti Ganes: Machines Making Us Think

Ignazio Licata

Dip. Scienze della Formazione e Scienze Cognitive,Univ. Messina
Via Concezione 6/8, 98100 Messina, Italy

Abstract. Turing proposed his game as an operational definition of intelligence. Some years before the Italian mathematician Bruno de Finetti had showed how the concept of "bet" could be formalized by the notion of subjective probability. We will show how both approaches exhibit strong analogies and hide behind the formalism the complex debate on subjectivity and the abductive facet of human cognition.

A. Chella et al. (Eds.): Biologically Inspired Cognitive Architectures 2012, AISC 196, p. 35.
springerlink.com

How to Simulate the Brain without a Computer

Karlheinz Meier

EPFL - BLUE BRAIN PROJECT
CH-1015 Lausanne - Switzerland

Abstract. The brain is fundamentally different from numerical information processing devices. On the system level it features very low power consumption, fault tolerance and the ability to learn. On the microscopic level it is composed of constituents with a high degree of diversity and adaptability forming a rather uniform fabric with universal computing capabilities.

Neuromorphic architectures attempt to build physical models of such neural circuits with the aim to capture the key features and exploit them for information processing. In the talk I will review recent work and discuss future developments.

A. Chella et al. (Eds.): Biologically Inspired Cognitive Architectures 2012, AISC 196, p. 37.
springerlink.com

Odor Perception through Network Self-organization: Large Scale Realistic Simulations of the Olfactory Bulb

Michele Migliore

Institute of Biophysics
National Research Council
Via U. La Malfa 153, 90146 Palermo, Italy

Abstract. Analysis of the neural basis for odor recognition may have significant impact not only for a better explanation of physiological and behavioral olfactory processes, but also for industrial applications in the development of odor-sensitive devices and applications to the fragrance industry. While the underlying mechanisms are still unknown, experimental findings have given important clues at the level of the activation patterns in the olfactory glomerular layer by comparing the responses to single odors and to mixtures. The processing rules supporting odor perception are controversial. A point of general importance in sensory physiology is whether sensory activity is analytic (or elemental) or synthetic (or configural). Analytic refers to the perception of individual components in a mixture, whereas synthetic refers to the creation of a unified percept. An example of analytic perception is taste in that sweet and sour can be combined in a dish and tasted individually. The mixture of blue and yellow to make green is an example of synthetic perception in vision. Which one of these properties applies to any given odor mixture is unclear and often confusing. In recent papers this problem has been intensely investigated experimentally but the underlying computational and functional processes are still poorly understood. The main reason is that the practical impossibility of recording simultaneously from a reasonable set of cells makes it nearly impossible to decipher and understand the emergent properties and behavior of the olfactory bulb network. We addressed this problem using a biological inspired cognitive architecture: a large-scale realistic model of the olfactory bulb. We directly used experimental imaging data on the activation of 74 glomeruli by 72 odors from 12 homologous series of chemically related molecules to drive a biophysical model of 500 mitral cells and 10,000 granule cells (1/100th of the real system), to analyze the operations of the olfactory bulb circuits. The model demonstrates how lateral inhibition modulates the evolving dynamics of the olfactory bulb network, generating mitral and granule cell responses that support/explain several experimental findings, and suggests how odor identity can be represented by a combination of temporal and spatial patterns, with both feedforward excitation and lateral inhibition via dendrodendric synapses as the underlying mechanisms facilitating network self-organization and the emergence of synchronized oscillations. More generally, our model provides the first physiologically

A. Chella et al. (Eds.): Biologically Inspired Cognitive Architectures 2012, AISC 196, pp. 39–40.
springerlink.com

plausible explanation of how lateral inhibition can act at the brain region level to form the neural basis of signal recognition. Through movies and real time simulations, it will be shown how and why the dynamical interactions between active mitral cells through the granule cell synaptic clusters can account for a variety of puzzling behavioral results on odor mixtures and on the emergence of synthetic or analytic perception.

Extending Cognitive Architectures

Alexei V. Samsonovich

Krasnow Institute for Advanced Study, George Mason University, Fairfax, VA 22030, USA
asamsono@gmu.edu

Abstract. New powerful approach in cognitive modeling and intelligent agent design, known as biologically inspired cognitive architectures (BICA), allows us to create in the near future general-purpose, real-life computational equivalents of the human mind, that can be used for a broad variety of practical applications. As a first step toward this goal, state-of-the-art BICA need to be extended to enable advanced (meta-)cognitive capabilities, including social and emotional intelligence, human-like episodic memory, imagery, self-awareness, teleological capabilities, to name just a few. Recent extensions of mainstream cognitive architectures claim having many of these features. Yet, their implementation remains limited, compared to the human mind. This work analyzes limitations of existing extensions of popular cognitive architectures, identifies specific challenges, and outlines an approach that allows achieving a "critical mass" of a human-level learner.

Keywords: BICA Challenge, human-level AI, learner critical mass, episodic memory, goal generation.

1 Introduction

Emergent new field of BICA[1] research brings together artificial intelligence, cognitive and neural modeling under a new umbrella: the overarching BICA Challenge to create a computational equivalent of the human mind [1, 2]. The challenge calls for an extension of cognitive architectures with new features that should bring them to the human level of cognition and learning. The list of these features includes episodic memory, theory-of-mind, a sense of self, autonomous goal setting, various forms of metacognition, self-regulated and meta-learning, emotional[2] and social intelligence, and more. Many recently extended popular cognitive architectures are claimed to have some or most of these features and capabilities. However, a critical question is whether the level of their implementation and usage is adequate to requirements set

---

[1] BICA stands for "biologically inspired cognitive architectures". The acronym was coined by DARPA in 2005 as the name of a program intended to develop psychologically and neurobiologically based computational models of human cognition.

[2] While the terms "emotional cognition" and "emotional intelligence" are highly overloaded in the literature with controversial semantics, they are used here generically to refer to cognitive representation and processing of emotions, moods, feelings, affects, appraisals, etc.

A. Chella et al. (Eds.): Biologically Inspired Cognitive Architectures 2012, AISC 196, pp. 41–49.
springerlink.com

by the challenge [2]. The present work addresses this question by examining particular examples, pointing to problems with existing implementations and setting specific challenges for future research.

Since the onset of cognitive modeling as a research paradigm, attempts are made to implement and study complete cognitive agents embedded in virtual or physical environments [3]. Computational frameworks used for designing these agents are known as cognitive architectures [4-8]. A cognitive architecture is considered "biologically inspired" when it is motivated by the organization and principles of biological intelligent systems, primarily, the human brain-mind. From this point of view, the majority of modern cognitive architectures belong to the BICA category. E.g., the most popular cognitive architectures, including ACT-R [9, 10] and Soar [11-14], originated from the Allen Newell's goal to model principles and mechanisms of human cognition [3], as opposed to the original goal of reproducing human intelligent capabilities in artificial intelligence [15] without necessarily replicating their mechanisms, or the goal in neuroscience – to understand how the brain works at the neuronal level.

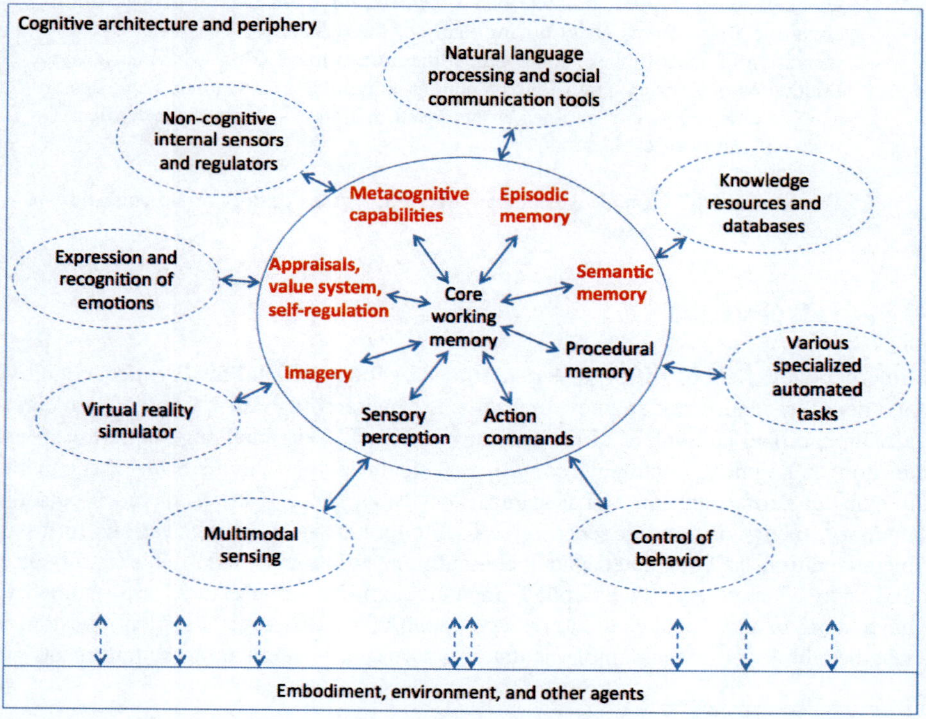

Fig. 1. Template for comparison of extensions of cognitive architectures. Only components within the solid circle belong to the cognitive architecture proper, of which the components shown in bold red are typically considered extensions. Virtually all circles may have direct connections to the environment (dashed vertical arrows).

A certain minimal set of elements including perception, cognition based on stored procedures, and action control, is common for all cognitive architectures due to the requirement of completeness. In these sense, other features can be regarded as extensions. The focus here is on extensions that are necessary for solving the BICA Challenge (Figure 1, red; [2]).

2 State of the Art and Limitations of Extensions: Examples

2.1 Limited Episodic Memory

As an example, let us consider the state of the art of episodic memory implementation and usage in cognitive architectures represented by the extended Soar [13, 14]. Episodic memory in Soar is stored as snapshots of contents of working memory taken together with contextual metadata. In principle, forms of episodic memory in Soar also include prospective memories of intents and plans. Retrieval is possible by activation of cues or by context. Usage may include many functions [14]: e.g., analysis of past episodes and retroactive learning, prediction and guidance in action selection, repetition avoidance.

This broad spectrum of functionality and usage of episodic memory looks much better than early implementations; yet many limitations still remain. Unlike human episodic memory, episodic memory in Soar does not remain plastic after its formation, and it is not modified or replicated every time when it is accessed (cf. [16]). Also, it mostly represents experiences of the actual past situations of the agent. The uniqueness of remembered episodes in many cases appears not critical for their usage: merging similar episodic memories may be allowed, which in psychological terms means mixing the notions of episodic and semantic memory.

The rich system of relations among remembered episodes characteristic of human memory is missing in these implementation, and as a consequence, strategic retrieval mechanisms with step-by-step contextual reinstatement [25, 26] are not implementable. Episodic memory of imagery is also missing, but is on the list in [14].

2.2 Limited Metacognition

Metacognition is a very broad notion, which in various forms is intimately interleaved virtually in all human cognition. It is impossible to address this topic here in detail. First, it is important to separate from metacognition anything that is not cognition on its own: e.g., internal sensing and autoregulation of the computational process (Figure 1, upper-left circle).

Speaking of metacognition as cognition about cognition, of particular interest are functions with known implementations in cognitive architectures, for example, Theory-of-Mind reasoning and autonomous goal generation [17, 18]. One general limitation of these implementations is that reasoning about goals is driven by a persistent meta-goal, and in this sense amounts to a sub-goal reasoning or planning.

Similarly, Theory-of-Mind in artificial intelligence is traditionally understood as a system of beliefs about beliefs of others processed from the same first-person mental perspective of the agent, in contrast with more rich mental simulations of others' mental perspectives performed by humans [20, 21].

2.3 Limited Affective Cognition

Extended Soar [13, 14] implements emotional intelligence by adding an appraisal detector as a separate module, which implements the theory of Scherer. The usage of tis module is that it generates one global characteristic (appraisal) of the current state of the agent, which can be used as a reward signal in reinforcement learning.

Limitations compared to human emotional cognition are innumerous; only a few examples can be named here. First and foremost, this approach does not allow for representation of social (complex) emotions. Secondly, it does not allow for simultaneous processing of multiple appraisals and appraisals of mental states of others.

3 Specific Challenges for BICA Designs: Examples

The above limitations suggest challenges for new cognitive architecture designs. The subset of challenges selected as examples below is not random. They contribute to an emergent coherent story that addresses the critical question of the BICA Challenge [2]: how to achieve the human learner critical mass? One specific approach is outlined in Section 4.

3.1 Plastic Prospective Episodic Memory

Episodic memory in humans is not limited to static snapshots of past experiences: it also stores imagined future or abstract situations, imagined experiences of others, and it is changed every time when it is accessed by mechanisms like reconsolidation and multiple trace formation. Many of these features will be critical for the believability of the agent and for its autonomous cognitive growth up to a human level. For example, remembered dreams of the future may help to generate new goals (see below); re-evaluating the past based on corrected beliefs may improve self-consistency of the agent cognition, and so on.

One specific challenge for future implementations is to have plastic prospective episodic memory of the imagined future scenario, in which not only the plan of achieving the goal, but also the understanding of the goal itself, as well as sub-goals, may change in response to new information in a more natural, human-like way. As a prerequisite for doing this, a significant first step, e.g., in Soar would be getting a goal situation represented in episodic memory as an imagined experience of the agent, thereby giving the agent new reasoning capabilities. This format of goal representation is already the standard for some existing frameworks: e.g., GMU BICA [24].

3.2 Creative Autonomous Goal Generation

The extension of episodic memory discussed above will allow the agent to reason about the goal as a perceived state of the world, questioning own beliefs, applying new knowledge and performing mental simulations in that state. The challenge is then to enable bootstrapped generation of higher and more complex goals that make sense in a given world, starting from a minimal set of innate primitive drives. A successful approach will combine many cognitive capabilities discussed in this article.

With multiple potential goal situations represented as plastic prospective episodic memories that are subject to metacognition, the agent will have possibilities of engineering and selection of goals. This process also requires metacognitive reasoning about goals. Processes of goal reasoning and goal selection can be automated using various approaches [17, 27], and in general rely on a system of values.

3.3 Human-Level Emotional Intelligence

Thus, a system of values appears necessary for the agent to be able to generate new goals. In order to be human-like, the goal selection process in an agent must be guided by a human-level system of values. This is only one aspect of the challenge of achieving human-level emotional intelligence in artifacts. Another aspect is the necessity for an agent to be integrated with human partners in a team, implying the ability to develop mutual relationships of trust, respect, subordination, etc. Complex social emotions are inevitably involved in the formation and maintenance of such relationships, which means that artifacts should be able to understand, generate, recognize and express social emotions. The "understand" part appears to be the hardest at present.

3.4 Human Learner Critical Mass

The human learner critical mass challenge is to identify, design and create a minimal cognitive architecture with minimal initial knowledge and skills, capable of human-like learning up to a level that a typical human can achieve in similar settings. The learning process does not need to be autonomous and may involve human instruction, mentoring, scaffolding, learning from examples, interaction with resources of various nature available to humans. The challenge can be further divided into a set of domain-specific human learner critical mass challenges and a general-purpose human learner critical mass challenge, the former being a precondition for the latter. The hypothesis is that a solution can be found in a simple form, as opposed to manual implementation and integration of all human intelligent capabilities. In this case, interactive learning, self-organization and evolution is expected to be among the main techniques used for creation of intelligent agents.

4 Toward Human-Like Intelligent Artifacts

The above consideration suggests that the challenge of cognitive growth of an artifact up to a human level of general intelligence (the human learner critical mass challenge) impinges on the evolvability of goals and values in an artificial cognitive system, which in turn requires artificial emotional intelligence. This section introduces a new approach to addressing the challenge, based on developing emotional intelligence in artifacts [22] and using it together with potential goal enumeration in order to generate a system of goals.

4.1 Emotional Extension of the Mental State Framework

The mental state framework [19] essentially relies on two building blocks: a mental state, that attributes specific content of awareness to a specific mental perspective of an agent, and a "schema": the term in this case refers to a specific structure that can be used to represent any concept or category. Instances of schemas populate mental states. Each instance has a standard set of attributes [24].

"Emotional" extension of this framework is based on the introduction of three new elements [22]: (i) an emotional state (an attribute of a mental state), (ii) an appraisal (an attribute of a schema), and (iii) higher-order appraisal schemas, that can be also called "moral schemas". As the name suggests, these schemas recognize patterns of appraisals and emotional states, and are intended to represent complex or social emotions and relationships, including pride, shame, trust, resentment, compassion, jealousy, sense of humor, etc. Available phenomenological data [28] can be used to define these schemas – or to map naturally emerging in the architecture new schemas onto familiar concepts.

4.2 Enumeration of Potential Goals

A useful enumeration of possible, virtually relevant, or potential goals in a given world or situation could be the key to goal generation. In order to enumerate possible goals in a useful way, one can use a semantic metalanguage [23, 29]: specifically, the shared by all languages lexical-conceptual core of semantic primes and their associated grammar. Examples of semantic primitives include very basic notions like "above", "big", "more", "have", "inside", "move", "see", "want", etc. From these fundamental notions, generic goal-like notions can be formed, e.g.: "survive", "satisfy desire", "possess", "secure", "dominate", "have freedom", "explore", etc. that can be applied to specific objects and situations in various combinations. Therefore, in a given setup, a conceptual lattice [31] of potential goals can be generated using the fundamental primitives. Then, classification and selection among potential goals should be done using a system of values organized in a Maslow hierarchy [30].

4.3 Generation of New Values

Thus, goal selection guided by the system of values requires a human-like system of values, and its natural development depends on the agent's ability to generate new values, which can be done by moral schemas, which therefore play the role of a "critical element" of the critical mass of a human-level learner. Moral schemas can be innate or emergent. Future studies will estimate this component of the critical mass in terms of a minimal subset of moral schemas that enable autonomous development of a human-compatible system of values and goals in a given environment.

5 Discussion

This paper presented a brief overview of cognitive architecture extensions with advanced, human-inspired cognitive capabilities, and pointed to the wide gap between existing implementations and the human mind. Several examples of specific challenges in bridging the gap were outlined, that allow us to decompose the BICA Challenge. Possible approaches to solving some of these challenges were discussed.

The key question is, which of these biologically inspired advanced features are critical, and which may be optional? "Critical" here means critical for acceptance as cognitively "equal" minds by humans, and for achieving a human-level learner critical mass. The analysis of the latter challenge presented here suggests that the critical set should include the above examples described as specific challenges, and more. Specifically, human-level emotional intelligence appears to be a necessary feature for the agent believability and for the sense of co-presence associated with the agent. It is also an essential component in self-regulated learning, which is one of the mechanisms required for achieving the critical mass: this aspect will be discussed elsewhere. As the consideration presented here illustrates, a general approach to solving the outlined challenges can be based on the formalism of multiple mental states simultaneously present in working memory [19], which therefore appears to be promising.

5.1 Conclusions

As a first step toward solving the BICA Challenge, state-of-the-art BICA need to be extended to enable advanced cognitive capabilities, including emotional intelligence, human-like episodic memory, and the ability to generate new goals and values. Existing extensions of mainstream cognitive architectures remain limited, compared to the human mind. Based on the analysis of these limitations and challenges, it is found here that human-level emotional intelligence is a critical component in the human-level learner critical mass. A specific approach to achieving the critical mass outlined here implies that more complex goals can be generated automatically based on an enumerated set of potential goals generated using universal cognitive elements like semantic primes, and the rules of goal selection can be based on an evolving system of values generated by moral schemas. These findings suggest tasks for future research.

References

1. Chella, A., Lebiere, C., Noelle, D.C., Samsonovich, A.V.: On a roadmap to biologically inspired cognitive agents. In: Samsonovich, A.V., Johannsdottir, K.R. (eds.) Biologically Inspired Cognitive Architectures 2011: Proceedings of the Second Annual Meeting of the BICA Society. Frontiers in Artificial Intelligence and Applications, vol. 233, pp. 453–460. IOS Press, Amsterdam (2011)

2. Samsonovich, A.V.: On the roadmap for the BICA Challenge. Biologically Inspired Cognitive Architectures 1(1), 100–107 (2012)
3. Newell, A.: Unified Theories of Cognition. Harward University Press, Cambridge (1990)
4. SIGArt, Special section on integrated cognitive architectures. Sigart Bulletin 2(4) (1991)
5. Pew, R.W., Mavor, A.S. (eds.): Modeling Human and Organizational Behavior: Application to Military Simulations. National Academy Press, Washington, DC (1998), `http://books.nap.edu/catalog/6173.html`
6. Ritter, F.E., Shadbolt, N.R., Elliman, D., Young, R.M., Gobet, F., Baxter, G.D.: Techniques for Modeling Human Performance in Synthetic Environments: A Supplementary Review. Human Systems Information Analysis Center (HSIAC), Wright-Patterson Air Force Base (2003)
7. Gluck, K.A., Pew, R.W. (eds.): Modeling Human Behavior with Integrated Cognitive Architectures: Comparison, Evaluation, and Validation. Erlbaum, Mahwah (2005)
8. Gray, W.D. (ed.): Integrated Models of Cognitive Systems. Series on Cognitive Models and Architectures. Oxford University Press, Oxford (2007)
9. Anderson, J.R., Lebiere, C.: The Atomic Components of Thought. Lawrence Erlbaum Associates, Mahwah (1998)
10. Anderson, J.R.: How Can the Human Mind Occur in the Physical Universe? Oxford University Press, New York (2007)
11. Laird, J.E., Rosenbloom, P.S., Newell, A.: Universal Subgoaling and Chunking: The Automatic Generation and Learning of Goal Hierarchies. Kluwer, Boston (1986)
12. Laird, J.E., Newell, A., Rosenbloom, P.S.: SOAR: An architecture for general intelligence. Artificial Intelligence 33, 1–64 (1987)
13. Laird, J.E.: Extending the Soar cognitive architecture. In: Wang, P., Goertzel, B., Franklin, S. (eds.) Artificial General Intelligence 2008: Proceedings of the First AGI Conference, pp. 224–235. IOS Press, Amsterdam (2008)
14. Laird, J.E.: The Soar Cognitive Architecture. MIT Press, Cambridge (2012)
15. McCarthy, J., Minsky, M.L., Rochester, N., Shannon, C.E.: A proposal for the Dartmouth summer research project on artificial intelligence. In: Chrisley, R., Begeer, S. (eds.) Artificial Intelligence: Critical Concepts, vol. 2, pp. 44–53. Routledge, London (1955)
16. Nadel, L., Samsonovich, A., Ryan, L., Moscovitch, M.: Multiple trace theory of human memory: Computational, neuroimaging, and neuropsychological results. Hippocampus 10(4), 352–368 (2000)
17. Molineaux, M., Klenk, M., Aha, D.W.: Goal-driven autonomy in a Navy strategy simulation. In: Proceedings of the National Conference on Artificial Intelligence, vol. 3, pp. 1548–1554 (2010)
18. Hiatt, L.M., Khemlani, S.S., Trafton, J.G.: An explanatory reasoning framework for embodied agents. Biologically Inspired Cognitive Architectures 1, 23–31 (2012)
19. Samsonovich, A.V., De Jong, K.A., Kitsantas, A.: The mental state formalism of GMU-BICA. International Journal of Machine Consciousness 1(1), 111–130 (2009)
20. Gallese, V., Goldman, A.: Mirror neurons and the simulation theory of mind-reading. Trends in Cognitive Science 2, 493–501 (1998)
21. Nichols, S., Stich, S.: Mindreading: An Intergrated Account of Pretence, Self-Awareness, and Understanding Other Minds. Oxford University Press, Oxford (2003)
22. Samsonovich, A.V.: An approach to building emotional intelligence in artifacts. In: Burgard, W., Konolige, K., Pagnucco, M., Vassos, S. (eds.) Cognitive Robotics: AAAI Technical Report WS-12-06, pp. 109–116. The AAAI Press, Menlo Park (2012)
23. Wierzbicka, A.: Semantic complexity: conceptual primitives and the principle of substitutability. Theoretical Linguistics 17(1-3), 75–97 (1991)

24. Samsonovich, A.V., De Jong, K.A.: Designing a self-aware neuromorphic hybrid. In: Thorisson, K.R., Vilhjalmsson, H., Marsela, S. (eds.) AAAI 2005 Workshop on Modular Construction of Human-Like Intelligence: AAAI Technical Report, pp. 71–78. AAAI Press, Menlo Park (2005)
25. Becker, S., Lim, J.: A computational model of prefrontal control in free recall: Strategic memory use in the California Verbal Learning task. Journal of Cognitive Neuroscience 15, 821–832 (2003)
26. Fletcher, P.C., Henson, R.N.A.: Frontal lobes and human memory—Insights from functional neuroimaging. Brain 124, 849–881 (2001)
27. Jaidee, U., Muñoz-Avila, H., Aha, D.W.: Integrated learning for goal-driven autonomy. In: Proceedings of IJCAI 2011, pp. 2450–2455 (2011)
28. Ortony, A., Clore, G., Collins, A.: The Cognitive Structure of Emotions. Cambridge University Press, Cambridge (1988)
29. Goddard, C.: Semantic primes, semantic molecules, semantic templates: Key concepts in the NSM approach to lexical typology. Linguistics 50(3), 711–743 (2012)
30. Maslow, A.H.: A theory of human motivation. Psychological Review 50(4), 370–396 (1943)
31. Ganter, B., Wille, R.: Formal Concept Analysis: Mathematical Foundations. Springer, Berlin (1999)

Babies and Baby-Humanoids to Study Cognition

Giulio Sandini

Robotics, Brain and Cognitive Sciences Department, Istituto Italiano di Tecnologia
Via Morego 30, 16163 Genova - Italy

Abstract. Simulating and getting inspiration from biology is certainly not a new endeavor in robotics. However, the use of humanoid robots as tools to study human cognitive skills it is a relatively new area of research which fully acknowledges the importance of embodiment and interaction (with the environment and with others) for the emergence of cognitive as well as motor, perceptual and social abilities.

The aim of this talk is to present our approach to investigating human cognitive abilities by explicitly addressing cognition as a developmental process through which the system becomes progressively more skilled and acquires the ability to understand events, contexts, and actions, initially dealing with immediate situations and increasingly acquiring a predictive capability.

During the talk I will also argue that, within this approach, robotics engineering and neuroscience research are mutually supportive by providing their own individual complementary investigation tools and methods: neuroscience from an "analytic" perspective and robotics from a "synthetic" one. In order to demonstrate the synergy between neuroscience and robotics I will start by briefly reviewing a few aspects of neuroscience and robotics research in the last 30 years to show that the two fields have indeed progressed in parallel even if, until recently, on almost independent tracks. I will then show that since the discovery of visuo-motor neurons our view of how the brain implements perceptual abilities has evolved, from a model where perception was linked to motor control for the only (yet fundamental) purpose of providing feedback signals, toward an integrated view where perception is mediated not only by our sensory inputs but also by out motor abilities. As a consequence the implementation and use of perceptive and complex humanoid robots has become not only a very powerful modeling tool of human behavior (and the underlying brain mechanisms) but also an important source of hypothesis regarding the mechanisms used by the nervous system to control our own actions and to predict the goals of someone elses actions. During the talk I will present results of recent psychophysical investigations in children and human adults as well as artificial implementations in our baby humanoid robot iCub.

A. Chella et al. (Eds.): Biologically Inspired Cognitive Architectures 2012, AISC 196, p. 51.
springerlink.com

Towards Architectural Foundations for Cognitive Self-aware Systems

Ricardo Sanz and Carlos Hernández

Autonomous Systems Laboratory
Universidad Politcnica de Madrid, Spain

Abstract. The BICA 2012 conference main purpose is to take a significant step forward towards the BICA Challenge -creating a real-life computational equivalent of the human mind. This challenge apparently calls for a global, multidisciplinary joint effort to develop biologically-inspired dependable agents that perform well enough as to to be fully accepted as autonomous agents by the human society. We say "apparently" because we think that "biologically-inspired" needs to be re-thought due to the mismatch between natural and artificial agent organization and their construction methods: the natural and artificial construction processes. Due to this constructive mismatch and the complexity of the operational requirements of world-deployable machines, the question of dependability becomes a guiding light in the search of the proper architectures of cognitive agents. Models of perception, cognition and action that render self-aware machines will become a critical asset that marks a concrete roadmap to the BICA challenge.

In this talk we will address a proposal concerning a methodology for extracting universal, domain neutral, architectural design patterns from the analysis of biological cognition. This will render a set of design principles and design patterns oriented towards the construction of better machines. Bio-inspiration cannot be a one step process if we we are going to to build robust, dependable autonomous agents; we must build solid theories first, departing from natural systems, and supporting our designs of artificial ones.

A. Chella et al. (Eds.): Biologically Inspired Cognitive Architectures 2012, AISC 196, p. 53.
springerlink.com

Achieving AGI within My Lifetime: Some Progress and Some Observations

Kristinn R. Thórisson

CADIA & School of Computer Science, Reykjavik University and Icelandic Institute for Intelligent Machines
Menntavegur 1, 101 Reykjavik, Iceland
thorisson@iiim.is, thorisson@hr.is

Abstract. The development of artificial intelligence (AI) systems has to date taken largely a constructionist approach, with manual programming playing a central role. After half a century of AI research, enormous gaps persist between artificial and natural intelligence. The differences in capabilities are readily apparent on virtually every scale we might want to compare them on, from adaptability to resilience, flexibility to robustness, to applicability. We believe the blame lies with a blind application of various constructionist methodologies building AI systems by hand. Taking a fundamentally different approach based on new constructivist principles we have developed a system that goes well beyond many of the limitations of present AI systems. Our system can automatically acquire complex skills through observation and imitation. Based on new programming principles supporting deep reflection and auto-catalytic principles for maintenance and self-construction of architectural operation, the system is domain-dependent and can be applied to a vast array of problem areas. We have tested the system on a challenging task: Learning a subset of socio-communicative skills by observing humans engaged in a simulated TV interview. This presentation introduces the core methodological ideas, architectural principles, and shows early test scenarios of the system in action.

A. Chella et al. (Eds.): Biologically Inspired Cognitive Architectures 2012, AISC 196, p. 55.
springerlink.com

Learning and Creativity in the Global Workspace

Geraint A. Wiggins

Professor of Computational Creativity
Queen Mary, University of London

Abstract. The key goal of cognitive science is to produce an account of the phenomenon of mind which is mechanistic, empirically supported, and credible from the perspective of evolution. In this talk, I will present a model based on Baars' [1] Global Workspace account of consciousness, that attempts to provide a general, uniform mechanism for information regulation. Key ideas involved are: information content and entropy [4,8], expectation [3,7], learning multi-dimensional, multi-level representations [2] and data [5], and data-driven segmentation [6].

The model was originally based in music, but can be generalised to language [9]. Most importantly, it can account for not only perception and action, but also for creativity, possibly serving as a model for original linguistic thought.

References

1. Baars, B.J.: A cognitive theory of consciousness. Cambridge University Press, Cambridge (1988)
2. Gärdenfors, P.: Conceptual Spaces: the geometry of thought. MIT Press, Cambridge (2000)
3. Huron, D.: Sweet Anticipation: Music and the Psychology of Expectation. Bradford Books, MIT Press, Cambridge, MA (2006)
4. MacKay, D.J.C.: Information Theory, Inference, and Learning Algorithms. Cambridge University Press, Cambridge (2003)
5. Pearce, M.T.: The Construction and Evaluation of Statistical Models of Melodic Structure in Music Perception and Composition. PhD thesis, Department of Computing, City University, London, UK (2005)
6. Pearce, M.T., Müllensiefen, D., Wiggins, G.A.: The role of expectation and probabilistic learning in auditory boundary perception: A model comparison. Perception 9, 1367–1391 (2010)
7. Pearce, M.T., Wiggins, G.A.: Auditory Expectation: The Information Dynamics of Music Perception and Cognition. Topics in Cogntive Science (in press)
8. Shannon, C.: A mathematical theory of communication. Bell System Technical Journal 27, 379–423, 623–656 (1948)
9. Wiggins, G.A.: "I let the music speak": cross-domain application of a cognitive model of musical learning. In: Rebuschat, P., Williams, J. (eds.) Statistical Learning and Language Acquisition. Mouton De Gruyter, Amsterdam, NL (2012)

A. Chella et al. (Eds.): Biologically Inspired Cognitive Architectures 2012, AISC 196, p. 57.
springerlink.com

Multimodal People Engagement with iCub

Salvatore M. Anzalone, Serena Ivaldi, Olivier Sigaud, and Mohamed Chetouani

Institut des Systemes Intelligents et de Robotique
CNRS UMR 7222 & Universite Pierre et Marie Curie, Paris, France

Abstract. In this paper we present an engagement system for the iCub robot that is able to arouse in human partners a sense of "co-presence" during human-robot interaction. This sensation is naturally triggered by simple reflexes of the robot, that speaks to the partners and gazes the current "active partner" (e.g. the talking partner) in interaction tasks. The active partner is perceived through a multimodal approach: a commercial rgb-d sensor is used to recognize the presence of humans in the environment, using both 3d information and sound source localization, while iCub's cameras are used to perceive his face.

Keywords: engagement, attention, personal robots.

1 Introduction

A sense of attention and engagement naturally emerges in humans during social interactions. This sensation of "co-presence" depends on different behaviors that show cognitive capabilities [1] [2]. In particular, establishing and maintaining eye contact between people involved in interactions is one of the key factors causing the "social engagement" [3] [4]. Thanks to this behavior we can communicate in a non-verbal way a variety of social information about attention and involvement in a conversation or a context. Harnessing people perception is thus fundamental for enhancing human-robot interaction. For example when robot and humans cooperate to perform a task or share a plan, if the robot expresses its engagement to the human partner by fixation, the latter is naturally prone to consider that the robot is following its commands. Social robots should be provided with this ability to improve the sense of social connection with humans [5] [6].

Fig. 1. People interacting with iCub (left). An sketch of a typical interaction scenario (center). A human partner catch the robot's attention to show a new object (right).

A. Chella et al. (Eds.): Biologically Inspired Cognitive Architectures 2012, AISC 196, pp. 59–64.
springerlink.com

In this paper we present a simple engagement system for the humanoid robot iCub [7], which will be foundational to more evolved HRI modules. The proposed system implements a multimodal approach to people detection and is able to let the robot identify and track the "active" partner (Figure 1).

2 System Overview

In typical human-robot interaction scenarii, one or more humans interact with the robot as partners sharing a plan or cooperating to fulfill a common task. The robotic system presented in this paper follows a multimodal approach to perceive and track these partners. A Microsoft Kinect is used to perceive and reconstruct the environment to retrieve the possible presence of humans. At the same time the 4 microphones array embedded in the Kinect is used to localize sound sources. It is assumed that a sound coming from the same direction of a human corresponds to his voice or to some noise that the partner intentionally produces to catch the robot's attention. Hence, when a sound is perceived, the system fuses the source localization information with the people location information and, accordingly, marks the "active partner", i.e. the current person talking or producing sounds. This information is exploited by a gaze controlling mechanism to move the robot's head and eyes so as to center the person in its field of view. Finally, the iCub camera stream is analyzed to detect human faces, to fine tune the person localization.

The proposed approach has been implemented as the interoperation of several software modules through different technologies. As shown in Figure 2, rgb-d data from the Kinect is elaborated in the "Skeleton Tracking" system, to extract and track the three-dimensional models of the human partners in the scene. At the same time, the "Sound Source Localization" system detects the presence of sounds and locates their directions. This information is collected by an "Attention System", which fuses the multisensory processed data so as to determine the current active partner. Lastly, this localization information is used by the "iCub Gaze Controller" to accordingly move the head and eyes (both are actuated) of the robot, so that the cameras are centered on the human partner. At the same time, the "Face Detection" module carries on the detection of humans based on their faces, collecting and analyzing image frames from the iCub's cameras.

The iCub control system takes advantage of the YARP framework [8], while the perception system that collects and elaborates data from the Kinect has been built using the ROS framework [9]. Both YARP and ROS support the development of several independent software modules that can be executed in a distributed environment and are able to communicate among different channels. Since the communication protocol of the two middlewares is different, a two ways YARP-ROS bridge has been built using standard Unix pipes to exchange the data between the two according to their own rules.

3 Multimodal Active Partner Tracking

The active partner tracking system relies on a multimodal approach that fuses different types of information: a 3d model of the human presents in the environment; the direction of the perceived sounds; the faces of the detected partners. The tracking of

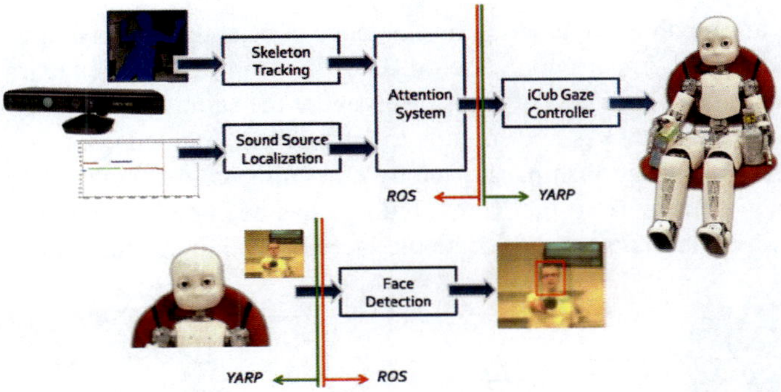

Fig. 2. The modules that composes the system

all the human partners interacting with the robot exploits the 3d sensor of the Kinect and a skeleton recognition algorithm from the OpenNi library[1]. After a calibration step, in which the person is encouraged to assume a standard pose in front of the Kinect (Figure 3a), the algorithm is able to detect and track the human partner in the environment. The points cloud collected from the sensor is analyzed to extract the presence of human bodies; then the single body parts (head, trunk, shoulders, elbows, knees, etc.) are recognized. Finally, a complete description of the skeleton is obtained in terms of positions in the environment of the body parts. Though having a detailed description of human pose, only the position of the head of each human partner is retained. Particularly, the head of the "active" partner is localized in order to make the robot gaze directly at it. The Kinect's 4-microphones array is used for sound source detection and localization. The processing relies on HARK library[2], which implements a version of the MUSIC algorithm [10], that is able to localize and segregate sounds assuming near-field range conditions (Figure 3b). In particular, the algorithm focuses on noisy dialog contexts between humans and a robot [11]. If the sound perceived comes from the same direction of a human partner, the sound is supposed to be his voice (or a noise intentionally produced by the human to catch the robot's attention). According to this simple rule, the speaking partner is marked as "active partner". A roto-translational matrix is applied to the position of the current active partner to convert his coordinates from the Kinect's to the iCub's reference frame in the Cartesian space; finally iCub is controlled to point its eyes towards this point, to gaze to the current active partner. In the gaze action, all the head joints are involved: neck (3 DOF: roll, pitch and yaw) and eyes (3 DOF: they have common tilt but can pan independently). It is worth noticing that the association between sound source and human body has some constraints: if the perceived sound is too far from any human, it is ignored; if a sound is perceived but no people is detected in the environment, iCub can point at the direction of the sound. These "automatic reflexes" have been chosen as human-like and easily readable behaviors. The cameras located in the iCub's eyes are also used. In particular, the Viola-Jones algorithm is used to recognize frontal faces [12]. Viola-Jones classifier is an efficient detection system that relies on a particular kind of features, called Haar-like.

[1] `http://www.openni.org/`

[2] `http://winnie.kuis.kyoto-u.ac.jp/HARK/`

These features are capable of encoding the existence of oriented contrast in different regions of the image [13]. In particular, a set of Haar-like features has been extracted from a training set of different pictures of faces taken under the same lighting conditions and normalized to line up the eyes and mouths to encode the oriented contrasts and their particular geometrical relationships showed by human faces. As shown in Figure 3c, , the image stream coming from the iCub's left eye has been elaborated by the trained classifier that was able to detect and locate the faces of the human partners.

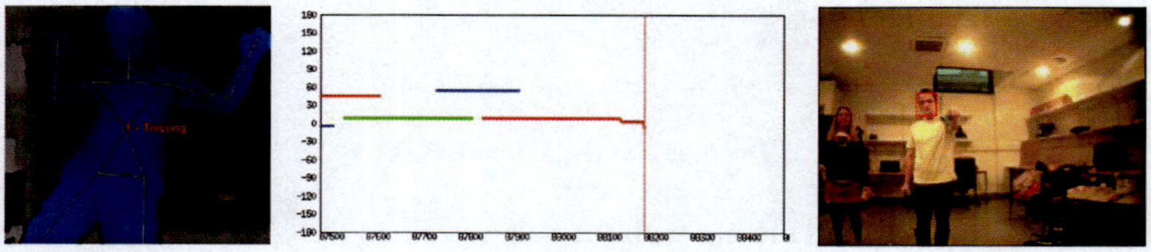

(a) (b) (c)

Fig. 3. The different perceptive sources. Figure 3a: a human skeleton extracted by the rbgd data. Figure 3b: the stream of sound sources detected by the Kinect (precisely, two sources). Figure 3c: a frame from iCub's camera is processed to detect the faces of the active human partner.

4 Experimental Results

First, in a developmental robotics scenario, where human partners interact with the robot to teach new objects. Two or more caregivers are in front of the robot, their 3D location being tracked as described in the previous section. As one of the people speaks, the produced sound is detected and matched with the person's location. In this case, the robot points his head towards the "active caregiver", i.e. the human partner that is currently trying to teach to the robot. As shown in Figure 4, the system is able to detect and track the humans in the environment, to mark the active partner, and to detect his face. If the caregiver speaks and moves at the same time, the robot tracks the human gazing at his head. If people talk alternatively, as during a conversation, the robot gaze at the "active" one, emulating a third "listening" to the conversation. The video showing some of the experiments can be found at: http://macsi.isir.upmc.fr.

From the point of view of the human partners, a sensation of "social cohesion", or "togetherness", grows up from the gazing of the robot: humans have the feeling of being perceived by the robot while they interact with it as in a real relation between human student and human teacher. In particular, the pointing behaviour is able to generate a feeling of social connection between the robot and the active partner: the human caregiver feels the attention of the robot as feedback to his teaching attempts. Remarkably, when partners were asked a feedback about the experiments, they admitted that the simple gaze of the robot was enough to give the impression that the robot was effectively listening to their conversation. In other words, co-presence has been felt by the partners as an increase of the mutual engagement during interaction.

As shown in Figure 5, the system has been used in real-life scenarios in which the robot joins a conversation between people. In this situation, the robot should be provided with advanced speech recognition and synthesis abilities to actively engage in

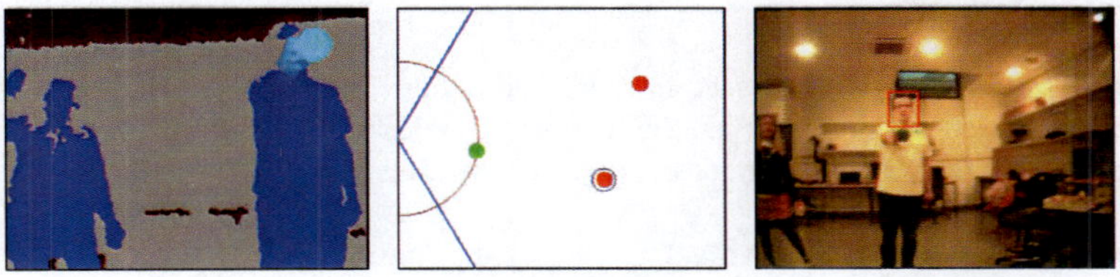

(a) Disparity map (b) People localized (c) Faces tracked

Fig. 4. Multimodal people localization and tracking. The rgb-d information (fig. 4a) is used to retrieve the presence of human partners in the environment. The "active partner" is identified matching this information with the sound sources localized by the microphones array. Figure 4b shows the information fusion: on the left the Kinect field of view, in blue; over the red semicircle, the green dot is the direction of the sound source perceived; red dots corresponds to the human partners perceived; the "active partner" is marked using a blue highlight. The face of the "active partner" is then extracted from iCub's camera frames (fig. 4c).

Fig. 5. iCub focuses on the "active partner", alternatively on the left and on the right

a conversation. However, our simple experiments suggest that the gazing behaviour arouse in humans to produce a feeling of social presence: even remaining silent, the robot is perceived as an active agent involved in the conversation.

5 Conclusions and Future Works

In this paper a multimodal people engagement system for the humanoid iCub has been presented. The system has been tested in experimental scenarios of close interaction with humans. Experiments show that by the use of gazing and faces engagement it is possible to establish a direct connection during interactions between humans and robots and to arouse in the human partners a sense of co-presence of the robot in the environment. Results encourage us to improve the mutual engagement adding new perception-action modalities, and further investigate the role of engagement in teaching scenarios, evaluating its effect in a qualitative way as well as in a quantitative way.

Acknowledgment. Authors would like to thank M. Guenaien for her efforts during the development of this project. This work has been partially supported by French National Research Agency (ANR) through TecSan program (project Robadom ANR-09-TECS-012) and by the MACSi Project (ANR 2010 BLAN 0216 01).

References

1. Durlach, N., Slater, M.: Presence in shared virtual environments and virtual togetherness. Presence: Teleoperators & Virtual Environments (2000)
2. Zhao, S.: Toward a taxonomy of copresence. Presence: Teleoperators & Virtual Environments (2003)
3. Breazeal, C.: Toward sociable robots. Robotics and Autonomous Systems (2003)
4. Breazeal, C.L.: Designing sociable robots. The MIT Press (2004)
5. Al Moubayed, S., Baklouti, M., Chetouani, M., Dutoit, T., Mahdhaoui, A., Martin, J.C., Ondas, S., Pelachaud, C., Urbain, J., Yilmaz, M.: Generating robot/agent backchannels during a storytelling experiment. In: IEEE International Conference on Robotics and Automation, ICRA 2009. IEEE (2009)
6. Rich, C., Ponsler, B., Holroyd, A., Sidner, C.L.: Recognizing engagement in human-robot interaction. In: 2010 5th ACM/IEEE International Conference on Human-Robot Interaction (HRI). IEEE (2010)
7. Metta, G., Sandini, G., Vernon, D., Natale, L., Nori, F.: The icub humanoid robot: an open platform for research in embodied cognition. In: Proceedings of the 8th Workshop on Performance Metrics for Intelligent Systems. ACM (2008)
8. Metta, G., Fitzpatrick, P., Natale, L.: Yarp: Yet another robot platform. International Journal on Advanced Robotics Systems (2006)
9. Quigley, M., Conley, K., Gerkey, B., Faust, J., Foote, T., Leibs, J., Wheeler, R., Ng, A.Y.: Ros: an open-source robot operating system. In: ICRA Workshop on Open Source Software (2009)
10. Nakadai, K., Okuno, H.G., Nakajima, H., Hasegawa, Y., Tsujino, H.: An open source software system for robot audition hark and its evaluation. In: 8th IEEE-RAS International Conference on Humanoid Robots, Humanoids 2008. IEEE (2008)
11. Asono, F., Asoh, H., Matsui, T.: Sound source localization and signal separation for office robot jijo-2. In: Proceedings of the 1999 IEEE/SICE/RSJ International Conference on Multisensor Fusion and Integration for Intelligent Systems, MFI 1999. IEEE (1999)
12. Viola, P., Jones, M.J.: Robust real-time face detection. International Journal of Computer Vision (2004)
13. Viola, P., Jones, M.: Rapid object detection using a boosted cascade of simple features. In: Proceedings of the 2001 IEEE Computer Society Conference on Computer Vision and Pattern Recognition, CVPR 2001. IEEE (2001)

Human Action Recognition from RGB-D Frames Based on Real-Time 3D Optical Flow Estimation

Gioia Ballin, Matteo Munaro, and Emanuele Menegatti

Department of Information Engineering of the University of Padova,
via Gradenigo 6B, 35131 - Padova, Italy
gioia.ballin@gmail.com, {munaro,emg}@dei.unipd.it

Abstract. Modern advances in the area of intelligent agents have led to the concept of cognitive robots. A cognitive robot is not only able to perceive complex stimuli from the environment, but also to reason about them and to act coherently. Computer vision-based recognition systems serve the perception task but they also go beyond it by finding challenging applications in other fields such as video surveillance, HCI, content-based video analysis and motion capture. In this context, we propose an automatic system for real-time human action recognition. We use the Kinect sensor and the tracking system in [1] to robustly detect and track people in the scene. Next, we estimate the 3D optical flow related to the tracked people from point cloud data only and we summarize it by means of a 3D grid-based descriptor. Finally, temporal sequences of descriptors are classified with the Nearest Neighbor technique and the overall application is tested on a newly created dataset. Experimental results show the effectiveness of the proposed approach.

Keywords: Action recognition, 3D optical flow, RGB-D data, Kinect.

1 Introduction

The challenge of endowing robotic agents with human-like capabilities is currently addressed by the cognitive robotics research field. In cognitive robotics, the aim is to create smart agents able to efficiently perform complex tasks in partially observable environments. In order to achieve real-world goals, a cognitive robot is equipped with a processing architecture that combine the perception, cognition and action modules in the most effective way. Then, a cognitive robot is not only able to perceive complex stimuli from the environment, but also to reason about them and to act coherently. Furthermore, a cognitive robot should also be able to safely interact and cooperate with humans.

The interaction with humans and the interpretation of human actions and activities have recently gained a central role in the researchers' community since the spread of new robotic devices has reached real-life environments such as offices, homes and urban environments. In this context, we propose an automatic system for real-time human action recognition. Human action recognition is an active research area in computer vision. First investigations about this topic began in the seventies with pioneering studies accomplished by Johanssons [2]. From then on, the interest in the field grew increasingly, motivated by a number of potential real-world applications such as video

A. Chella et al. (Eds.): Biologically Inspired Cognitive Architectures 2012, AISC 196, pp. 65–74.
springerlink.com

surveillance, HCI, content-based video analysis and retrieval. Moreover, in recent years the task of recognizing human actions has gained increasingly popularity thanks to the emergence of modern applications such as motion capture and animation, video editing and service robotics.

Our system relies on the acquisition of RGB-D data and exploits the Robot Operating System [3] as a framework. We use the Microsoft Kinect sensor and the tracking system described in [4] and [1] to robustly detect and track people in the scene. Next, we estimate the 3D optical flow of the points relative to each person. For this purpose, we propose a novel technique that estimates 3D velocity vectors from point cloud data only, thus obtaining a real-time calculus of the flow. Then, we compute a 3D grid-based descriptor for representing the flow information within a temporal sequence and we recognize actions by means of the Nearest Neighbor classifier. We tested this technique on a RGB-D video dataset which contains six actions performed by six different actors. Our system is able to recognize the actions in the dataset with a 80% accuracy while running at a medium frame rate of 23 frames per second.

The remainder of the paper is organized as follows: Section 2 provides a complete review about the recent advances in human action recognition systems. Section 3 describes the proposed real-time computation of 3D optical flow, while Section 4 outlines the data structure used to summarize the estimated flow information. Experimental results are reported in Section 5 and Section 6 concludes the paper and outlines the future work.

2 Related Work

Most of the works on human action recognition rely on information extracted from 2D images and videos. These approaches mostly differ in the features representation. Popular global representations are edges [5], silhouettes of the human body [6] [7] [8], 2D optical flow [9] [10] [11] and 3D spatio-temporal volumes [6] [7] [8] [12] [13] [14]. Conversely, effective local representations mainly refer to [15] [16] [17] [18] [19] [20] and [21]. The recent spread of inexpensive RGB-D sensors has paved the way to new studies in this direction. Recognition systems that rely on the acquisition of 3D data could potentially outperform their 2D counterparts, but they still need to be investigated. The first work related to RGB-D action recognition is signed by Microsoft Research [22]. In [22], a sequence of depth maps is given as input to the system. Next, the relevant postures for each action are extracted and represented as a bag of 3D points. The motion dynamics are modeled by means of an action graph and a Gaussian Mixture Model is used to robustly capture the statistical distribution of the points. Recognition results state the superiority of the 3D silhouettes with respect to their 2D analogous.

Subsequent studies mainly refer to the use of two different technologies: Time of Flight cameras [23][24] and active matricial triangulation systems, in particular the Microsoft Kinect [25], [26], [27], [28], [29], [30]. [25][26] represent the first attempt to exploit skeleton tracking information to recognize human actions. This information is used to compute a set of features related to the human body pose and motion. Then, the proposed approach is tested on different classifiers: the SVM classification is compared with both the one-layer and the hierarchical Maximum Entropy Markov Model classification. Finally, the dataset collected for testing purposes has been made publicly available by the authors. As for [25][26], the recently published work by Yang *et al.* [27] relies on the acquisition of skeleton body joints. The 3D position differences of body joints are exploited to characterize the posture and the motion of the observed human body. Finally, Principal Component Analysis is applied to compute the so-called *EigenJoints* and the Näives-Bayes-Nearest-Neighbor technique is used to classify these descriptors. This work consistently outperforms that of Li *et al.* [22] while using about a half their number of frames. A different approach is followed in [28] by Zhang and Parker, where the popular 2D spatio-temporal features are extended to the third dimension. The new features are called 4D spatio-temporal features, where the "4D" is justified by the 3D spatial components given by the sensor plus the time dimension. The descriptor computed is a 4D hyper cuboid, while Latent Dirichlet Allocation with Gibbs sampling is used as classifier. Another work in which typical 2D representations are extended to 3D is [29]. The authors extend the existing definitions of spatio-temporal interest points and motion history images to incorporate also the depth information. They also proposes a new publicly available dataset as test bed. For the classification purpose, SVMs with different kernels are used.

From the application point of view [25], [26], and [29] are targeted to applications in the personal robotics field, while [22] and [27] are addressed to HCI and gaming applications. Finally, [28] and [30] are primarily addressed to applications in the field of video surveillance. In [30], Popa *et al.* propose a system able to continuously analyze customers' shopping behaviours in malls. By means of the Microsoft Kinect sensor, Popa *et al.* extract silhouette data for each person in the scene and then compute moment invariants to summarize the features.

In [23][24], a kind of 3D optical flow is exploited for the gesture recognition task. Unlike our approach, Holte *et al.* compute the 2D optical flow using the traditional Lukas-Kanade method and then extend the 2D velocity vectors to incorporate also the depth dimension. At the end of this process, the 3D velocity vectors are used to create an annotated velocity cloud. 3D Motion Context and Harmonic Motion Context serve the task of representing the extracted motion vector field in a view-invariant way. With regard to the classification task, [23] and [24] do not follow a learning-based approach, instead a probabilistic Edit Distance classifier is used in order to identify which gesture best describes a string of primitives. [24] differs from [23] because the optical flow is estimated from each view of a multi-camera system and is then combined into a unique 3D motion vector field.

3 3D Optical Flow

In this section, we propose a novel approach to compute the 3D optical flow as an extension to the third dimension of the traditional 2D optical flow [9] [10] [11]. Computing the optical flow with real-time performances is really a challenging problem: traditional 2D approaches involve several computations on every pixel of the input images thus leading to poor temporal performances. On the contrary, we compute 3D optical flow only for relevant portions of the overall 3D scene.

In details, we associate a cluster to each tracked person by means of the underlying tracking-by-detection system [1]. Such a cluster is defined as a 4D point cloud representing an individual in the 3D world. The four dimensions represent the 3D geometric coordinates and the RGB color component of each point. With this information, we estimate the 3D optical flow associated to each identified cluster frame-by-frame. The first step of this process concerns storing the appropriate information. Indeed, at each frame F and for each track k, we store two elements: the cluster associated to k at frames F and $F-1$. The second step involves matching the two point clouds stored at each frame in order to find correspondences between points.

3.1 Points Matching

Matching cluster points relative to different time instants represents the true insight of this work. Since we deal with human motion, we cannot assume the whole person cluster to undergo a rigid transformation. For this reason, we exploit local matching techniques.

For each track k, let $A_k(F-1)$ be the cluster associated to k at the frame $F-1$ and let $A_k(F)$ the cluster associated to k at the frame F. Let P be a generic point in $A_k(F-1)$. If we can find the spatial location of P in $A_k(F)$, then we can also estimate the actual 3D velocity vector representing the movement of P. In order to find a match between the points in $A_k(F-1)$ and in $A_k(F)$, a two-way matching algorithm is applied. First, $A_k(F)$ is kept fixed and for each point in $A_k(F-1)$ a 1-nearest neighbor search is performed in order to find a matching point in $A_k(F)$. Next, the same pattern is repeated with $A_k(F-1)$ fixed instead of $A_k(F)$. This way, two different vectors of correspondences are returned, those vectors are then intersected and a final vector of correspondences is returned. Since clusters have an average of 300 points, kdtrees with FLANN searches are used to speed-up the computations. Furthermore, searches are driven by both the 3D geometric coordinates of the points and the RGB color information. In Fig. 1, we show the correspondences estimated while a person is hand waving. In Fig. 1(a) two consecutive images are shown, while in Fig. 1 (b) the corresponding cluster points are reported and the points correspondences are drawn in red. In this work, the 3D optical flow is seen as a set of estimated 3D velocity vectors that are obtained in constant time from the vector of estimated correspondences by dividing the spatial difference by the temporal difference. At an implementation level, the computed optical flow can be stored as an additional field for each point of $A_k(F)$, thus creating an annotated point cloud.

4 3D Grid-Based Descriptor

In order to recognize human actions from the 3D optical flow estimated in Section 3, first a suitable description for the flow is required. Indeed, the proposed approach generally returns a different number of velocity vectors in each frame. To achieve a fixed size descriptor we compute a 3D grid surrounding each cluster in the scene. The 3D grid provides an effective spatial partition of the cluster points. Furthermore, since each of the 3D velocity vectors is associated to a cluster point by means of an annotated

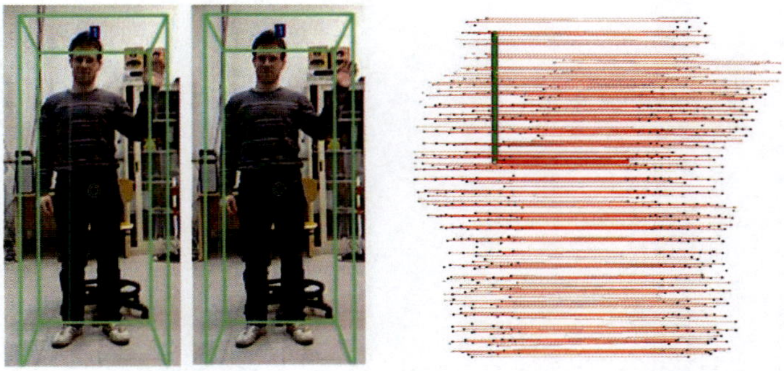

(a) RGB frames (b) Cluster points and estimated correspondences

Fig. 1. Display of two consecutive RGB frames where the tracking output is drawn as a 3D bounding box (a) and the cluster points relative to the tracked person, together with the estimated correspondences (b) with regard to the hand waving action

point cloud, the grid provides also a 3D optical flow partition. In this work, the 3D grid represents the baseline data structure of the flow descriptor and it is defined by a fixed number of spatial partitions along the three geometric axes. The grid computation involves three basic steps. The first step is concerned with defining the minimum and maximum grid bounds along the three dimensions. Bounds are set so that the current cluster is centered in the grid, even if movements of the upper limbs occur. In the second step, the bounds of the grid are combined with the spatial divisions in order to define the minimum and maximum grid ranges associated to each 3D cube of the grid. The last step is devoted to place the right points into the right 3D cube. In particular, the current cluster is scanned and each point is put into the right cube based on its 3D geometric coordinates. We finally choose to have four partitions along the x, y, and z axis. This choice is justified by our will of keeping separate the right side of the human body from the left side, while also keeping limited the size of the descriptor. Such a 3D grid is shown in Fig. 2. The final descriptor is obtained by summarizing the flow information contained in each grid cube: the 3D average velocity vector is computed for each cube and all these vectors are concatenated in a column vector.

The 3D grid-based descriptor is calculated frame by frame, for each track k and frame F. For the classification purpose, we collect a sequence of n 3D grid-based descriptors and this sequence represents the final descriptor for the action at issue. Since an action actually represents a sequence of movements over time, considering multiple frames could potentially provides more discriminant information to the recognition task with respect to approaches in which only a single-frame classification is performed. In this work, we set the constant n to 10 and [31] provides a justification to our choice. Since we do not have the a priori knowledge about the action duration and since each different action is characterized by a different temporal duration, interpolation and sampling techniques are used to create the final descriptor. At the end, the final descriptor is normalized and used for training and testing purposes. Normalization enables to partially discard the presence of noise in the raw data.

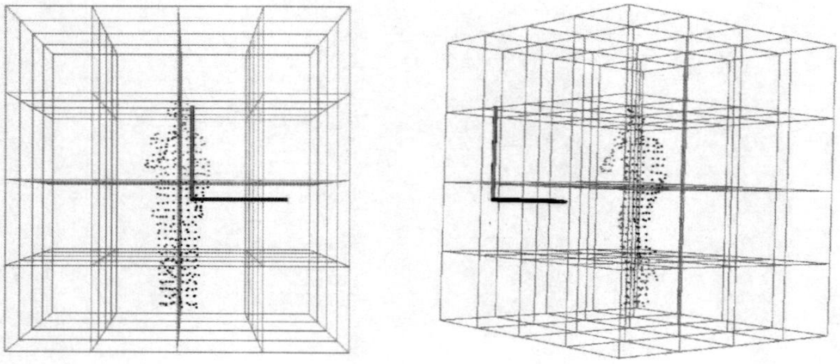

Fig. 2. Two different views of the computed 3D grid: 4 partions along the x, y and z axis are used

5 Experiments

In order to test the recognition performances of our descriptor we exploited a simple 1-Nearest Neighbor classifier. The test phase involves four main steps. In the first step we collect single-frame descriptors in the so called final descriptor until 10 frames are reached. When the 10 frames are exceeded, a window sampling is applied to obtain a 10-frames final descriptor. As a second step, the final descriptor is normalized, while in the third step we compute a distance between the test descriptor and all the training examples in the dataset. We used two types of distance function: the Euclidean distance function and the Mahalanobis distance function. Finally, the label related to the training descriptor that has the shortest distance to the test descriptor is chosen as predicted class.

5.1 Dataset and Setup

Our work is mainly targeted to video surveillance applications. Since no public RGB-D dataset devoted to recognize typical video surveillance actions is currently available, we collected a new RGB-D dataset in a lab environment in order to test our recognition system. The dataset contains six types of human actions: *standing*, *hand waving*, *sitting down*, *getting up*, *pointing*, *walking*. Each action is performed once by six different actors and recorded from the same point of view. We invited the six volunteers to naturally execute the actions, and we gave no indication to them about how to accomplish movements. Each of the segmented video samples spans from about 1 second to 7 seconds.

5.2 Results

This section discusses the experimental results achieved by performing the 1-Nearest Neighbor classification on 10-frames final descriptors. Tests have been executed by following the leave-one-out approach: first we chose an actor from the dataset to use its recordings as test bed, then we trained the classifier with the examples related to the other five subjects. We collected the classification results related to the unseen actor with respect to the training samples. Finally, the process has been performed for each actor in the dataset. Results are provided in the form of confusion matrices and they are shown in Table 1 and Table 2. Table 1 is related to a Nearest Neighbor classification obtained by using the Mahalanobis distance, while Table 2 refers to a Nearest Neighbor classification based on the Euclidean distance computation. We are able to achieve an accuracy of 80% and a precision of 74% for the Euclidean-based classification, while we obtain an accuracy of 78% and a precision of 72% for the Mahalanobis-based classification. We can notice that the Mahalanobis distance led to worse results with respect to the Euclidean distance, suggesting that the number of training examples was not enough for computing reliable means and variances. Moreover, the computation of the covariance matrix and its inverse is costly when dealing with many dimensions. Experiments also show that our application is able to achieve good recognition results for those actions in which the movement of the entire human body is involved (e.g. *getting up* and *walking*), while fairly good performances are obtained from the recognition of actions characterized by the upper limbs motion only (e.g. *hand waving* and *sitting*).

Table 1. Confusion matrix related to a Nearest Neighbor classification obtained by using the Mahalanobis distance. In the matrix: **STA** stands for *standing*, **HAW** stands for *hand waving*, **SIT** stands for *sitting down*, **GET** stands for *getting up*, **POI** stands for *pointing* and finally **WAL** stands for *walking*.

	STA	**HAW**	**SIT**	**GET**	**POI**	**WAL**
STA	**0.67**	0.17			0.17	
HAW	0.17	**0.50**			0.17	
SIT			**0.83**			
GET		0.17		**1.00**		0.17
POI	0.17	0.17	0.17		**0.50**	
WAL						**0.83**

Table 2. Confusion matrix related to a Nearest Neighbor classification obtained by using the Euclidean distance. In the matrix: **STA** stands for *standing*, **HAW** stands for *hand waving*, **SIT** stands for *sitting down*, **GET** stands for *getting up*, **POI** stands for *pointing* and finally **WAL** stands for *walking*.

	STA	**HAW**	**SIT**	**GET**	**POI**	**WAL**
STA	**0.83**	0.33	0.17		0.33	
HAW		**0.50**				
SIT			**0.83**			
GET		0.17		**1.00**		
POI	0.17				**0.67**	
WAL						**1.00**

With regards to temporal performances, we run the application on a notebook equipped with a 2nd generation Intel Core i5 processor characterized by a processor speed that ranges from 2.4 GHz to 3 GHz if the Intel Turbo Boost technology is enabled. On this working station, the application runs in real-time with a medium frame rate of 23 frames per second.

6 Conclusions and Future Work

In this paper, we proposed a method for real-time human action recognition for a cognitive robot endowed with a RGB-D sensor. We focused on the features extraction step and in particular we exploited 3D optical flow information directly extracted from people point clouds to obtain a suitable representation of human actions. To this aim we also proposed a 3D grid-based descriptor to encode the 3D flow information into a single vector. The estimation of the 3D optical flow field proved to be effective to the recognition task with a Nearest Neighbor classifier: we achieved an accuracy of 80% and a precision of 74% on six basic actions performed of a newly collected RGB-D dataset. Furthermore, the application runs in real-time at a medium frame rate of 23 fps.

As future works, we envision to make the descriptor more discriminant by using histograms of 3D flow orientation instead of mean flow orientations. Moreover, we plan to use more sophisticated classifiers and to extend our dataset in order to include more actions, even in presence of partial occlusion.

References

1. Munaro, M., Basso, F., Menegatti, E.: Tracking people withing groups with rgb-d data. In: Proc. of the International Conference on Intelligent Robots and Systems (IROS), Vilamoura, Portugal (2012)
2. Johansson, G.: Visual perception of biological motion and a model for its analysis. Attention, Perception, & Psychophysics 14, 201–211 (1973), 10.3758/BF03212378
3. Quigley, M., Gerkey, B., Conley, K., Faust, J., Foote, T., Leibs, J., Berger, E., Wheeler, R., Ng, A.: Ros: an open-source robot operating system. In: Proceedings of the IEEE International Conference on Robotics and Automation, ICRA (2009)
4. Basso, F., Munaro, M., Michieletto, S., Pagello, E., Menegatti, E.: Fast and Robust Multi-People Tracking from RGB-D Data for a Mobile Robot. In: Lee, S., Cho, H., Yoon, K.-J., Lee, J. (eds.) Intelligent Autonomous Systems 12. AISC, vol. 193, pp. 269–281. Springer, Heidelberg (2012)
5. Carlsson, S., Sullivan, J.: Action recognition by shape matching to key frames. In: IEEE Computer Society Workshop on Models versus Exemplars in Computer Vision (2001)
6. Yilmaz, A., Shah, M.: Actions sketch: a novel action representation. In: IEEE Computer Society Conference on Computer Vision and Pattern Recognition, CVPR 2005, vol. 1, pp. 984–989 (June 2005)
7. Blank, M., Gorelick, L., Shechtman, E., Irani, M., Basri, R.: Actions as space-time shapes. In: Proc. Tenth IEEE Int. Conf. Computer Vision ICCV 2005, vol. 2, pp. 1395–1402 (2005)
8. Rusu, R.B., Bandouch, J., Meier, F., Essa, I.A., Beetz, M.: Human action recognition using global point feature histograms and action shapes. Advanced Robotics 23(14), 1873–1908 (2009)

9. Efros, A.A., Berg, A.C., Mori, G., Malik, J.: Recognizing action at a distance. In: Proceedings of the Ninth IEEE International Conference on Computer Vision, vol. 2, pp. 726–733 (October 2003)
10. Yacoob, Y., Black, M.J.: Parameterized modeling and recognition of activities. In: Sixth International Conference on Computer Vision, pp. 120–127 (January 1998)
11. Ali, S., Shah, M.: Human action recognition in videos using kinematic features and multiple instance learning. IEEE Transactions on Pattern Analysis and Machine Intelligence 32(2), 288–303 (2010)
12. Ke, Y., Sukthankar, R., Hebert, M.: Efficient visual event detection using volumetric features. In: Proc. Tenth IEEE Int. Conf. Computer Vision ICCV 2005, vol. 1, pp. 166–173 (2005)
13. Liu, J., Ali, S., Shah, M.: Recognizing human actions using multiple features. In: IEEE Conference on Computer Vision and Pattern Recognition, CVPR 2008, pp. 1–8 (June 2008)
14. Scovanner, P., Ali, S., Shah, M.: A 3-dimensional sift descriptor and its application to action recognition. In: Proceedings of the 15th International Conference on Multimedia, MULTIMEDIA 2007, pp. 357–360. ACM, New York (2007)
15. Laptev, I., Lindeberg, T.: Space-time interest points. In: Proc. Ninth IEEE Int. Computer Vision Conf., pp. 432–439 (2003)
16. Laptev, I., Marszalek, M., Schmid, C., Rozenfeld, B.: Learning realistic human actions from movies. In: Proc. IEEE Conf. Computer Vision and Pattern Recognition CVPR 2008, pp. 1–8 (2008)
17. Dollar, P., Rabaud, V., Cottrell, G., Belongie, S.: Behavior recognition via sparse spatio-temporal features. In: Proc. 2nd Joint IEEE Int. Visual Surveillance and Performance Evaluation of Tracking and Surveillance Workshop, pp. 65–72 (2005)
18. Schuldt, C., Laptev, I., Caputo, B.: Recognizing human actions: a local svm approach. In: Proc. 17th Int. Conf. Pattern Recognition ICPR 2004, vol. 3, pp. 32–36 (2004)
19. Niebles, J.C., Wang, H., Fei-Fei, L.: Unsupervised learning of human action categories using spatial-temporal words. Int. J. Comput. Vision 79, 299–318 (2008)
20. Kläser, A., Marszałek, M., Schmid, C.: A spatio-temporal descriptor based on 3d-gradients. In: British Machine Vision Conference, pp. 995–1004 (September 2008)
21. Weinland, D., Ronfard, R., Boyer, E.: Free viewpoint action recognition using motion history volumes. Comput. Vis. Image Underst. 104(2), 249–257 (2006)
22. Li, W., Zhang, Z., Liu, Z.: Action recognition based on a bag of 3d points. In: 2010 IEEE Computer Society Conference on Computer Vision and Pattern Recognition Workshops (CVPRW), pp. 9–14 (June 2010)
23. Holte, M.B., Moeslund, T.B.: View invariant gesture recognition using 3d motion primitives. In: IEEE International Conference on Acoustics, Speech and Signal Processing, ICASSP 2008, March 31-April 4, pp. 797–800 (2008)
24. Holte, M.B., Moeslund, T.B., Nikolaidis, N., Pitas, I.: 3d human action recognition for multi-view camera systems. In: 2011 International Conference on 3D Imaging, Modeling, Processing, Visualization and Transmission (3DIMPVT), pp. 342–349 (May 2011)
25. Sung, J., Ponce, C., Selman, B., Saxena, A.: Human activity detection from rgbd images. In: Plan, Activity, and Intent Recognition. AAAI Workshops, vol. WS-11-16. AAAI (2011)
26. Sung, J., Ponce, C., Selman, B., Saxena, A.: Unstructured human activity detection from rgbd images. In: International Conference on Robotics and Automation, ICRA (2012)
27. Yang, X., Tian, Y.: Eigenjoints-based action recognition using naive-bayes-nearest-neighbor. In: IEEE Workshop on CVPR for Human Activity Understanding from 3D Data (2012)
28. Zhang, H., Parker, L.E.: 4-dimensional local spatio-temporal features for human activity recognition. In: 2011 IEEE/RSJ International Conference on Intelligent Robots and Systems (IROS), pp. 2044–2049 (September 2011)

29. Ni, P.B., Wang, G., Moulin, P.: Rgbd-hudaact: A color-depth video database for human daily activity recognition. In: 2011 IEEE International Conference on Computer Vision Workshops (ICCV Workshops), pp. 1147–1153 (November 2011)
30. Popa, M., Koc, A.K., Rothkrantz, L.J.M., Shan, C., Wiggers, P.: Kinect Sensing of Shopping Related Actions. In: Wichert, R., Van Laerhoven, K., Gelissen, J. (eds.) AmI 2011. CCIS, vol. 277, pp. 91–100. Springer, Heidelberg (2012)
31. Schindler, K., van Gool, L.: Action snippets: How many frames does human action recognition require? In: IEEE Conference on Computer Vision and Pattern Recognition, CVPR 2008, pp. 1–8 (June 2008)

Modality in the MGLAIR Architecture

Jonathan P. Bona and Stuart C. Shapiro

University at Buffalo Computer Science and Engineering,
Buffalo, NY, USA
{jpbona,shapiro}@buffalo.edu

Abstract. The MGLAIR cognitive agent architecture includes a general model of modality and support for concurrent multimodal perception and action. It provides afferent and efferent modalities as instantiable objects used in agent implementations. Each modality is defined by a set of properties that govern its use and its integration with reasoning and acting. This paper presents the MGLAIR model of modalities and mechanisms for their use in computational cognitive agents.

1 Introduction

The MGLAIR (Multimodal Grounded Layered Architecture with Integrated Reasoning) cognitive agent architecture extends the GLAIR architecture [10] to include a model of concurrent multimodal perception and action.

Humans constantly sense and act in multiple modalities simultaneously. MGLAIR aims to replicate this functionality in computational agents by including modular afferent and efferent modalities as part of the architecture – without necessarily mimicking the actual implementation of the human mind.

As a cognitive architecture [7], MGLAIR provides a high-level specification of a system in terms of its parts, their functionality, and the interactions between parts. This system may be used as a platform to implement and test models of cognition.

MGLAIR is a layered architecture: the Knowledge Layer (KL) and its subsystems perform conscious reasoning, planning, and acting. Each agent's Sensori-Actuator Layer (SAL) is embodiment-specific and includes low-level controls for its sensori-motor capabilities. The Perceptuo-Motor Layer (PML) connects the mind (KL) to the body (SAL), grounding conscious symbolic representations through perceptual structures. The PML is itself divided into sub-layers.

MGLAIR's Knowledge Layer is implemented in SNePS, which is a logic-based Knowledge Representation and Reasoning system [11] [12]. SNeRE, the SNePS subsystem that handles planning and acting [5], is a key component of MGLAIR. It expands agents' capabilities by connecting logic-based reasoning with acting. The plans formed and actions taken by an agent at any time depend in part on the agent's beliefs (including beliefs about the world based on its perceptions) at that time.

SNeRE includes mental acts (*believing* or *disbelieving* a proposition; *adopting* or *unadopting* a policy), control acts (control structures such as act *sequence* and

A. Chella et al. (Eds.): Biologically Inspired Cognitive Architectures 2012, AISC 196, pp. 75–81.
springerlink.com

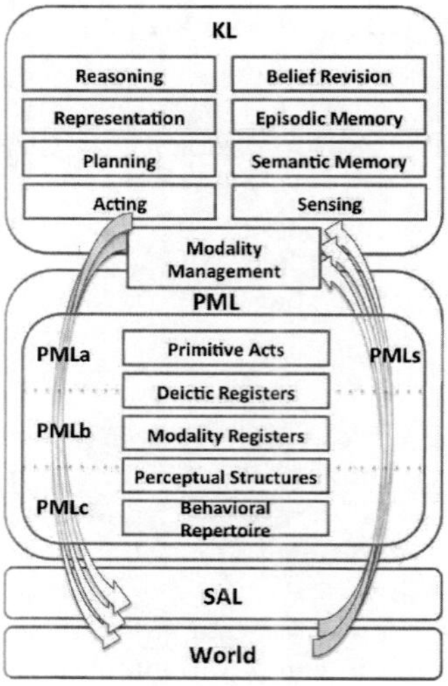

Fig. 1. MGLAIR

selection), and primitive, agent-specific external acts such as moving, sensing, or other interactions with the world. Sensing and acting within MGLAIR involve the flow of data between the agent's conscious mind and its body. When, in the course of reasoning and interacting with the environment, an agent decides to move, it does so by performing actions that utilize the appropriate modality.

Though some existing GLAIR agents use multiple sensory and motor capabilities [8][9], and were initially considered to be "MGLAIR agents," their multiple modalities are implemented *ad hoc* in software and lack an underlying structure and support from, or integration with, the architecture or the agent as a whole.

Of central importance to MGLAIR is its treatment of afferent and efferent *modalities* as instantiable objects that are part of agent implementations. The architecture specifies how modalities are defined and managed, what properties they posses, and how their use is integrated with the rest of the system. Agents using these modalities deal independently with sense data and acts that correspond to distinct capabilities.

2 Modality

Each modality corresponds to a single afferent or efferent capability of an agent: a limited resource capable of implementing only a limited number of related activities.

The set of modalities available to an agent necessarily depends on its embodiment: an agent with motorized wheels and a camera-based vision system moves by consciously performing actions that use its locomotion modality, and consciously perceives the world through the use of its visual modality.

Other architectures (e.g. ACT-R [1], Soar [6], CERA [2]) include processing modules for sensory and motor capabilities/modalities. These are usually fixed in number and in nature, even in architectures such as EPIC [4], which focuses on support for multimodal embodied tasks. While MGLAIR provides default modality definitions for commonly-used modalities, its general model allows agent designers to extend, exclude, or modify these, and to define as many additional modalities as are needed for different agents and embodiments. It provides sensible, customizable, default procedures for the integration of percepts within the knowledge layer.

In MGLAIR, all of an agent's conscious thought takes place in the agent's symbolic language of thought at the KL. MGLAIR's PML structures, with which symbolic knowledge is aligned, constitute unconscious multi-modal representations.

Any action an agent consciously performs is available as an abstract symbolic representation in the knowledge layer, corresponding to low-level motor control commands in the SAL, which it is connected to through alignment with the intermediate PML structures. The flow of distinct types of sensory and motor impulses between the agent's mind and body occur independently, each in its own modality.

An MGLAIR modality possesses a directional data channel that connects the mind to the body. The modality itself handles the transmission and transformation of data in the channel: raw sense data within a modality at the SAL is passed up to the PML and converted to perceptual structures. These structures in the PML are converted to conscious symbolic representations of percepts as they are passed upwards to the KL. The exact nature and implementation of that transformation depends on the type of modality in question and on the agent's embodiment.

The SNePS KL and its connections to the PML are implemented in Common Lisp. MGLAIR does not require the PML itself or the SAL to be implemented using any particular technology: the software implementations of an agent's KL, PML, and SAL may even run on different physical machines.

2.1 Modality Properties

An MGLAIR modality definition is a set of **required** and *optional* properties:

A unique name, by which the agent may refer to the modality;
A modality type, a subtype of either `afferent` or `efferent`;
Knowledge layer predicates for symbolic representation of percepts or acts;
Data channel specification, including connection type and parameters, as well as a **modality buffer specification**, discussed more in Section 2.2;
Conscious access permissions granting the agent knowledge of the modality;
Focus permissions governing focus for the modality (see §2.2);
A conflict handling specification governing conflicting access attempts (§2.4);
A human-readable description of the modality;
Relations to other modalities – e.g. for proprioception, discussed in §2.5.

Each modality's definition is shared among the layers of the architecture.

2.2 Modality Buffers

Each afferent modality data channel includes a buffer of sense data arriving from the SAL via the lower levels of the PML. Sense data are added to the buffer as they are generated, and removed in the same order for processing by the upper levels of the PML. Sense data in the modality buffers are marked with timestamps. Neither the raw sense data nor their timestamps are consciously accessible to the agent, though both can be used within the PML to construct percepts.

To prevent agents from becoming overwhelmed by sense data generated faster than they can be perceived, modality buffers have an optional fixed capacity, which is specified when the modality is defined. A modality with a fixed capacity may handle incoming sense data that would otherwise be placed in its (full) buffer either by refusing and discarding arriving sense data when the buffer is full, or by pushing out old data to make room for the new. Modality buffers allow for either of these possibilities.

Modality buffers without a fixed capacity must specify an expiration interval for data. Old data in a modality buffer that reaches expiration are discarded without being processed/perceived. A buffer may have both a capacity and an expiration interval.

2.3 Modality Focus

An MGLAIR agent may elect to focus more on a particular afferent modality than on others, or to ignore a particular modality altogether. Each distinct modality has its own processes in the PML that operate continually and concurrently to manage the flow and transformation of data/impulses in the modality. Altering a modality's focus increases or decreases the frequency with which its internal processes run.

The ability to govern modality focus is useful when an agent is carrying out a task that relies heavily on perceiving with a subset of its sensory modalities, or when percepts in some modalities may be more urgent than percepts in others. An agent embodied in a robot with cliff detection and speech recognition might focus more on perceiving with its cliff detection modality than on its speech recognition. In this case, and in many others, the less urgent modality may have a much higher volume of sense data than the more urgent one. Without a focusing mechanism, such an agent would dedicate its attention to the modality producing the most percepts.

An agent may alter its focus on a modality by performing one of the following acts with the modality name as its argument: `focus` increases the agent's focus on the modality; `unfocus` decreases focus; `ignore` causes the agent to ignore the modality completely; `attend` causes the agent to stop ignoring the modality, restoring it to the focus level it held before being ignored; and `restore-focus` resets the modality's focus to its starting level.

In order for an agent to consciously act to modify its focus on a modality, the agent must be aware of the modality, and must have a way of referring to it. Not all modalities must be consciously accessible to their agents. Whether a modality is accessible to an agent is a property of the modality, specified in its definition.

By default, all of an agent's modalities have equal focus levels. Each modality definition may include an initial level of focus for the modality. A modality definition may also prohibit the agent from changing a modality's level of focus.

2.4 Sensing and Acting

Rules in the upper layer of each afferent modality transform perceptual data from the PML into conscious knowledge within the KL. This involves converting perceptual structures into propositional terms for what is being perceived. These rules and structures are implemented in PMLs: the PML, sub-layer *s*. Each such rule is specific to its modality and uses predicates and terms that may be unique to the modality to generate the propositional terms that correspond to percepts in that modality.

Though it is up to the agent designer to specify how complex sense data are are translated into conscious percepts, the architecture includes default translations for each afferent modality. These defaults are activated by instantiating and using a modality that does not redefine them. Separate processes for each modality connect the PML and KL, converting sense data into knowledge concurrently as it arrives. This allows the agent to perceive via all of its afferent modalities simultaneously.

Percepts arriving at the KL can be linked to the modality that produced them, allowing the agent to reflect on how it came to believe that something is the case ("I saw it"), and to perform meta-reasoning about its own perceptual capabilities.

When a primitive act is defined it is associated with an efferent modality. A single modality may have associated with it multiple acts that use the same basic bodily capability or resource. For instance, an agent with a grasping effector might have a modality dedicated to its use. Distinct acts that use the modality, such as *positioning the grasper*, *grasping an object*, and *releasing an object*, have distinct conscious representations. When the agent performs acts, the PML and SAL decompose them from abstract symbols (e.g. `grasp`) into low-level commands (e.g. `apply voltage to motor m for n seconds`), and executes those commands.

Each efferent modality can carry out only a single activity at a time. Any desired activity that seems to violate this principle should be broken into component acts in separate modalities for concurrent use. Plans for complex acts are formed by combining other acts, either primitive or complex, via control acts.

If an agent attempts to use a single efferent modality while it is already in use, the conflict is handled in one of several ways: The new impulse may be blocked and discarded; the new impulse may interrupt and discard the current activity, then proceed as normal; the new impulse may interrupt and suspend the current activity, then proceed as normal, in which case the interrupted activity resumes once the interrupter finishes; or the new impulse may be placed in an act queue for that modality and be carried out when the modality is next available. The manner in which such conflicts are handled is a property of each modality, and part of its definition.

2.5 Relations between Modalities

We have implemented a simple model of proprioception for MGLAIR in which an efferent modality is associated with a related afferent modality, allowing the agent to perceive its own actions based on sense data generated in the SAL. For example, an agent that is embodied in a wheeled robot with motors that include a tachometer can use this feature to sense its use of its motors.

The modalities for *locomotion* and *locomotive proprioception* are connected at the PML, where a locomotive proprioception modality register stores perceptual data pertaining to the current state of locomotion. This may include a simple sense corresponding to its just being in use, or may be a more advanced sense based on the position of motors, the duration of the current use, etc. Depending on the application and design of the agent, the agent may be given conscious access to this register. In this case, the agent may simply know whenever its motors are in use because it is able to directly perceive that fact. Alternatively, proprioceptive modality registers may remain hidden in the PML with their contents inaccessible to conscious reasoning. In this case, the sense information they represent can still be used to manage act impulses for that modality within the PML, for instance by blocking new acts in the associated efferent modality while it is still in use at the SAL.

This example of an efferent modality tied to a corresponding afferent modality may serve as a model for the development of more complex inter-modal interactions. A similar approach might join senses from modalities before they are perceived. E.g., an agent with vision and sonar-based object detection might merge sensory data from those two modalities into joint percepts in the PML. This approach might be used to model many of the forms of cross-modal interactions observed in nature [13] [3].

3 Conclusions

The MGLAIR architecture provides a model of afferent and efferent modalities for computational cognitive agents. MGLAIR agents that instantiate these modality objects can use them to sense and act - and make inferences and plans about percepts and actions - concurrently in different modalities.

By dividing agents' capabilities into modular modalities, each with its own properties governing its use, MGLAIR allows agents to sense and act simultaneously using different resources with minimal interference, and to consciously decide which resources to focus on for particular tasks. The resulting agents are more useful than counterparts that can be implemented without such modalities, and more accurately mimic the capabilities of cognitive agents observed in nature.

References

1. Anderson, J.: ACT: A simple theory of complex cognition. American Psychologist 51, 355–365 (1996)
2. Arrabales, R., Ledezma, A., Sanchis, A.: A Cognitive Approach to Multimodal Attention. Journal of Physical Agents 3(1), 53 (2009)
3. Calvert, G., Spence, C., Stein, B. (eds.): The Handbook of Multisensory Processes. MIT Press (2004)
4. Kieras, D., Meyer, D.: An overview of the EPIC architecture for cognition and performance with application to human-computer interaction. Human-Computer Interaction 12(4), 391–438 (1997)
5. Kumar, D.: A unified model of acting and inference. In: Nunamaker Jr., J.F., Sprague Jr., R.H. (eds.) Proceedings of the Twenty-Sixth Hawaii International Conference on System Sciences, vol. 3, pp. 483–492. IEEE Computer Society Press, Los Alamitos (1993)
6. Laird, J.E., Newell, A., Rosenbloom, P.S.: Soar: an architecture for general intelligence. Artif. Intell. 33(1), 1–64 (1987)
7. Langley, P., Laird, J., Rogers, S.: Cognitive architectures: Research issues and challenges. Cognitive Systems Research 10(2), 141–160 (2009)
8. Shapiro, S., Anstey, J., Pape, D., Nayak, T., Kandefer, M., Telhan, O.: MGLAIR agents in virtual and other graphical environments. In: Proceedings of the National Conference on Artificial Intelligence, p. 1704. AAAI Press, MIT Press, Menlo Park, Cambridge (1999/2005)
9. Shapiro, S.C., Anstey, J., Pape, D.E., Nayak, T.D., Kandefer, M., Telhan, O.: The Trial The Trail, Act 3: A virtual reality drama using intelligent agents. In: Young, R.M., Laird, J. (eds.) Proceedings of the First Annual Artificial Intelligence and Interactive Digital Entertainment Conference (AIIDE 2005), pp. 157–158. AAAI Press, Menlo Park (2005)
10. Shapiro, S.C., Bona, J.P.: The GLAIR Cognitive Architecture. International Journal of Machine Consciousness 2(2), 307–332 (2010)
11. Shapiro, S.C., Rapaport, W.J.: The SNePS family. Computers & Mathematics with Applications 23(2-5), 243–275 (1992)
12. S. C. Shapiro and The SNePS Implementation Group. SNePS 2.7 User's Manual. Department of Computer Science and Engineering, University at Buffalo, The State University of New York, Buffalo, NY (2007), `http://www.cse.buffalo.edu/sneps/Manuals/manual27.pdf`
13. Stein, B., Meredith, M.: The Merging of the Senses. The MIT Press (1993)

Robotics and Virtual Worlds: An Experiential Learning Lab

Barbara Caci[1], Antonella D'Amico[1], and Giuseppe Chiazzese[2]

[1] Dipartimento di Psicologia, Università degli Studi di Palermo, Viale delle Scienze, Ed. 15. 90128 Palermo - Italy
[2] Istituto per le Tecnologie Didattiche di Palermo, Consiglio Nazionale delle Ricerche, Via Ugo La Malfa, 153, 90146 Palermo - Italy
{barbara.caci,antonella.damico}@unipa.it,
giuseppe.chiazzese@itd.cnr.it

Abstract. Aim of the study was to investigate the cognitive processes involved and stimulated by educational robotics (LEGO® robots and Kodu Game Lab) in lower secondary school students. Results showed that LEGO® and KGL artifacts involve specific cognitive and academic skills. In particular the use of LEGO® is related to deductive reasoning, speed of processing visual targets, reading comprehension and geometrical problem solving; the use of KGL is related to visual-spatial working memory, updating skills and reading comprehension. Both technologies, moreover, are effective in the improvement of visual-spatial working memory. Implications for Human-Robot Interaction and BICA challenge are discussed.

Keywords: Educational robotics, cognitive skills, academic performance, human-robot interaction.

1 Introduction

Educational robotics is one of the most challenging research area in the domain of Human-Robot Interaction (HRI), and is eligible as one of the best practical domain for the BICA challenge [1; 2]. In the constructivist framework [3], robotic and virtual interfaces are considered powerful tools for learning concepts about Mathematics, Computer programming, and Physics [4], for improving visual-constructive abilities, reasoning and problem-solving skills [5; 6] and for enhancing narrative and paradigmatic thinking [7]. Both LEGO® robotic kits and KGL allow children to build robots or agents able to act goal-oriented behaviors, such as to direct themselves toward a light, avoid obstacles or move inside a maze, and so on.

Using LEGO® robotic kits, children build the body of the robot assembling motors and sensors (e.g., touch, ultrasonic, or light) with the brain of robot, a programmable microcontroller-based brick (NTX). Then, they may program the mind of the robot using an object-based interface inspired to Logo, and based on *if-then* rules. However, LEGO® kits are limited in the kind of behaviors that subjects can design and program.

A. Chella et al. (Eds.): Biologically Inspired Cognitive Architectures 2012, AISC 196, pp. 83–87.
springerlink.com

A more extensive programming practice is offered by the recent Kodu Game Lab (KGL) [8], a 3D virtual stage for the creation of virtual worlds, in which children can define more complex behaviors, movements and interactions with characters and objects using similar programming rules [9]. Using KGL children may design virtual environments, defining terrain scenarios and adding mountain, volcanoes, lakes, and so on. Then, they may enrich the environment by adding funny characters (e.g. Kodu, cycle, boat, fish), environmental elements (trees, apples, rocks clouds), paths (streets, enclosures, walls) and city elements (houses, factories, huts, castles) and, finally, they may assign specific behaviors to some of the elements, generating an interactive virtual world. Although the literature about KGL is quite recent, first experiments on use of Kodu have shown that its visual programming language is particularly easy to use also for novice students [10], compared to textual language models used by Alice [11], Greenfoot [11] and Scratch [13].

Starting from this theoretical framework, an experiential learning laboratory (32-hours) involving children was designed with a twofold goal: to study some of the cognitive and academic abilities involved in building and programming robots/agents; to measure the effectiveness of the laboratory in the enhancement of the same cognitive and academic skills.

2 Method

2.1 Participants

The study involved 59 students of 11 years of age, attending an Italian Secondary School, that were casually assigned to experimental (F =14, M=22) and control condition (F=15, M=18). Children of the experimental group (EG) followed the laboratory described below; children of the control group (CG) followed the regular school activities.

2.2 Materials and Procedures

A pre-post test design was adopted. During the pre-post test phases the cognitive abilities and academic performances of EG and CG were measured using:

- eight syllogistic and conditional reasoning tasks, drawn from the Automated System for the Evaluation of Operator Intelligence [14].
- the PML working memory battery [15], aimed at measuring phonological and visual-spatial working memory, executive functions (shifting, updating, inhibition), rate of access to long term memory and speed of processing.
- the MT Reading Comprehension tasks [16] requiring children to read a narrative and an informative passage and to answer to 15 multiple-choice questions for each passage.
- two arithmetical problem solving tasks and one geometrical problem solving task.

Moreover, during laboratory the programming skills of children were assessed using two questionnaires (LEGO® -Q and KGL-Q) respectively aimed at examining the acquisition of concepts about fundamentals of LEGO® and KGL interfaces, and their programming rules.

2.3 The Laboratory

The laboratory consisted of eight four–hours sessions, performed during the curricular school time, and described below:

1. *Familiarization with Robots* – movies describing different kind of fantasy and real robot (e.g. Transformers, AIBO, RunBot) were presented to children, ad used as a frame to start a circle time discussion about scientific concepts related to Biology, Physics and Mathematics (e.g., the notion of brain, mind, sensory-motor apparatus, velocity, light spectrum).
2. *Construction of LEGO® robots* – In order to let children familiarize with *LEGO®* robotic kit, and to build a small-mobile robot following the instruction provided by the manual.
3. *Programming training with LEGO® and creation of the arena* – The programming interface was introduced to children. Then, they created a narrative scenario (e.g. a script) for the robot behavior and built a physical environment (i.e. the arena) using pasteboard, colors, modeling paste and other materials.
4. *Programming session with LEGO® and verification* - Children realized the programming algorithms for adapting the robot to the created environment. Finally, they completed the *LEGO®*-Q questionnaire about the acquisition of *LEGO®* programming skills.
5. *Familiarization with KGL environment* – Children acquired the functionality of the Xbox controller for interacting with KGL interface and explored some illustrative virtual worlds. Next, they designed and realized their own environment, choosing terrains, mountains, lakes, and specific characters.
6. *Familiarization with tile-based visual programming of KGL* - Children were trained to program the virtual world using appropriate rules (a rule in KGL is composed by a *when-do* tile).
7. *Construction of the virtual world and programming of virtual agents*- Children reproduced the narrative scenario previously realized with LEGO® (session 3), enriching it with all the elements and additional features available in KGL.
8. *Programming session with KGL and verification* – Children were involved in collaborative programming sessions, and were then presented with the KLG-Q questionnaire aimed at verifying the acquisition of KLG programming skills

3 Results

A correlation analysis was performed with the aim to explore the relationships among the performance in cognitive or academic tests used in the pretest phase, and the results obtained by children in *LEGO®*-Q and KLG-Q. Results at Pearson's product-moment correlation tests showed that deductive reasoning skills ($r=.72$; $p<.01$) and speed of processing visual targets are related with *LEGO*®-Q scores ($r=.45$; $p<.05$). Scores in Visual-spatial working memory ($r=.49$; $p<.05$) and in Updating executive function ($r=.46$; $p<.05$) are associated with KGL-Q scores. Reading Comprehension scores are correlated both with *LEGO*®-Q ($r=.66$; $p<.01$) and KGL-Q scores ($r=.48$; $p<.05$), whereas Geometry problem-solving scores are related with *LEGO*®-Q scores ($r=.60$; $p<.01$).

In order to examine the effectiveness of laboratorial activities for the improvement of cognitive and academic skills, a series of 2X2 repeated univariate ANOVA with two levels of the *between-subject* factor Group (EG and CG) and two levels of the *within-subject* factor Time (Pre-test and Post-test) was performed on the individual scores obtained by children in experimental and control group, in each of the cognitive and academic tests. An interaction Group x Time revealed that children in EG improved significantly their visual-spatial working memory score compared to the CG ($F_{2\text{-}43}=4.05$; $p<.005$).

4 Discussion and Conclusion

Results corroborate the importance of educational robotics in the framework of HRI. Indeed, as already suggested in other studies, we demonstrated that LEGO® and KGL involve specific cognitive and academic skills and may be effective in the improvement of visual-spatial working memory. Moreover, using KGL software, we gave children the opportunity to create a more complex and rich virtual world, overcoming some of the limits of using concrete artifacts and real environments. By this way, children metaphorically experienced the passage from a concrete phase of thinking-with-objects to an abstract phase of thinking-with-virtual-objects or representations.

Experiential robotic labs like this show advantages both for educational purposes and for enhancing HRI: by discovering the way robots work, children reason about and stimulate their own basic level cognitive abilities. By contextualizing the simple and narrow behaviors of robots in a narrative context, children start to consider robot as parts of human life, as agents experiencing emotions and motivations. In this sense, such educational labs may represent a step on the road to the BICA challenge.

Acknowledgments. The research is included in the Project "Edutainment: education and entertainment in experiential learning labs" funded by Assessorato Regionale Istruzione e Formazione Professionale, Sicilia and promoted by MIUR, Agenzia Nazionale per lo Sviluppo dell'Autonomia Scolastica, Sicilia (ANSAS). Others partner are: Istituto Comprensivo Buonarroti, Palermo; Children's museum BIMPA, Palermo.

References

1. Samsonovich, A.V.: On a road map for BICA challenges. In: Biologically Inspired Cognitive Architectures, vol. 1, pp. 100–107 (July 2012)
2. Chella, A., Lebiere, C., Noelle, D.C., Samsonovich, A.V.: On a roadmap to biologically inspired cognitive agents. In: Samsonovich, A.V., Jóhannsdóttir, K. (eds.) Biologically Inspired Cognitive Architectures 2011. Frontiers in Artificial Intelligence and Applications, vol. 233, pp. 453–460. IOS Press, Amsterdam (2011)

3. Harel, I., Papert, S.: Constructionism. Ablex Publishing Corporation, Norwood (1991)
4. Resnick, M., Martin, F., Sargent, R., Silverman, B.: Programmable bricks: Toys to think with. IBM Systems Journal 35(3&4) (1996)
5. Caci, B., D'Amico, A.: Children's cognitive abilities in construction and programming robots. In: Proceedings of Roman 2002, Berlino, Settembre 23-26 (2002)
6. Caci, B., D'Amico, A., Cardaci, M.: New frontiers for psychology and education: robotics. Psychological Reports 94, 1327–1374 (2004)
7. Caci, B., D'Amico, A., Cardaci, M.: La robotica educativa come strumento di apprendimento e creatività. Form@re (53) (2007)
8. Stolee, K.T., Fristoe, T.: Expressing computer science concepts through Kodu game lab. In: Proceedings of the 42nd ACM Technical Symposium on Computer Science Education-SIGCSE 2011, Dallas, TX, USA (2011)
9. Fowler, A., Fristoe, T., MacLauren, M.: Kodu Game Lab: a programming environment. The Computer Games Journal 3(1), 17–28 (2011)
10. Chiazzese, G., Laganà, M.R.: Online Learning with Virtual Puppetry. Journal of e-Learning and Knowledge Society 7(3), 121–129 (2010)
11. Cooper, S., Dann, W., Pausch, R.: Alice: a 3-d tool for introductory programming concepts. In: Northeastern Conference on the Journal of Computing in Small Colleges, pp. 107–116 (2000)
12. Greenfoot, `http://www.greenfoot.org/` (retrieved July 1, 2012)
13. Maloney, J., Burd, L., Kafai, Y., Rusk, N., Silverman, B., Resnick, M.: Scratch: A sneak preview. In: C5 2004: Conference on Creating, Connecting and Collaborating through Computing, pp. 104–109 (2004)
14. D'Amico, A., Cardaci, M., Guarnera, M.: S.A.V.I.O.: Sistema Automatizzato di Valutazione dell'Intelligenza Operatoria. Descrizione e Sperimentazione pilota dello strumento. Bollettino di Psicologia Applicata 235 (2002)
15. D'Amico, A., Lipari, C.: La batteria PML per la misurazione della memoria di Lavoro. Firera e Liuzzo editore (2012)
16. Cornoldi, C., Colpo, G.: Nuove Prove di Lettura MT per la Scuola Media Inferiore. O.S. Organizzazioni Speciali, Firenze (1995)

Comprehensive Uncertainty Management in MDPs

Vincenzo Cannella, Roberto Pirrone, and Antonio Chella

Dipartimento di Ingegneria Chimica, Gestionale, Informatica, Meccanica,
University of Palermo, Viale delle Scienze, Edificio 6, Palermo, Italy
{vincenzo.cannella26,pirrone,chella}@unipa.it

Abstract. Multistage decision-making in robots involved in real-world tasks is a process affected by uncertainty. The effects of the agent's actions in a physical environment cannot be always predicted deterministically and in a precise manner. Moreover, observing the environment can be a too onerous for a robot, hence not continuos. Markov Decision Processes (MDPs) are a well-known solution inspired to the classic probabilistic approach for managing uncertainty. On the other hand, including fuzzy logics and possibility theory has widened uncertainty representation. Probability, possibility, fuzzy logics, and epistemic belief allow treating different and not always superimposable facets of uncertainty. This paper presents a new extended version of MDP, designed for managing all these kinds of uncertainty together to describe transitions between multi-valued fuzzy states. The motivation of this work is the design of robots that can be used to make decisions over time in an unpredictable environment. The model is described in detail along with its computational solution.

1 Introduction

Observing the environment can be a too onerous for a robot. A common example is offered by vision, a task usually carried out by very expensive routines. In these cases, the perception can not be continous. On the contrary, it is carried out at regular intervals, during which the perception of the environment is absent. In these cases, the robot has to interact with the world without an adequate perception of the environemt. This problem is often treated through a stochastic approach. An example is found in [21]. This paper introduces an extended version of the classical Markov Decision processes able to manage delays in perceptions. As a matter of fact, the robot has to take into account the inherently uncertain nature of this task. Moreover, an agent interacting with the world could suffer the limitations due to the imprecisions of its actuators. This can be true in particular for robots. In general, robots deal with uncertainty at different levels. The environment or the goal can be imprecise; the agent can not foresee the effects of its actions on the environment; agent's resources may be inadequate to compute a good strategy. Traditionally, uncertainty has been modeled through the probability theory. In this paper we propose a wider definition of the concept

A. Chella et al. (Eds.): Biologically Inspired Cognitive Architectures 2012, AISC 196, pp. 89–94.
springerlink.com

of uncertainty. After the first formulations based on probability, the concept of uncertainty has been widened including possibility. Fuzzy logics was the first theoretical background for the definition of possibility given by Zadeh. According to a classical view uncertainty was treated through the Probability Theory. At this time, uncertainty can be treated with the Possibility Theory and Fuzzy Logics. The term "Possibility Theory" [8] was coined by Zadeh to express the intrinsic fuzziness of natural languages as well as uncertainty information. This was the premise for defining the concept of *possibility*, which can be related to fuzzy sets. According to Zadeh, associating an uncertain quantity to a fuzzy set induces for this quantity a possibility distribution, representing the information about the values this quantity can assume. Another theory proposed to express uncertainty is the so-called Evidence Theory (or Dempster-Shafer Theory), or the theory of belief functions. It is a generalization of the Bayesian theory of subjective probability. Evidence Theory is a mathematical tool for combining empirical evidences to construct a coherent picture of reality [15]. Spohns theory of epistemic beliefs (EBs) [17], also referred to as kappa calculus, is viewed as a qualitative counterpart of Bayesian probability theory. In same cases, it is also referred as rank-based system and qualitative probabilitiy. This theory is a calculus designed to represent and reason with plain human beliefs. It dexcribes a formalism to represent plain EBs and procedures for revising beliefs when new information is obtained. According to it an epistemic state is represented functionally through the so-called disbelief function. This function can be revised in light of new information on the basis of ad hoc rules. A disbelief function, very similar to a probability distribution function, is specified by its values for the singleton subsets of configurations of the variable. In turn, the theoretical findings outlined above allowed researchers to devise new models for agents dealing with uncertainty. In general, these are possibilistic or fuzzy extensions of their previous versions inspired to the probabilistic approach. Examples of fuzzy or possibilistic (PO)MDPs can be found in [1] and in [5]. In this work an extended version of MDP is presented, which manages the different uncertainty models in a unified framework. Transitions values can be instantiated seamlessly in terms of either probability, possibility or EB. Moreover the model allows to describe the states as multi-valued fuzzy variables. The action strategy is computed using a reformulation of the backward induction algorithm that takes into account *believability* values i.e. both probability, possibility, and EB. The rest of the paper is arranged as follows. In the next section we propose a unified management of uncertainty. Next, the proposed model is detailed. Finally, the last section reports conclusions, and future work.

2 A Unified Approach to Uncertain Reward

The three kinds of uncertainty mentioned above share many different aspects. Many authors have proposed a unified approach in many different manners. For this reason we introduce the notion of believability, as a generalization of all the three different concepts. In our model, we called *believability* the generic uncertain quantity u that can be unified with either probability, possibility, or EB. This quantity can be referred to as a fuzzy value $a \in A$. Beleivability can be joined to a believability distribution $u(a)$, whose definition is similar to the definition of the probability, or possibility, distribution. We introduce also the conditional believability distribution $u(b|a)$ [3] for all fuzzy values $a \in A$ and $b \in B$, as the degree to which it is certain that the element $b \in B$ appears, knowing that the element a is true. A and B are independent if and only if $u(b|a) = u(b)$. The joint believability distribution $u(a, b)$ for all fuzzy values $a \in A$ and $b \in B$, is the degree to which it is certain that the element $a \in A$ and $b \in B$ are both true. In all these definitions the u function is a measure of the believability degree of its arguments. These functions can be manipulated through a set of operators. In this work we focus on $\oplus_b$, the believability accumulation operator, and $\odot_b$, the believability combination operator. The first operator sums the effects of new information to the previous knowledge. The latter one represents how new pieces of informations interact together. For compactness, we define the operator

$$\bigoplus_{a \in A} {}_b\, u(a) = u(a_1) \oplus_b \cdots \oplus_b u(a_n) \tag{1}$$

with $A = \{a_1, \cdots, a_n\}$ Given the joint believability distribution $u(a, b)$, the marginal believability distribution over A is defined by

$$u_1(a) = \underset{\oplus_b b \in B}{u}(a, b) \forall a \in A \tag{2}$$

and the marginal believability distribution over B is given by

$$u_2(b) = \underset{\oplus a \in A}{u}(a, b) \forall b \in B \tag{3}$$

To solve a MDP based on believability, we must be able to manage the uncertainty of the expected reward from a policy. Reward can also be managed in a unified manner through an ad hoc gain accumulation operator $\oplus_g$. The gain associated to the agent's next state is not sure, because the agent cannot be sure of the state itself. The expected reward over all next states $s \in S$ can be computed as a weighted combination of their believabilities $ub(s)$. To this aim we introduce the gain combination operator $\odot_g$. The expected reward ER in a state s can be computed with the equation:

$$ER = \underset{\oplus_g s \in S}{[} \odot_g (r(s, a), ub(s))] \tag{4}$$

Here S is the state space and $r(s, a)$ is the reward obtained if the agents performs the action a when being in the state s. As proposed in [18], defined operators can be instanziated for each distinct kind of managed uncertainty. In table 1 there is a collection of the most common operators. To combine two or more kinds of uncertainty together, mixed operators have to be defined. In particular, possibility has been deeply investigated in the past, exploiting the relation that exist between possibility, probability, and fuzzy. Many approaches and criteria to convert probability into possibility and vice versa have been defined [8], [12], [2], [4],[10], [20].

Table 1. Operators defined for different common kinds of uncertainties: probability, possibility, and Spohn's epistemic belief

$\oplus_b$	$\odot_b$	$\oplus_g$	$\odot_g$
$+$	$*$	$+$	$*$
max	min	min	$max(1 - believability, gain)$
min	$+$	$+$	$+$

3 A Unified Model of MDP

In this paper we present a unified model of MDP able to manage at the same time many kinds of uncertainty togetehr. It is defined as a tuple $\{S, A, T, r, \oplus_b, \odot_b, \oplus_g, \odot_g\}$ where:

- S represents a finite set of states
- A represents a finite set of actions
- $B_T : S \times A \times S \Rightarrow \Pi(S)$ is the state transition function where $B_T(s' \mid s, a)$ is the conditional believability of moving to state s' when the action a has been executed in the state s.
- $r : S \times A \Rightarrow R$ is the reward function and $r(s, a)$ is the expected reward for taking an action a when the agent is in the state s.
- $\oplus_b, \odot_b, \oplus_g, \odot_g$: the operators used to manage the uncertainty and reward.

Procedure 1. Calculate an optimal policy

Input: $S, B_T, ub, \oplus_b, \odot_b, \oplus_g, \odot_g$
  pol = []; {define the policy var}
  for all $s \in S$ **do**
    $V^*_{N+1}(s) = 0$
  end for
  for $i = N \rightarrow 1$ **do**
    for all $s \in S$ **do**
      $V^*_t(s) = max_{a \in A} \oplus_g [r(s, a), \oplus_g[\underset{\odot_g s' \in S}{(}\; ub_a(s), V^*_{t+1}(s'))]]$ {$V^*_t(s)$ is the Bellman equation. At each step, the partial reward is computed from the previous computed one over all the states and all the actions.}
      $a^*_t(s) = \arg\max_{a \in A} \oplus_g[r(s, a), \oplus_g[\underset{\odot_g s' \in S}{(}\; ub_a(s), V^*_{t+1}(s'))]]$ {At each step the best action $a^*_t(s)$ in the state s maximizes the partial reward.}
    end for
  end for
Output: pol, V^*

Each state $s \in S$ can be joined to an array of F_s fuzzy features $A_1, A_2, \ldots, A_{F_s}$. Each feature is characterized by a membership function $\mu_i(A_i)$. Uncertainty in the state is modeled by the *uncertain belief state* $Ub = [ub(s_1), \ldots, ub(s_{|S|})]^T$ that is the believability vector on the state space S. At each step, the system updates $ub_a(s)$ that is the believability of reaching state s after acting a:

$$Ub_a(s') = \underset{\oplus_b s \in S}{[} \odot_b (B_T(s'|s,a), Ub(s))] \quad (5)$$

The agent plans its actions on the basis of a reward function. At each step, the amount of the total gain is updated by adding the new gain to the previous one but also other solutions are possible. A plan can be found by adopting the backward induction algorithm, one of the most known algorithm for MDP. The algorith is described in the Procedure 1.

4 Conclusions and Future Work

The presented extended MDP model is able to deal with different uncertainty representations in a unified framework. It can treat all those cases of uncertainty suffered by robots for an inadequate perception of the environment, or imprecise actuators. The MDP can be instantiated seamlessly using all different kinds of transition uncertainties together. The backward induction algorithm has been reformulated also to cope with such uncertainty representations. Finally, states are described as arrays of fuzzy values. At this moment, an intense experimentation of the model is carried out. By now, in the proposed formulation the agent access to the environment without errors. Future work will be devoted to overcome this limitation, and to propose an extended version of POMDPs. In this way, the model will cover the uncertainty related to perceptual aspects in the interaction with a physical environment. Suitable strategies will be studied for enabling the agent either to choice the best description of the environment or to adopt a mixed model.

References

1. Pardo, M.J., de la Fuente, D.: Design of a fuzzy finite capacity queuing model based on the degree of customer satisfaction: Analysis and fuzzy optimization. Fuzzy Sets and Syst. 159(24), 3313–3332 (2008)
2. Klir, G.J., Parviz, B.: Probability-possibility transformations: A comparison. Int. J. Gen. Syst. 21(3), 291–310 (1992)
3. Gwét, H.: Normalized conditional possibility distributions and informational connection between fuzzy variables. Int. J. Uncertain, Fuzziness and Know-Based Syst. 5(2), 177–198 (1997)
4. Jumarie, G.: Possibility-probability transformation: A new result via information theory of deterministic functions. Kybernetes: Int. J. Syst. Cybernetics 23(5), 56–59 (1994)
5. Sabbadin, R., Fargier, H., Lang, J.: Towards qualitative approaches to multi-stage decision making. Int. J. Approx. Reason. 19(3-4), 441–471 (1998)
6. Zhou, Z.J., Hu, C.H., Zhang, B.C., Chen, L.: An improved fuzzy kalman filter for state estimation of nonlinear systems. J. Phys: Conference Series 96(1), 012130 (2008)

7. Klir, G.: Principles of uncertainty: What are they? why do we need them? Fuzzy Sets Syst. 74(1), 15–31 (1995)
8. Zadeh, L.A.: Fuzzy sets as a basis for a theory of possibility. Fuzzy Sets Syst. 100(suppl. 1(0)), 9–34 (1999)
9. Matía, F., Jiménez, A., Al-Hadithi, B.M., Rodríguez-Losada, D., Galán, R.: The fuzzy kalman filter: State estimation using possibilistic techniques. Fuzzy Sets Syst. 157(16), 2145–2170 (2006)
10. Wonneberger, S.: Generalization of an invertible mapping between probability and possibility. Fuzzy Sets Syst. 64, 229–240 (1994)
11. Oussalah, M., De Schutter, J.: Possibilistic kalman filtering for radar 2d tracking. Inf. Sci. 130, 85–107 (2000)
12. Dubois, D., Prade, H.: Unfair coins and necessity measures: Towards a possibilistic interpretation of histograms. Fuzzy Sets Syst. 10, 15–20 (1983)
13. Chen, G., Xie, Q., Shieh, L.S.: Fuzzy kalman filtering. Inf. Sci. 109, 197–209 (1998)
14. Klir, G., Folger, T.: Fuzzy sets, uncertainty, and information. Prentice Hall (1988)
15. Shafer, G.: A Mathematical Theory of Evidence. Princeton University Press, Princeton (1976)
16. Mohamed, M., Gader, P.: Generalized hidden markov models. part i. theoretical frameworks. IEEE Transactions on Fuzzy Syst. 8(1), 67–81 (2000)
17. Giang, P.H., Shenoy, P.P.: A qualitative linear utility theory for Spohns theory of epistemic beliefs. In: UAI, pp. 220–229. Morgan Kaufmann (2000)
18. Dubois, D., Prade, H.: Possibility theory as a basis for qualitative decision theory. In: Proceedings of the 14th International Joint Conference on Artificial Intelligence, vol. 2, pp. 1924–1930. Morgan Kaufmann Publishers Inc., San Francisco (1995)
19. Ghorbani, A.A., Xu, X.: A fuzzy markov model approach for predicting user navigation. In: IEEE/WIC/ACM International Conference on Web Intelligence, pp. 307–311 (2007)
20. Yamada, K.: Probability-possibility transformation based on evidence theory. In: Joint 9th IFSA World Congress and 20th NAFIPS International Conference, vol. 1, pp. 70–75 (2001)
21. Hsu, K., Marcus, S.I.: Decentralized control of finite state Markov processes. IEEE Trans. Auto. Control AC-27(2), 426–431 (1982)

A New Humanoid Architecture for Social Interaction between Human and a Robot Expressing Human-Like Emotions Using an Android Mobile Device as Interface

Antonio Chella[1], Rosario Sorbello[1], Giovanni Pilato[2], Giorgio Vassallo[1], and Marcello Giardina[1]

[1] DICGIM, RoboticsLab, Universitá degli Studi di Palermo, Viale delle Scienze, 90128 Palermo, Italy
[2] ICAR, CNR Viale delle Scienze, 90128, Palermo, Italy
rosario.sorbello@unipa.it

Abstract. In this paper we illustrate a humanoid robot able to interact socially and naturally with a human by expressing human-like body emotions. The emotional architecture of this robot is based on an emotional conceptual space generated using the paradigm of Latent Semantic Analysis. The robot generates its overall affective behavior (Latent Semantic Behavior) taking into account the visual and phrasal stimuli of human user, the environment and its personality, all encoded in his emotional conceptual space. The robot determines its emotion according by all these parameters that influence and orient the generation of his behavior not predictable from the user. The goal of this approach is to obtain an affinity matching with humans. The robot exhibit a smoothly natural transition in his emotion changes during the interaction with humans taking also into account the previous generated emotions. To validate the system, we implemented the distribute system on an Aldebaran NAO small humanoid robot and on a Android Phone HTC and we tested this social emotional interaction using the phone device as intelligent interface between human and robot in a complex scenario.

Keywords: Humanoid Robot, Cognitive Architecture, Emotions, Google Android.

1 Introduction

The Emotional competence is one of main key of the minimal agency of a robot where the Biologically Inspired Cognitive Architecture (BICA) community wants to invest their efforts [1].

An emotional BICA agent need to exibit the capability [2] to capture and feels about human emotions during interaction with a person if we want to accept

A. Chella et al. (Eds.): Biologically Inspired Cognitive Architectures 2012, AISC 196, pp. 95–103.
springerlink.com

him in people society[3][4]. Emotional equipment is nowadays a key condition of minimal agency for an artificial agent like a humanoid robot[5].

Emotions [6],[7] improves the capability of humanoid robots[8] and allow their behaviors [9]to be effective in the interaction with humans[10].

We aim at equipping a humanoid robot Nao with emotional competence[11] during their interaction with humans. Many studies have shown the importance of emotion [12],[13] in the field of human-humanoid interaction (HHI)[14],[15],[16]. The proposed emotional cognitive architecture of humanoid robot is based on the creation of a probabilistic emotional conceptual space automatically induced from data. The approach is based on the application of the paradigm of Latent Semantic Analysis whose procedure it has been widely explained in Chella et al. 2008 [17]. The architecture of the presented system is inspired to the layered approach illustrated in Chella et al. 1998 [18]. This layered architecture is organized in three main areas: the sub-conceptual area, the emotional area and the behavioral area.

We have been conducted different and etherogenus experiments using the humanoid robot Nao in order to test the cognitive emotional capabilities of the robot by reproducing various emotive situations: storyteller robot [19], robot steward[20], entertainment robot[21] and emotional robot [22] using a mobile phone.

The paper is structured as follows: in section 2 the cognitive architecture of the humanoid robot is described; in section 3 the coding of emotions events and personality of the robot is illustrated; finally in last section 4 the experimental results conducted with the humanoid robot Nao Aldebaran using the new architecture are shown.

2 The Cognitive Architecture

The paper illustrates an approach which is the evolution of what presented in [23],[22]. The key ideas are to code in some manner a "personality" to a robot in order to make the interaction between humans and robots amusing and neither mechanical nor deterministic, avoiding a predictable and trivial behavior. The personality of the robot is coded in a semantic space which is built starting form a matrix which encodes occurrences of "events". An event is anything that can happen in a defined time-slot, including feeling something. The matrix encodes the temporal sequence of events: if event "i" temporally precedes event "j", then the $ij-th$ element of the matrix is increased. This makes the event matrix non-symmetric and by construction. The way the event matrix is built determines the robot personality an its reaction to events. The event matrix is decomposed by using the TSVD algorithm. This determines the set of vectors representing the $i-th$ event as two vectors: the $i-th$ event as happening before another event "j" or the $i-th$ event as a consequence of an event "j". In order to take into account the past history of events, an average vector that is resulting form a time frame is considered. This choice makes it possible to determine the most probable emotion that is triggered in the robot by his past history according to its personality "hardwired" in his event matrix which constitutes its reaction to events.

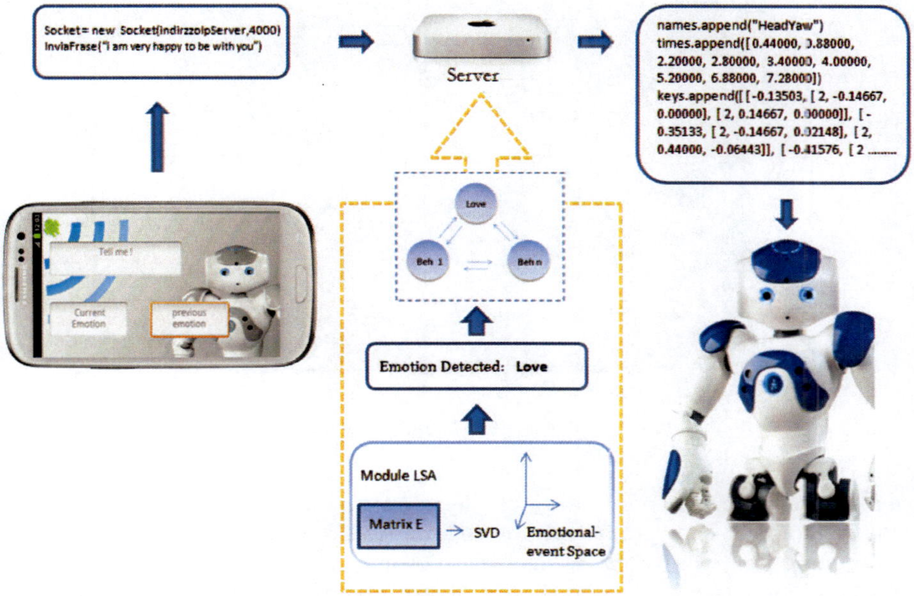

Fig. 1. The overall schema of the proposed architecture

The Architecture, as shown in Fig. 1 uses an Android mobile phone as user-friendly interface available to human for the emotional interaction with robot. The Architecture as described in details in [22] is organized in three main areas: perceptual, emotional, and behavioral. The first area processes perceptual data coming from the sensors. The second area is the "conceptual space of events - emotional states" which constitutes the sub-symbolic representation of emotions, events, personality. The third area activates a latent semantic behavior (LSB) related to the humanoid emotional state.

3 Sub-symbolic Personality and Event Reaction Coding

Let W the set of all M events that can occur in our scenario, with the term event we mean anything that can happen during a period of time in the scenario. This includes also words that can be pronounced or words that describe things or that express a feeling. Let W_{em} the subset of W which consists of particular events that represent emotions. We have selected the following emotional expressions: *sadness, fear, anger, joy, surprise, love.* We create an *experience matrix* **E** that relates events with events, including emotions with events and vice-versa for construction, in a static manner through a training phase.

E is a $M*M$ matrix whose *(i,j)-th* entry is the square root of the sample probability of the *i-th* event occurring before the *j-th* event, given a time-slot. The manner in which the experience matrix **E** is created during the training

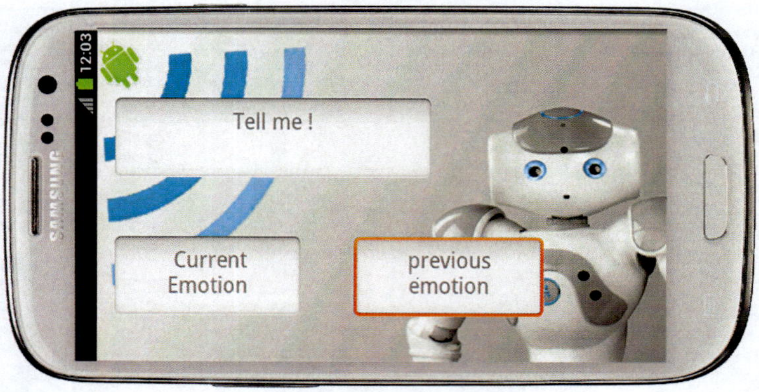

Fig. 2. A snapshot of the Android Phone Interface

phase determines the personality of the robot: the fact that an event causes an emotion, or that en event usually precedes another event is implicitly hardwired in the experience matrix **E**. We consider not only the co-occurrences of events or events and emotions, but also their temporal sequence: if the event A occurs in a timeframe before the event B, the e_{AB} item of the matrix **E** will be increased; otherwise, if the event B occurs in a timeframe before than the event A, the e_{BA} item of the matrix **E** will be increased instead. The Singular Value Decomposition of the matrix **E** is then performed, so that **E** is decomposed in the product of three matrices: a column-orthonormal M∗M matrix **U**, a column-orthonormal M∗M matrix **V** and a M∗M diagonal matrix Σ, whose elements are the singular values of **E**.

$$E = U\Sigma V^T \tag{1}$$

Let us suppose that $\mathbf{E}_s$ singular values are ranked in decreasing order. Let R be a positive integer with $R < M$, and let $\mathbf{U}_R$ be the $M{*}R$ matrix obtained from **U** by suppressing the last $M{-}R$ columns,Σ_R the matrix obtained from Σ by suppressing the last $M{-}R$ rows and the last $M{-}R$ columns and $\mathbf{V}_R$ be the $N{*}R$ matrix obtained from **V** by suppressing the last $M{-}R$ columns. Then

$$E_R = U_R \Sigma_R V_R^T \tag{2}$$

is an approximation of **E**. Each event is represented by two vectors: the rows of $\mathbf{U}_R$, that we generically indicate as u_i and the rows of $\mathbf{V}_R$, that we generically indicate as $\mathbf{v}_i^T$. The $\mathbf{u}_i$ vector represents the sub-symbolic coding of the event i for the past, while the $\mathbf{v}_i\ ^T$ vector represents the sub-symbolic coding of the event i for the future. Let us transform the $\mathbf{u}_i$ and $\mathbf{v}_i$ vectors in order to take into account the contribution of the Σ matrix, so that we obtain:

$$u_i \leftarrow u_i\sqrt{\Sigma} \tag{3}$$

$$v_i^T \leftarrow \sqrt{\Sigma v_i^T} \tag{4}$$

Let us suppose that in a given time slot of N events, we have a set P of events occurring in the timeframe, which represents the near past of our scene, and their sub-symbolic coding given by the set of vectors $\mathbf{u}_i$ associated to the events belonging to P. We want to determine the most probable emotion that is triggered in the robot by this past history. In our approach we compute the average vector that represents the past, i.e.

$$\bar{u} = \frac{1}{N} \sum_{u_i : e_i \in P} u_i \tag{5}$$

where N is the number of events that we consider in the timeframe. In order to find the emotion triggered by this history we compute the dot product between the average vector representing the past and all the vectors representing the future emotions, i.e. the $\mathbf{v}_i{}^T$ vectors associated to the events e_i belonging to the subset W_{em}

$$v_i^* = \underset{v_i^T : e_i \in Wem}{\arg\max} \; f(\bar{u} \cdot v_i^T) \tag{6}$$

v_i* is the sub-symbolic coding of the emotion e_i* that is caused by the past history.

4 Experimental Results

4.1 Scenario

We have coded the event matrix considering a set of events, among which: *User is sad; User is happy; User does not find her doll; User asks something to the robot; User recognizes object; User does not recognizes object; Robot is happy; Robot is sad; Robot asks something to User; Robot angry; Robot finds something; Robot brings something to User; Robot made a mistake; Robot makes two subsequent mistakes;Robot makes more of two subsequent mistakes; User thanks Robot.* As an example, an high value in the sequence *User is sad, Robot is sad* makes the robot empathetic; the sequence *Robot is sad, User is sad* makes no sense if we do not want that the user is empathetic with the robot, so we have a low value of occurrences for this last sequence.

In the Fig. 3 we can see a schema of the finite State Machine representing the switching emotional behaviors interested in the scenario determined by the most probable emotion that is triggered in the robot by his past history according to its personality.

The scenario designed for the experimental test conducted was setted in a simulated Living room. The actors that take part in the simulated human - robot interaction was: Nao (the robot), Sara (a child) and Mary (the doll), used

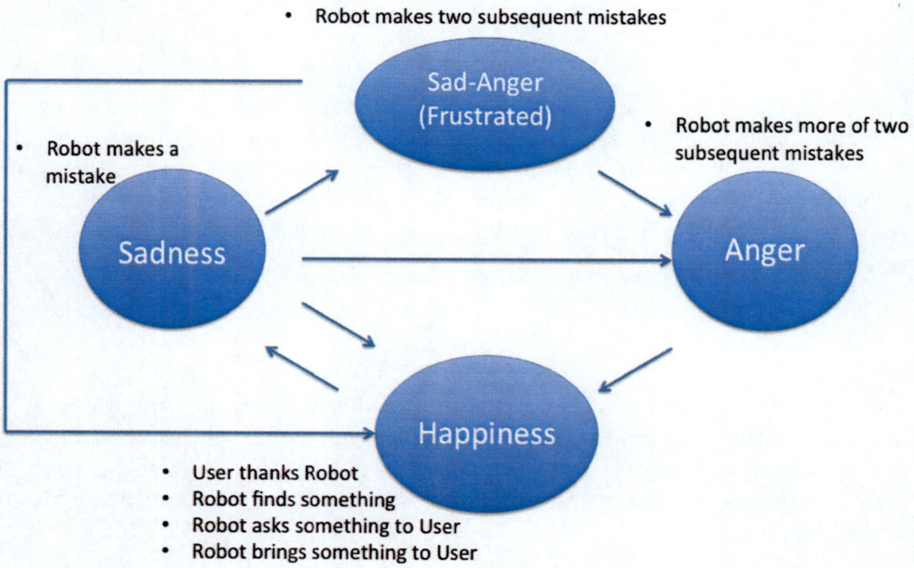

Fig. 3. The finite State Machine representing the switching emotional behaviors interested in the scenario

for make a task for the robot. The interested emotion evoked in the scenario were: the Sadness, the Anger, the frustration and the happines.

Below we are describing the introduction of Story and the Flows of events between Actor Child and Actor Robot.

It tells the story of a child Sara , who is very sad because she lost her doll Mary. Just a little robot can help here to find Mery and make happy Sara:

1. Sara is **sad** because she no finds longer her doll.
2. Sara asks to the robot to find her lost doll.
3. The robot is **sad** because the child is *sad*.
4. The robot asks to child if he can help her by going to looking for of her doll.
5. Sara tells to robot that is very **happy** of his proposal and accepts his help.
6. The robot asks to child to describe her doll.
7. Sara describes her doll.
8. The robot goes to looking for the doll and takes the first doll that is in its path. (but the doll taken isn't Mary).
9. The robot brings the doll to the child.
10. Sara doesn't recognize in the doll, Mary and she becomes sad because the doll isn't her.
11. The robot is **sad** because the child is **sad** and because he made a mistake.
12. The robot comes back to looking for the doll and he takes the first doll that is in its path (but again the doll taken isn't Mary).
13. The robot brings the doll to the child.

Fig. 4. A snapshot of the proposed scenario where the robot find the doll requested from the Child Sara

14. Sara doesnt recognize in the doll, Mary and she remains sad because the doll isn't her.
15. The robot gets **sad-angry (frustrated)** because he made a second mistake.
16. The robot returns to looking for the doll and he takes the first doll that is in its path (but again the doll taken isn't Mary).
17. The robot brings the doll to the child.
18. Sara doesn't recognize in the doll, Mary and she remains **sad** because the doll isn't her.
19. The robot gets **angry** because he committed a third mistake.
20. The robot comes back to looking for the doll and he takes the first doll (Fig. 4) that is in its path (in this case the doll is the right one).
21. The robot brings the doll to the child.
22. Sara, recognize Mary and, now, she is **happy** because he finally found her doll.
23. Sara say thank you to the robot for the hard work he had done.
24. The robot (not more angry) is **happy** because he has absolved his task.

5 Conclusion and Future Works

A new release of Emotional Social NAO Robot has been described. The proposed version of the Cognitive Architecture introduce the concept of event for the construction of the emotional conceptual space. The Scenario presented in the Experimental results show the effectiveness of the approach in a complex scenario during a multi-events HHI. Future work will be oriented to the improvement of the connection between events, emotions, and heterogenous stimuli coming from user with the goal of having an integration of the robots in the multi-events daily activities of humans. This evolution will represent a step forward for a BICA robot to be considered more "human-level intelligent" [1].

References

1. Chella, A., Lebiere, C., Noelle, D., Samsonovich, A.: On a roadmap to biologically inspired cognitive agents. In: Frontiers in Artificial Intelligence and Applications, vol. 233, pp. 453–460. IOS Press (2011)
2. Breazeal, C.: Emotion and sociable humanoid robots. International Journal of Human-Computer Studies 59(1), 119–155 (2003)
3. Oztop, E., Franklin, D., Chaminade, T., Cheng, G.: Human-humanoid interaction: is a humanoid robot perceived as a human? International Journal of Humanoid Robotics 2(4), 537 (2005)
4. Fellous, J.: From human emotions to robot emotions. In: Architectures for Modeling Emotion: Cross-Disciplinary Foundations, pp. 39–46. American Association for Artificial Intelligence (2004)
5. Menegatti, E., Silvestri, G., Pagello, E., Greggio, N., Cisternino, A., Mazzanti, F., Sorbello, R., Chella, A.: 3d models of humanoid soccer robot in usarsim and robotics studio simulators. International Journal of Hr: Humanoid Robotics 5(3), 523 (2009)
6. Thagard, P., Shelley, C.: Emotional analogies and analogical inference. The Analogical Mind: Perspectives from Cognitive Science, 335–362 (2001)
7. Arnold, M.: Emotion and personality. Psychological aspects, vol. i (1960)
8. Monceaux, J., Becker, J., Boudier, C., Mazel, A.: Demonstration: first steps in emotional expression of the humanoid robot nao. In: Proceedings of the 2009 International Conference on Multimodal Interfaces, pp. 235–236. ACM (2009)
9. Xie, L., Wang, Z., Wang, W., Yu, G.: Emotional gait generation for a humanoid robot. International Journal of Automation and Computing 7(1), 64–69 (2010)
10. Hudlicka, E.: Reasons for emotions. Integrated Models of Cognition Systems 1, 263 (2007)
11. Chella, A., Sorbello, R., Pilato, G., Vassallo, G., Balistreri, G., Giardina, M.: An Architecture with a Mobile Phone Interface for the Interaction of a Human with a Humanoid Robot Expressing Emotions and Personality. In: Pirrone, R., Sorbello, F. (eds.) AI*IA 2011. LNCS, vol. 6934, pp. 117–126. Springer, Heidelberg (2011)
12. Bruce, A., Nourbakhsh, I., Simmons, R.: The role of expressiveness and attention in human-robot interaction. In: Proceedings of the IEEE International Conference on Robotics and Automation, ICRA 2002, vol. 4, pp. 4138–4142. IEEE (2002)
13. Arkin, R., Fujita, M., Takagi, T., Hasegawa, R.: Ethological modeling and architecture for an entertainment robot. In: IEEE International Conference on Robotics and Automation, Proceedings 2001 ICRA, vol. 1, pp. 453–458. IEEE (2001)
14. Le, Q., Hanoune, S., Pelachaud, C.: Design and implementation of an expressive gesture model for a humanoid robot. In: 2011 11th IEEE-RAS International Conference on Humanoid Robots (Humanoids), pp. 134–140. IEEE (2011)
15. Zecca, M., Mizoguchi, Y., Endo, K., Iida, F., Kawabata, Y., Endo, N., Itoh, K., Takanishi, A.: Whole body emotion expressions for kobian humanoid robot - preliminary experiments with different emotional patterns. In: The 18th IEEE International Symposium on Robot and Human Interactive Communication, RO-MAN 2009, September 27-October 2, pp. 381–386 (2009)

16. Miwa, H., Itoh, K., Matsumoto, M., Zecca, M., Takanobu, H., Rocella, S., Carrozza, M., Dario, P., Takanishi, A.: Effective emotional expressions with expression humanoid robot we-4rii: integration of humanoid robot hand rch-1. In: Proceedings of the 2004 IEEE/RSJ International Conference on Intelligent Robots and Systems (IROS 2004), vol. 3, pp. 2203–2208. IEEE (2004)
17. Chella, A., Pilato, G., Sorbello, R., Vassallo, G., Cinquegrani, F., Anzalone, S.M.: An Emphatic Humanoid Robot with Emotional Latent Semantic Behavior. In: Carpin, S., Noda, I., Pagello, E., Reggiani, M., von Stryk, O. (eds.) SIMPAR 2008. LNCS (LNAI), vol. 5325, pp. 234–245. Springer, Heidelberg (2008)
18. Chella, A., Frixione, M., Gaglio, S.: An architecture for autonomous agents exploiting conceptual representations. Robotics and Autonomous Systems 25(3), 231–240 (1998)
19. Chella, A., Barone, R., Pilato, G., Sorbello, R.: An emotional storyteller robot. Emotion, Personality, and Social Behavior (2008)
20. Anzalone, S., Nuzzo, A., Patti, N., Sorbello, R., Chella, A.: Emo-dramatic robotic stewards. Social Robotics, 382–391 (2010)
21. Anzalone, S., Balistreri, G., Sorbello, R., Chella, A.: An emotional robotic partner for entertainment purposes. International Journal of Computational Linguistics Research 1, 94–104 (2010)
22. Chella, A., Sorbello, R., Pilato, G., Balistreri, G., Anzalone, S., Giardina, M.: An innovative mobile phone based system for humanoid robot expressing emotions and personality. In: International Conference of Biologically Inspired Cognitive Architectures, pp. 57–63 (2011)
23. Chella, A., Sorbello, R., Vassallo, G., Pilato, G.: An architecture for humanoid robot expressing emotions and personality. In: Proceedings of the 2010 Conference on Biologically Inspired Cognitive Architectures 2010: Proceedings of the First Annual Meeting of the BICA Society, pp. 33–39. IOS Press (2010)

The Concepts of Intuition and Logic within the Frame of Cognitive Process Modeling

O.D. Chernavskaya[1], A.P. Nikitin[2], and J.A. Rozhilo[1]

[1] Lebedev Physical Institute, Lebedev, RAS, Moscow, Russia
{olgadmitcher,yarikas}@gmail.com
[2] General Physics Institute, RAS, Moscow, Russia
apnikitin@nsc.gpi.ru

Abstract. Possible interpretation of intuition and logic is discussed in terms of neurocomputing, dynamic information theory, and pattern recognition. The thinking is treated as a self-organizing process of recording, storing, processing, generation, propagation of information under no external control. Information levels, being formed successively due to this process are described, with the transition between each level is shown to be accompanied with reduction of information. The hidden information could be interpreted as a basis for intuition, whereas verbalized abstract concepts and relations refer to the logical thinking.

Keywords: image, symbol, neural processor, connections, lamina.

The concepts of intuition and logic (as well as consciousness and subconsciousness), despite of their popularity, are still controversial, so mechanisms of these types of thinking continue to be at the center of scientific interest. There are many descriptions of these concepts with no clear and commonly accepted definitions. We'll rely on, first, Kant's approach, where intuition is the *direct discretion of truth*. Second, following the etymology of λογος (word, thought), we accept "average" understanding of logic as "a verbal argument, based on conventional cause-and-effect relations".

In this paper, we discuss possible interpretation of these concepts within the model based on neurocomputing, dynamic information theory, and pattern recognition (see [1-2] and refs. therein). An intellectual system is represented as a series of coupled neuroprocessors, that is, laminas (plates) being populated by formal neurons — bistable elements which can exist either in active or passive stationary state. We consider two types of processors: Hopfield's type — linear additive network (***H***) for operations with image information; Grossberg's type — nonlinear competitive interaction of neurons for the formation of internal symbols. Information is stored in a form of trained (modified due to the system evolution) interconnections between neurons.

Below, the thinking is treated as the ***self-organizing process of recording, storing, processing, generation, and propagation of information under no external control.*** Earlier [2] we have shown that the functions of recording and storage (as well as generation and reception) of information are *complementary*, so that they are to be shared

A. Chella et al. (Eds.): Biologically Inspired Cognitive Architectures 2012, AISC 196, pp. 105–107.
springerlink.com

between *two different subsystems*. The first one is to contain *noise* (occasional effects) to provide a *mixing layer* necessary to generate information; it corresponds to the ***right hemisphere*** (***RH***) of human brain, traditionally considered as responsible for intuition. The other subsystem is to be free of occasional elements; it corresponds to the ***left hemisphere*** (***LH***) responsible for processing information and logic.

The self-organization process provides several levels of *different type* information.

1. *Primary images* ***I*** include any available imaginary information: all signals from receptors are written as chains of activated neurons on lamina ***H*** forming the ***images***. The inter-plane connections between neurons are modified from weak ("grey") to strong ("black") ones upon presentation of objects. This level carries out the function of recording the "sensual" information and refers to ***RH***.
2. *Imaginary information* — ***typical images TI*** are presented at a special ***H***-type lamina (related to ***LH***) perceiving only the images recorded by strong enough "black" connections. Its function is to store useful and filter out unnecessary information.
3. *Symbolic* (*semantic*) *information* — ***symbols S*** correspond to *typical* images and are formed in ***RH*** (with the noise participation). Each symbol possesses a semantic content, i.e., awareness of the fact that this chain of active neurons describes a real particular object. At the same level one can find a ***standard symbol*** (symbol-word ***SW*** that is presented in ***LH*** mainly) to indicate the same specific object. This level provides interaction between the laminas, i.e., processing of sensible information.
4. *Abstract* (*verbalized*) *information* — infrastructure (in both subsystems) of symbols ***S***, standard symbols ***SW***, and their interrelationships, which are not connected with neurons-progenitors on lamina ***H***, and thus, are not associated with any imaginary object, but emerges in a well-trained system due to interaction of all the laminas (the "*inferential*" knowledge). Its function is to implement a communication with other systems ("*to explain by words*").

Thus, there are four basic elements: ***I***, ***TI***, ***S***, ***SW*** and connections between them. The first three levels represent the latent (hidden) individual information of the system, "thing-in-itself". Only the last one, verbalized information is sensible in the common sense ("*to bring on the level of consciousness*").

The emergence of each subsequent level is accompanied by a *reduction* of information. For example, images being recorded by weak "grey" connections (their role is to store *accidental* information that may become important in the future) are not transferred to ***TI*** level, and thus, cannot be associated with any symbol — this information turns out to be not conscious and out of control. This chain can be activated only by noise ("suddenly see by the mind's eye"). This act can be interpreted as an *inspiration* (*insight*), and the "grey" images themselves could be treated as the *subconsciousness*.

Within the model presented, the *hidden* information (that was ignored in course of emergence of next level) has its own "levels of depth", with the bulk being stored in ***RH*** subsystem; this very information could be interpreted as a basis for the *intuition*. The *logical* thinking is to be related to *verbalized concepts* and *abstract relations*, but those that are common in a given society. It refers to ***LH*** subsystem only.

References

1. Chernavskaya, O.D., Nikitin, A.P., Chernavskii, D.S.: The concept of intuitive and logical in neurocomputing. Biophysics 54(6), 1103 (2009)
2. Chernavskaya, O.D., et al.: On the role of "image" and 'symbol" concepts in modeling of thinking by means of neurocomputing. Appl. Nonlinear Dynamics 19(6), 5 (2011) (in Russian)

Do Humanoid Robots Need a Body Schema?

Dalia De Santis, Vishwanathan Mohan, Pietro Morasso, and Jacopo Zenzeri

Robotics, Brain and Cognitive Sciences Department,
Istituto Italiano di Tecnologia, Via Morego 30, Genova, Italy
{dalia.desantis,vishwanathan.mohan,
pietro.morasso,jacopo.zenzeri}@iit.it

Abstract. The concept of body schema is analysed, comparing the biological and artificial domains and emphasizing its logical necessity for the efficient coordination of redundant degrees of freedom. The implementation of the body schema by means of the passive motion paradigm is summarised, suggesting that a well-defined body schema may be the basic building block for developing a powerful cognitive architecture.

Keywords: body schema, equilibrium point hypothesis, passive motion paradigm.

1 Introduction

Although the concept of body schema is not clearly defined, it is generally accepted that humans have it and use it in skilled actions. The neuronal basis of this concept has been the topic of a number of studies (e.g. [12, 26, 31]), focusing on subtle distinctions such as the differences between motor body schema and perceptual body schema which are associated with different neuronal subsystems, such as the ventral and the dorsal pathways. Although such fine distinctions are somehow irrelevant from a more functional point of view, focused on the artificial physiology of a humanoid robot, they immediately point out at the distributed nature of such concept and also suggest that multi-sensory and motor aspects are intermingled in a very strong manner.

An interesting opening towards an understanding of the problem comes from clinical studies of strange or rare neurological disorders and associated sensorimotor illusions: for example, the phantom limb illusion [27], the Capgras syndrome (a recognition disorder of the body image of close acquaintances: [9]), and apotemnophilia/somatoparaphrenia syndromes (body integrity image disorders: [1, 4, 19]), which all emphasize the intimate relationship between the brain, the body and the 'body image'.

Conscious intentions and Motor awareness identify another relevant topic for addressing the computational background of the body schema problem. The recent review by Desmurget and Sirigu [8] is quite illuminating in this respect. They show that the subjective feelings of conscious intention and movement awareness are mediated by a neural sensorimotor network involving the posterior parietal cortex, supplementary motor area and premotor cortex. Intentions to carry out an action are necessary in order to be aware of the action and such motor awareness/conscious intention is likely to use the same cortical circuitry that is responsible for movement planning and

A. Chella et al. (Eds.): Biologically Inspired Cognitive Architectures 2012, AISC 196, pp. 109–115.
springerlink.com

control. Thus, consciousness of our movements is not an abstract mental process, separated from the action, but an integral part of the action generation mechanism. Moreover, as previously reported, the clinics of neurological disorders shows many examples of illusion of movement without movement or real movement without awareness of it, which can be explained in terms of dysfunctions of the parieto-frontal circuitry considered above.

Imitation is a form of social interaction related to motor awareness. In particular, Rizzolatti and Sinigaglia [28] point out that the parieto-frontal 'mirror' circuitry is the only mechanism that 'allows an individual to understand the action of others 'from the inside' and gives the observer a first-person grasp of the motor goals and intentions of other individuals'. In other words, the body schema is a kind of suit that can be worn by different owners. Although the studies on motor awareness and imitation via parieto-frontal circuitry are not viewed by their authors as relevant from the point of view of defining the neural substrate of the body schema concept, we suggest that this is just the other way around, in the framework of a biomimetic approach to humanoid robot cognition.

Additional inspiration for this line of reasoning comes from experiments on motor imagery. A growing amount of evidence in neuroimaging (in terms of EEG, fMRI, PET, NIRS) generally supports the idea of common underlying functional networks sub-serving both the execution and imagination of movements ([16, 17, 24]). The neural activation patterns include not only premotor, motor, and parietal cortical areas but also subcortical areas of the cerebellum and the basal ganglia. In particular, the presence of activity in the typically motor regions suggests that covert actions are in the same 'motor format' that is required by overt actions, with the only difference that during covert actions descending motor commands are inhibited at some point and reafferent signals (but not anticipated sensory consequences of planned actions) are missing. In fact, the concurrent activation of descending motor pathways might be involved in the generation of efference copies that propagate upstream to parietal and premotor cortex, thus predicting the potential consequences of the planned action. Marc Jeannerod [14] went a step forward by formulating the Mental Simulation Theory, which posits that cognitive motor processes such as motor imagery, movement observation, action planning & verbalization share the same representations with motor execution. Jeannerod interprets this brain activity as an internal simulation of a detailed representation of action and uses the term S-state for describing the corresponding time-varying mental states. The crucial point, in our view, is that if S-states occurring during covert actions are to a great extent quite similar to the states occurring during overt actions, then it is not unreasonable to posit that also real, overt actions are the results of the same internal simulation process. Running such internal simulations on an interconnected set of neuronal networks is, in our view, the main function of the body schema. Therefore, the body schema must not be considered a static structure, like the Penfield's homunculus, but a dynamical systems that generates goal-oriented, spatio-temporal, sensorimotor patterns. In agreement with the evidence reviewed above we may suggest that the distributed neural structure implementing such dynamical process is a set of parieto-frontal circuits.

Although the body schema concept has some overlap with the general issue of consciousness and "inner life" [29] we think it makes sense to separate them for several reasons, including an empirical/engineering one: a body schema is necessary, urgent, and doable for humanoid robots, whereas mechanisms of consciousness are still far away from engineering reality.

2 Humanoid Robots in Search of a Body Schema

The issue of body-schema is not very popular in humanoid robotics. The concept of embodiment is certainly more popular but it is not the same thing. If you have a body schema you also have embodiment but not the other way around. Vernon, von Hofsten and Fadiga [30] in their discussion on a roadmap for cognitive development in humanoid robots present a catalogue of cognitive architectures but in none of them the concept of body schema is a key element. Hoffmann et al [13] review the concept in robotics, emphasizing the gap between the concept and its computational implementations. The fundamental question is, in our view, the following one: Why does a humanoid robot need a body schema? The tentative answer, which we put forward as a working hypothesis, can be formulated as follows: humanoid robots need a body schema for the same reason for which a human or a chimp needs it, namely because humans/chimps/humanoids all have a complex body and without a suitable body schema they would be unable to use such body, take advantage of it, and ultimately survive. Roughly speaking, the complexity of a body can be measured by the number of controllable degrees of freedom (DoFs) which bring with them not only complex coordination/control problems but also complex multimodal perceptual problems. The human body has hundreds of DoFs, humanoid robots have several tens, much less insects or "simpler" biological organisms. Such simple organisms do not have a brain and do not need any body schema. Their behaviour is essentially reactive, with very simple neural circuits, connecting sensors to actuators via a few neuronal clusters.

Braitenberg's Vehicles [5] are examples of how purely reactive artificial agents – agents that have no internal states but only direct connections between sensors and motors – can exhibit complex, adaptive behaviors. Adaptive behavior without any central representation is also present in mollusks, like the "famous" Aplysia depilans [15]. In the realm of Artificial Intelligence, Rodney Brooks [6] introduced a computational framework (subsumption architecture) based on decentralized control and generalized the concept by proposing the idea of "Intelligence without representation" [7].

Taking this into account we like to speculate that a complex body without a brain capable to support a body schema would be a drawback instead of an improvement in natural or artificial phylogenesis. A purely reactive architecture is quick and efficient for low levels of complexity but does not scale up well: with increasing number of DoFs it would clearly suffer the curse of dimensionality because the number of useful/necessary coordination mechanisms among DoFs or operational modules, typical of 'subsumption architectures', would inevitably grow up exponentially with increasing difficulty in action selection and potential conflicts between modules. On the other hand, the explosive growth in the complexity of the body in a limited number of species, culminating in homo sapiens, is typically concentrated in two body parts (the hand and the vocal tract) that do not have a specific/specialized function from the Darwinian point of view, like the sharp teeth and claws for predators or the long beaks of hummingbirds. The hand and the vocal tract are indeed general-purpose tools (for manipulation and communication, respectively) to be employed in infinite numbers of possible manners and purposes. The emergence, in phylogenesis, of such complicated bodily tools implied the parallel evolution of a supporting hardware, capable to deal with an 'internal representation' of the complex body, i.e. a body schema. In a way this is just another version of the chicken or egg dilemma: a complex body 'causes' the evolution of a complex brain or the other way around? In any case we believe that no complex body can 'live' without a complex brain and no brain has a chance to display its power without a complex body. The critical change in evolution, from a reactive neural architecture to a brain with a clear distinction between central and peripheral neural circuits, is that the quick, causal link between sensory stimuli and motor responses is broken. Actions can be goal-oriented, not necessarily stimulus-oriented, can occur in anticipation of events/stimuli or in learned cycles. Real/overt actions can alternate with covert/mental actions in order to optimize the chance of success in a game or social interaction. In general, we suggest that for an organism with a complex body, acting is not limited to reactive, albeit adaptive, stimulus-response mechanisms. As a consequence, overt actions are just the tip of an iceberg: under the surface it is hidden a vast territory of actions without movements (covert actions) which are the essence of motor cognition.

3 Linking the Body to the Body Schema

We propose that the body schema is an internal model shared by real and imagined actions, in such a way to unify the computational background of the different aspects of purposive actions: execution, planning, reasoning, observation, imitation etc. There are several important consequences of this situation:

1. *Time* is not an independent variable but is intrinsic in the dynamics of the body schema.
2. The body schema must be a closed-loop system which includes an *inverse module* (mapping motor goals and task constraints into coordinated movements of the DoFs) and a *direct module* (mapping coordinated movements into the corresponding sensory consequences).
3. The body-schema must be *multi-referential*, in the sense of containing and integrating internal representations of the natural Frames of Reference (FoRs) related to sensors, actuators, tools, etc. This includes multiple *coordinate transformations* that must be carried out in an implicit, distributed way.
4. The body schema must be *adaptable*, in the sense of being able to self-calibrate for modifications of anatomical parameters, and *extendable,* in the sense of rapidly including internal representation of external tools that a skilled user learns to master [18].

In a sense, the logical necessity of the body schema is related to the so called Degrees of Freedom Problem [2]: the solution of this problem is a computational process by which the brain coordinates the action of a high-dimensional set of motor variables for carrying out the tasks of everyday life, typically described and learnt in a task-space of much lower dimensionality. Such dimensionality imbalance is usually expressed by the term motor redundancy.

A powerful approach for addressing Degrees of Freedom Problem is EPH (Equilibrium Point Hypothesis: [3, 10, 11]), which posits that body posture is not directly controlled by the brain in a detailed way but is the 'biomechanical consequence' of the equilibrium among a large set of muscular and environmental forces. In this view, 'movement' is a symmetry-breaking phenomenon, i.e. the transition from an equilibrium state to another. A generalization of this concept, which allows to model in the same manner overt actions (that include muscle activations) and covert actions (that are limited to imagined movements) is provided by PMP (Passive Motion Paradigm: [25]). The PMP formalism is based on a network of computational modules, somehow mirroring the parieto-frontal neural networks of the body schema, which are organized in such a way to display multi-referential attractor dynamics, driven by task-dependent force fields that represent task goals and constraints. Each motor space consists of a generalized displacement node and a generalized force node; Jacobian operators relate motion in different spaces (from higher to lower dimensionality) and transpose Jacobians map force fields from a space of lower dimensionality to a space of higher dimensionality; stiffness or admittance operators link displacement and force nodes. Force fields propagate activation throughout the network, animating it and ultimately moving from one equilibrium state to another. In this sense the body schema has no analogy with a lifeless geographic map but is a living, animated internal body model. Different tasks can recruit different body parts, possibly with the addition of manipulated tools. The developed formalism allows a real-time reconfiguration of the network, without the need of ad-hoc body models for each task.

We already applied the PMP formalism for representing the body schema of the iCub humanoid robot [20] in several EU projects [21-23]. The future challenge is to use it as a basic building block for a global cognitive architecture of the robot. It can be conceived as a synergy formation middleware between the neuromuscular layer and the abstract layers of the cognitive architecture. The former layer includes actuators, sensors, and spinal reactive mechanisms. This part is well developed in the iCub system, allowing the real-time processing in an asynchronous manner of a massive flow of information and control. The latter set of layers is still a work in progress and is the main target for an on-going EU project named DARWIN. The core of the system is a set of modules that aim at the acquisition of causal knowledge, relations and associated values of different objects, situations and actions (in different environmental contexts and with high-level goals) through explorative interventions in the world and observation of outcomes. This includes a reasoning system that on one hand is supposed to drive the explorative acquisition of conceptual knowledge and on the other hand can exploit the acquired knowledge in order to generate intelligent goal directed behaviours in different real-world scenarios. The reasoning system is also integrated with a mechanism of motor sequence learning, affordance learning, and an observational learning architecture formulated in terms of a 'mental simulation loop'. What is important is that such high-level cognitive functions can be grounded on a robust body model that allows, at the same time, to integrate the different cognitive requirements and generate coordinated actions, dominating the complexity of the real body.

Acknowledgments. This paper is partly supported by the EU project DARWIN (Grant No: FP7- 270138).

References

1. Bayne, T., Levy, N.: Amputees by choice: body integrity identity disorder and the ethics of amputation. J. Appl. Philos. 22, 75–86 (2005)
2. Bernstein, N.: The Coordination and Regulation of Movements. Pergamon Press (1967)
3. Bizzi, E., Hogan, N., Mussa Ivaldi, F.A., Giszter, S.: Does the nervous system use equilibrium-point control to guide single and multiple joint movements? Behav. Brain Sci. 15, 603 (1992)
4. Brang, D., McGeoch, P.D., Ramachandran, V.S.: Apotemnophilia: a neurological disorder. NeuroReport 19, 1305–1306 (2008)
5. Braitenberg, V.: Vehicles—Experiments in Synthetic Psychology. MIT Press (1986)
6. Brooks, R.A.: A robust layered control system for a mobile robot. IEEE J. Robotic Autom. 2, 14–23 (1986)
7. Brooks, R.A.: Intelligence without representation. Artif. Intell. J. 47, 139–159 (1991)
8. Desmurget, M., Sirigu, A.: A parietal-premotor network for movement intention and motor awareness. Trends Cogn. Science 13, 411–419 (2009)
9. Capgras, J., Reboul-Lachaux, J.: L'illusion des 'sosies' dans un délire systématise chronique. Bull. Soc. Clinique Med. Mentale 2, 6–16 (1923)
10. Feldman, A.G.: Functional tuning of the nervous system with control of movement or maintenance of a steady posture. Biophysics 11, 565–578 (1966)
11. Feldman, A.G., Levin, A.F.: The origin and use of positional frames of reference in motor control. Behav. Brain Sci. 18, 723 (1995)
12. Gallagher, S.: How the Body Shapes the Mind. Oxford Univ. Press, London (2005)
13. Hoffmann, M., Gravato Marques, H., et al.: Body Schema in Robotics: A Review. IEEE Trans. on Auton. Mental Development 2, 304–324 (2010)
14. Jeannerod, M.: Neural simulation of action: a unifying mechanism for motor cognition. Neuroimage 14, 103–109 (2001)
15. Kandel, E.R., Tauc, L.: Heterosynaptic facilitation in neurones of the abdominal ganglion of Aplysia depilans. J. Physiol. 181, 1–27 (1965)
16. Kranczioch, C., Mathews, S., et al.: On the equivalence of executed and imagined movements. Hum. Brain Mapping 30, 3275–3286 (2009)
17. Lotze, M., Montoya, P., et al.: Activation of cortical and cerebellar motor areas during executed and imagined hand movements: an fMRI study. J. Cogn. Neurosci. 11, 491–501 (1999)
18. Maravita, A., Iriki, A.: Tools for the body (schema). Trends Cogn. Sci. 8, 79–86 (2004)
19. McGeoch, P.D., Brang, D., et al.: Xenomelia: a new right parietal lobe sindrome. J. Neurol. Neurosurg. Psychiatry 82, 1314–1319 (2011)

20. Metta, G., Natale, L., Nori, F., et al.: The iCub humanoid robot: An open-systems platform for research in cognitive development. Neural Networks 23, 1125–1134 (2010)
21. Mohan, V., Morasso, P.: Towards reasoning and coordinating action in the mental space. Int. J. Neural Syst. 17, 1–13 (2007)
22. Mohan, V., Morasso, P., et al.: A biomimetic, force-field based computational model for motion planning and bimanual coordination in humanoid robots. Aut. Rob. 27, 291–301 (2009)
23. Mohan, V., Morasso, P., et al.: Teaching a humanoid robot to draw 'Shapes'. Aut. Rob. 31, 21–53 (2011)
24. Munzert, J., Lorey, B., Zentgraf, K.: Cognitive motor processes: the role of motor imagery in the study of motor representations. Brain Res. Rev. 60, 306–326 (2009)
25. Mussa Ivaldi, F.A., Morasso, P., Zaccaria, R.: Kinematic Networks. A Distributed Model for Representing and Regularizing Motor Redundancy. Biol. Cybernetics 60, 1–16 (1988)
26. Paillard, J.: Body schema and body image—A double dissociation in deafferented patients. In: Gantchev, Mori, Massion (eds.) Motor Control, Today and Tomorrow (1999)
27. Ramachandran, V.S., Blakeslee, S.: Phantoms in the brain: Probing the mysteries of the human mind. William Morrow & Company (1998)
28. Rizzolatti, G., Sinigaglia, C.: The functional role of the parieto-frontal mirror circuit: interpretations and misinterpretations. Nat. Rev. Neurosci. 11, 264–274 (2010)
29. Shanahan, M.: Embodiment and the inner life. Oxford University Press (2010)
30. Vernon, D., von Hofsten, C., Fadiga, L.: A roadmap for cognitive development in humanoid robots. Springer (2010)
31. de Vignemont, F.: Body schema and body image—Pros and cons. Neuropsychologia 48, 669–680 (2010)

Simulation and Anticipation as Tools for Coordinating with the Future

Haris Dindo[1], Giuseppe La Tona[1], Eric Nivel[2], Giovanni Pezzulo[3], Antonio Chella[1], and Kristinn R. Thórisson[2,4]

[1] Computer Science Engineering, University of Palermo (UNIPA/DICGIM), Viale delle Scienze, Ed. 6, 90100 Palermo, Italy
{haris.dindo,chella}@unipa.it, latona@dinfo.unipa.it

[2] Center for Analysis & Design of Intelligent Agents and School of Computer Science (CADIA), Reykjavik University, Menntavegur 1, IS-101 Reykjavik, Iceland
nivel@ru.is

[3] Consiglio Nazionale delle Ricerche, ILC-CNR and ISTC-CNR, Via S. M. della Battaglia, 44, Roma, Italy
giovanni.pezzulo@istc.cnr.it

[4] Icelandic Institute for Intelligent Machines, Uranus 2. h., Menntavegur 1, IS-101 Reykjavik, Iceland
thorisson@iiim.is

Abstract. A key goal in designing an artificial intelligence capable of performing complex tasks is a mechanism that allows it to efficiently choose appropriate and relevant actions in a variety of situations and contexts. Nowhere is this more obvious than in the case of building a *general* intelligence, where the contextual choice and application of actions must be done in the presence of large numbers of alternatives, both subtly and obviously distinct from each other. We present a framework for action selection based on the concurrent activity of multiple forward and inverse models. A key characteristic of the proposed system is the use of simulation to choose an action: the system continuously simulates the external states of the world (proximal and distal) by internally emulating the activity of its sensors, adopting the same decision process as if it were actually operating in the world, and basing subsequent choice of action on the outcome of such simulations. The work is part of our larger effort to create new observation-based machine learning techniques. We describe our approach, an early implementation, and an evaluation in a classical AI problem-solving domain: the Sokoban puzzle.

1 Introduction

In cognitive science and neuroscience, most research so far has focused on *action-outcome* mechanisms that give access to the proximal (i.e. immediate) effects of actions - named "internal forward models" (see e.g. [12]). The ability to predict the long-term effects of actions, rather than only their proximal outcomes, should

A. Chella et al. (Eds.): Biologically Inspired Cognitive Architectures 2012, AISC 196, pp. 117–125.
springerlink.com

in principle result in improved potential for adaptivity, as it allows considering distal events such as future dangers and opportunities. This feature has been recognized as one of the major ingredients of "true" intelligence and inserted in the roadmap for building biologically-inspired architectures [3]. Indeed, recent evidence indicates that humans and other animals are able to consider distal (not only proximal) action effects in their choice [17]. While this ability has been believed to be almost exclusive to humans, some evidences of its presence in other animals have been observed [16]. Certainly dogs, sea lions, and seals do a good job of predicting where a moving ball is headed.

It is still unclear how the human (and animal) brain might implement long-term predictions and use them for action selection. One theory that is receiving increasing empirical support is the idea of *action simulation*. According to this theory, multiple short-term predictions (as produced for instance by forward models) can be chained to produce long-term predictions [10]. In his *emulation theory of representation* Grush suggests a computational framework synthesizing theories of motor control, imagery and perception that is based on emulation [8]. In this view, internal representations of the world allow for prediction and simulation of future sensor stimuli. In a similar vein, the simulation hypothesis [9] argues that inner rehearsal and simulation can be used for better choice and constitutes a link between the domains of sensorimotor action and cognition. This representation is believed to be an enabling mechanism for higher level cognitive abilities. In prior work we have called this "coordinating with the future" [13]; we argue that the ability to represent non-existent, future or not yet perceivable states must be part of the ability of any general intelligence [18], as a key mechanism to realize distal goals and avoid dangers. Goal-directed action selection is based on the prediction and comparison of action outcomes, rather than on fixed stimulus-response mappings [1]. In goal-directed systems the effects of actions can be predicted at multiple levels, proximal or distal.

Some approaches have already been proposed to implement simulation and anticipation mechanisms in artificial agents. The cognitive architecture Polyscheme [2] uses simulation as a way to integrate different representations and algorithms, but simulation is not used for action selection. A few studies [7,21] use simulated sensory input to blindly control robot navigation, but they do not use simulation mechanisms for planning or long term decisions. Toussaint [19] presents a neural mechanism for stimulus anticipation, where simulation is used to enhance reinforcement learning, but only short-term prediction mechanism are proposed in this work. Pezzulo [14] implements action simulation by chaining multiple short-term predictions, and uses them for planning and predicting future dangers, but not as part of a more general mechanism of decision-making.

In this paper we present a framework for action selection that exploits simulation and anticipation mechanisms. Based on Thórisson's call for a new constructivist A.I. framework [18] the work is part of our larger effort to produce observation-based machine learning capabilities for cognitive architectures.

In the rest of the paper we refer to *simulation* as the ability to imagine or predict future courses of actions, and to *anticipation* as the ability to take actions, set up and pursue goals that are not dictated by currently available perceptions (see also [15] for conceptual clarifications). We first give an overview of our action selection framework in Sec. 2. We then describe how we implemented the simulation and anticipation processes in Sec. 3 and Sec. 4 respectively. We discuss the application of our framework to a problem solving domain in Sec. 5.

2 Action Selection Framework

In our framework knowledge is represented via internal models which encode operational knowledge about the system itself and other entities of the world. We distinguish between two types of internal models: *forward* (known as predictors) and *inverse* (known as controllers) [20]. Models operate on the stream of data, encoded as *messages*, collected from the environment and from the system's own inner activity. A message is a key-value pair; the semantics of the message is intrinsic to the key and interpreted by models. The set of messages available at a certain time t constitutes the *state* S_t of the system.

In our architecture each model possess a list of patterns on messages, meaning that we restrict the applicability of a particular model only to states that match the pattern of that model. A model also possesses a *production* that can be executed when its patterns are satisfied. Productions can be predictions of future states, in case of forward models, or controls, in case of inverse once. Thus, a forward model is defined as $M_f = \{Precondition, command, Production\}$ and an inverse model as $M_i = \{Precondition, Goal, Production\}$.

The system possesses a set of *multiple paired forward and inverse models*, each specialized for a specific situation. It decides whether a model can be executed or not through *model activation values*, which quantify the specificity of a model to the current situation. Multiple models can have their patterns satisfied, but only those with activation value higher than a threshold can be executed. Figure 1 (left) pictures the interaction of system processes that leads to action selection through simulation and anticipation. In brief: perceptual inputs and internal goals influence the activation value of models; the decision maker component then preforms look-ahead simulations and looks for candidate actions to anticipate before deciding which production to execute - if any (see Fig. 1 (b)).

3 Simulation

The system avoids explicit planning by simulating future consequences of its choices and choosing the most promising course of actions. This means that when more than one inverse model is active and competes to be executed, the system simulates the execution of each active model and finally commits to the one having the most promising simulated outcome - as measured by a domain-dependent cost function. The remaining of this section provides a computational account of the simulation processes in the architecture.

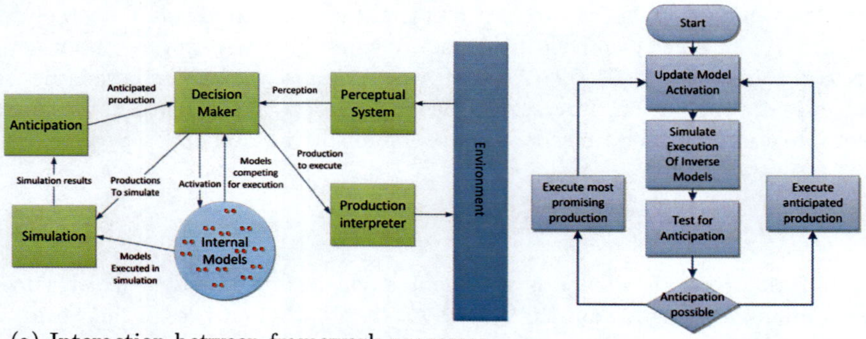

(a) Interaction between framework processes

(b) Flow chart of action selection algorithm

Fig. 1. (a) The "decision maker" process receives messages from the perceptual system and sends productions to execute to the "production interpreter" (directly connected to the environment via sensors and actuators) according to its internal models after taking into account the results of simulation and anticipation processes; (b) A schematic view of the process of simulation and anticipation

Forward models allow the system to predict the response of the system (and the environment) as a consequence of a command execution. In this respect, the forward models than can be viewed as producing faux sensory signals or "mock" sensors as in [8]. These mock sensors are the enabling mechanism for simulating or imagining future states of the system and the environment. Therefore, the system can manipulate its representation of future states in a way which is detached from situated action.

We want the system to simulate not only its proprioception or the evolution of important entities of the world, we also want the system to be able to *simulate its decisions*. We believe that this mechanism not only is efficient, but is also similar to the way humans simulate or imagine future courses of actions: humans do not usually consider every alternative when imagining a future course of actions, but instead implicitly assume the decisions that would be made in future situations.

To simulate future decisions we propose the mechanism of "*process virtualization*". By maintaining the same interface, the Decision Maker process sends its commands to fake production interpreters and receive fake sensory inputs from mock sensors. Therefore, to simulate future decisions, we can clone the Decision Maker process (i.e. its internal status) and link it to fake architectural processes that emulate the behavior of the real ones. We can view the Decision Maker as an operating system that can be installed on the real hardware or on a virtual machine whose hardware is emulated through forward models – the hardware being in this case the motor and the perceptual systems.

To simulate the outcome of a production with a limited time horizon n, the Simulation process spans the execution of a cloned Decision Maker process and the simulated processes that it needs to communicate with. The internal states of the system and the Decision Maker process are explicit and monitored; in particular, these are collected for each simulated step and concur to determine the final result of the simulation. The output of a simulation S is a list of simulation steps, each of which consists of the predicted status at that step, the predicted executed model and a set of flags that describe the predicted internal status of the Decision Maker process.

In a decision step, the execution of each active model is simulated. To establish which simulation result is the most promising - or the least dangerous - one we need a bias similar to a *somatic marker* [4], defined as a domain-specific heuristic function (other processes being domain-independent). The Decision Maker commits to the simulation trace with the lowest total bias - as computed by the said heuristics.

However, blindly committing to one course of action - albeit the most convenient one - prevents the agent from considering actions that could bring more value in the future if executed in advance, even if not strictly necessary for the goal at hand. How such an anticipative behavior can be achieved is depicted in the following section.

4 Anticipation

The ability of the system to coordinate with the future heavily depends on the mechanism of anticipation. The system not only uses simulation to choose the most promising action or to avoid possibly dangerous situations, but it anticipates needs that it currently does not have. The agent overcomes the "here and now" limitation by anticipating goals and affordances that will only be available at future times. This ability is considered almost exclusive for humans and a hallmark for intelligent behavior; people do grocery shopping even if they are actually not thirsty or hungry – they anticipate their future needs and act accordingly.

The simulation trace the decision maker commits to is called the *baseline simulation*. The Anticipation process analyzes its trace to search for possible productions that - if anticipated - would bring value in the future or would help avoiding future dangers. This implies two necessary steps for the Anticipation process:

1. Searching for productions eligible to be anticipated
2. Evaluating the efficacy of anticipating candidate productions.

Generally speaking, by analyzing the simulation trace the system looks for interesting situations that can be usefully anticipated. Obviously this statement is highly vague, the system needs a set of *criteria* that can label a simulation step as "interesting" for anticipation. We propose to define these criteria in a way that can easily be extended and modified by plugging in or removing mechanisms.

Definition 1. *An anticipation criterion C is a function that takes as input a simulation step and tests a logic condition on it (see below); if it is satisfied it outputs the production of the step; alternatively it outputs $false$.*

The system is endorsed with a set of anticipation criteria. An example anticipation criterion looks for failures of predictions or injections of new sub-goals. The anticipation process applies all the criteria to each step of the simulation, then it considers as *anticipation candidates* the productions that satisfy at least one criterion. Currently, the set of anticipation criteria is hand-coded; however, future works will explore the possibility to learn new anticipation criteria from experience.

Once the most promising anticipation candidates are selected, the anticipation process must test if these are useful productions to anticipate, and in affirmative case select one of them for execution. In order to carry out this evaluation of anticipation candidates, the anticipation process compares the course of action resulting from the execution of candidate productions to the baseline simulation. By doing so it can test whether anticipating a candidate production leads to a situation similar to the baseline simulation or brings the agent away from that course of actions. We want the system to anticipate a production exploiting current affordances and then continue acting to reach its original goals. If that was not the case the system could anticipate a future production conflicting, for example, with one of the agent's distal (i.e. future) goals.

This evaluation calls for the process to simulate the execution of production candidates and compare the result with the baseline simulation. To compare two simulations we cannot just compare the list of predicted states. Instead, we propose the use of a *criterion of similarity* between the simulations that is based on the activated models at each step of the simulation trace.

Definition 2. *A simulation S_a is similar to another simulation S_o if there exist an index i and an index j so that the sub-list of S_a that starts from i and the sub-list of S_o that starts from j have the same model for each step and the initial patterns of those models match the same state (i.e. message).*

Therefore, we test anticipation candidates by evaluating if the simulated anticipation is *similar* to the baseline simulation. The process tests the candidates in reverse chronological order (oldest first) and selects the first successful candidate. This choice is dictated by the computational costs of testing each anticipation candidate. Future work will explore the possibility of ranking all candidates according to a heuristic function.

5 Case Study: Game of Sokoban

To test our ideas in a simple yet challenging domain, we have chosen the famous problem solving puzzle Sokoban. In a grid containing a number of boxes, an agent has to push these boxes one at the time towards their target positions. Sokoban is classified as a PSPACE-complete motion planning problem and as an NP-HARD space problem [6]. Previous approaches explore either classical state-space search algorithms augmented with carefully selected heuristics [11], or adopt a reactive planning approach in which promising state-action pairs are learned via observation and imitation [5].

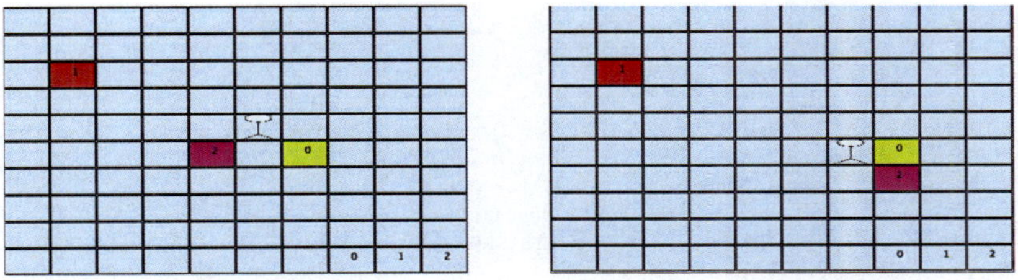

(a) Equally affordable objects (b) Possible anticipation

Fig. 2. (left) A situation that benefits from simulation; in fact simulation will predict that moving the box 0 first will result in a more convenient outcome since it is nearer to the desired position; (right) While the agent's goal is to push the box 0 downward, anticipation allows it to first push the box 1 in order to free the box 0 along its path to the container, even if it is not one of the system's current goals

However, Sokoban requires requires a great deal of anticipatory thinking in order to efficiently solve the puzzle without ending in a deadlock state. In fact, moving a box could result in a situation the agent cannot recover from; for instance, pushing a box to one of the borders leads to a deadlock as box can only be pushed and not pulled.

We have performed a preliminary investigation of our anticipatory mechanism with promising results. In situations similar to that of Fig. 2 we observed a behavior similar to that of a human player in which the agent moved a box exploiting its adjacency to it (an affordance) to free the future passage of another block. Indeed, simulation is extremely useful in situations in which the agent has to make a decision between boxes that present the same affordances (in our example, both boxes are adjacent to the agent; Fig. 2). By simulating the outcome of pushing first one box and then another, the agent can evaluate, through the domain-specific heuristic function, which choice is the most promising one. By comparing game traces between the system and humans[1] starting from the same initial state, we have observed that our systems adopts the same anticipative decisions as humans in 83% of situations.

We plan to further detach the simulation from situated action, by implementing the representation of counterfactuals. We also plan to investigate how to implement different levels of abstractions for simulations, for example using different time scales. We are also studying the adoption of reinforcement learning-based algorithms to learn domain-dependent anticipation criteria through experience.

Acknowledgments. This work has been supported in part by the EU funded project HUMANOBS: Humanoids That Learn Socio-Communicative Skills Through Observation, contract no. FP7-STREP-231453 (www.humanobs.org), and by a research grant from Rannis, Iceland. The authors would like to thank the HUMANOBS Consortium for valuable discussions and ideas, which have greatly benefited this work.

[1] 10 players selected from the population of Ph.D. students playing 6 games each. Each participant had never played Sokoban before.

References

1. Balleine, B.W., Dickinson, A.: Goal-directed instrumental action: contingency and incentive learning and their cortical substrates. Neuropharmacology 37(4-5), 407–419 (1998)
2. Cassimatis, N.L., Trafton, J.G., Bugajska, M.D., Schultz, A.C.: Integrating cognition, perception and action through mental simulation in robots. Robotics and Autonomous Systems 49(1-2), 13–23 (2004)
3. Chella, A., Lebiere, C., Noelle, D., Samsonovich, A.: T on a roadmap to biologically inspired cognitive agents. In: Biologically Inspired Cognitive Architectures. Frontiers in Artificial Intelligence and Applications, vol. 233, pp. 453–460 (2011)
4. Damasio, A.R., Everitt, B.J., Bishop, D.: The somatic marker hypothesis and the possible functions of the prefrontal cortex [and discussion]. Philosophical Transactions of the Royal Society of London. Series B: Biological Sciences 351(1346), 1413–1420 (1996)
5. Dindo, H., Chella, A., La Tona, G., Vitali, M., Nivel, E., Thórisson, K.R.: Learning Problem Solving Skills from Demonstration: An Architectural Approach. In: Schmidhuber, J., Thórisson, K.R., Looks, M. (eds.) AGI 2011. LNCS, vol. 6830, pp. 194–203. Springer, Heidelberg (2011)
6. Dor, D., Zwick, U.: Sokoban and other motion planning problems. Computational Geometry 13(4), 215–228 (1999)
7. Gigliotta, O., Pezzulo, G., Nolfi, S.: Evolution of a predictive internal model in an embodied and situated agent. Theory in Biosciences (2011)
8. Grush, R.: The emulation theory of representation: Motor control, imagery, and perception. Behavioral and Brain Sciences 27(03), 377–396 (2004)
9. Hesslow, G.: Conscious thought as simulation of behaviour and perception. Trends in Cognitive Sciences 6(6), 242–247 (2002)
10. Jeannerod, M.: Neural simulation of action: A unifying mechanism for motor cognition. NeuroImage 14, S103–S109 (2001)
11. Junghanns, A., Schaeffer, J.: Sokoban: Enhancing general single-agent search methods using domain knowledge. Artificial Intelligence 129(1-2), 219–251 (2001)
12. Kawato, M.: Internal models for motor control and trajectory planning. Current Opinion in Neurobiology 9, 718–727 (1999)
13. Pezzulo, G.: Coordinating with the future: the anticipatory nature of representation. Minds and Machines 18(2), 179–225 (2008)
14. Pezzulo, G.: A Study of Off-Line Uses of Anticipation. In: Asada, M., Hallam, J.C.T., Meyer, J.-A., Tani, J. (eds.) SAB 2008. LNCS (LNAI), vol. 5040, pp. 372–382. Springer, Heidelberg (2008)
15. Pezzulo, G., Butz, M.V., Castelfranchi, C.: The Anticipatory Approach: Definitions and Taxonomies. In: Pezzulo, G., Butz, M.V., Castelfranchi, C., Falcone, R. (eds.) The Challenge of Anticipation. LNCS (LNAI), vol. 5225, pp. 23–43. Springer, Heidelberg (2008)
16. Suddendorf, T., Corballis, M.C.: The evolution of foresight: What is mental time travel, and is it unique to humans? Behavioral and Brain Sciences 30(03), 299–313 (2007)
17. Szpunar, K.: Episodic future thought an emerging concept. Perspectives on Psychological Science 5(2), 142–162 (2010)

18. Thórisson, K.R.: From constructionist to constructivist A.I. Keynote. In: AAAI Fall Symposium Series: Biologically Inspired Cognitive Architectures, Washington D.C. Also available as AAAI Tech. Report FS-09-01, pp. 175–183. AAAI Press, Menlo Park (2009)
19. Toussaint, M.: A sensorimotor map: Modulating lateral interactions for anticipation and planning. Neural Computation 18(5), 1132–1155 (2006)
20. Wolpert, D., Kawato, M.: Multiple paired forward and inverse models for motor control. Neural Networks 11(7-8), 1317–1329 (1998)
21. Ziemke, T., Jirenhed, D.A., Hesslow, G.: Internal simulation of perception: a minimal neuro-robotic model. Neurocomputing 68, 85–104 (2005)

Solutions for a Robot Brain

Walter Fritz

Ex ITBA, Buenos Aires
Walter_1927@yahoo.com.ar

Abstract. Some problems exist when trying to build a cognitive architecture. It should be able to learn any activity, understand human words, learn from its own experiences and manage a bigger memory without slowing down. Here we show some possible solutions.

Keywords: Cognitive architecture, biologically inspired brain, artificial brain, stimulus and response, learning, intelligent system.

1 Introduction

As far as I know, some unsolved problems exist when trying to build a biologically inspired cognitive architecture (an intelligent system, a universally applicable robot brain). Here are the solutions.

2 Problems and Solutions

2.1 The Brain Has to Understand

Understanding comes from connecting a sense input to an action output. If you hear a Chinese word, you cannot understand it. But if they show you an apple and then say a word, you will understand the meaning of the word. In the same way, the robot brain reads a word, and creates a new concept and puts it into the concept "present situation". Then it sees an apple and creates another concept. Again it puts the concept into a new present situation. Now we have two consecutive "present situations". Now the brain creates a "response rule" that starts with the first situation and ends with the second situation. Now it **understands** the word apple and can use it.

This is called symbol grounding, connecting a symbol to a sense information or an action.

2.2 Start Simple

In trying to build an artificial brain, we have to start with a concrete and relatively simple task. To start by trying to build an artificial brain with all the functions of the natural one, is just as unreasonable as is if the Wright brothers tried to build a Boeing 707. Further functions of the brain will be added later.

A. Chella et al. (Eds.): Biologically Inspired Cognitive Architectures 2012, AISC 196, pp. 127–129.
springerlink.com

2.3 Stimulus and Response

The proposed brain is biologically inspired. There is a stimulus and response arc in a biological brain. Even if we look at the brain with a microscope, we observe neurons with a stimulus by a number of other neurons, and a response. Again, when we observe many related neurons (a neural column), we find a stimulus at the input and a response at the output.

All higher level activities of the brain are based on the stimulus and response of neural fields. The proposal is to use stimulus and response at an intermediate level, roughly equivalent to the neural column. The situation in the exterior environment, or the situation inside the brain, is the stimulus, expressed by concepts in the first part of the response rule. The last part of the response rule, also represented by concepts, is the output of a convenient action. The brain uses these response rules for activity in the environment and also for activity within the brain.

2.4 The Brain Reviews Its Memory of Experiences

When the brain is not occupied with external actions, it looks for patterns in its memory of the response rules just used, and thus learns generalized situations, abstract concepts, and creates total concepts from a group of part concepts. Here is how:

There are some elementary response rules. They exist at the first start of the program. They have computer code attached to them. Prolonged actions are composed from these elementary response rules. There are also some response rules, with code attached, that are not very elementary at present. They are used to perform actions on the situation inside the brain. With them the brain reviews its memory of recently used response rulers.

For instance, when the brain reads several response rules and finds that two response rules are identical, except that one has a "3" and the other has a "three", then it concludes that "3" and "three" are two concrete examples of a more abstract concept. It then creates a new concept and stores the number labels of these two concrete concepts as its content.

2.5 This Brain Has Excellent Scalability

The look-up of an adequate response rule for a given situation is independent of the size of the memory. Since, the present situation is composed of concepts, and these concepts include the labels of the applicable response rules, look-up is immediate.

3 Conclusion

We have shown solutions to some problems. I hope that you have found ideas that you can use in your own research.

The above shown solutions are based on the following experimental programs: General Learner [1], General Learner Three [2], and the proposed program Robot Brain [3]. If you need detailed information, there you can find it.

References

1. Fritz, W.: General Learner (1995), http://www.intelligent-systems.com.ar/intsyst/genlearn.html
2. Fritz, W.: General Learner Three (2006), http://www.intelligent-systems.com.ar/intsyst/genLearnThree.html
3. Fritz, W.: Robot Brain (2012), http://www.intelligent-systems.com.ar/intsyst/robotBrain.html

Exemplars, Prototypes and Conceptual Spaces

Marcello Frixione[1] and Antonio Lieto[2]

[1] DAFIST - University of Genoa, Italy
[2] Dept. of Computer Science - University of Turin, Italy
{frix@dist.unige.it, lieto@di.unito.it}

Abstract. This paper deals with the problem of the computational representation of "non classical" concepts, i.e. concepts that do not admit a definition in terms of necessary and sufficient conditions (sect. 1). We review some empirical evidence from the field of cognitive psychology, suggesting that concept representation is not an unitary phenomenon. In particular, it seems likely that human beings employ (among other things) both prototypes and exemplar based representations in order to deal with non classical concepts (sect. 2). We suggest that a cognitively in-spired, hybrid prototype-exemplar based approach could be useful also in the field of artificial computational systems (sect. 3). In sect. 4, we take into consideration conceptual spaces as a suitable framework for developing some aspects of such a hybrid approach (sect. 5). Some conclusion follows (sect. 6).

Keywords: Knowledge representation, Concept representation, Prototypes, Exemplar-Based Representations, Conceptual Spaces.

1 Representing Non Classical Concepts

Starting from the philosophical analyses by Ludwig Wittgenstein [1], and from Eleanor Rosch's [2] seminal work in the field of empirical psychology, the so-called "classical theory of concepts" has undergone severe criticisms. According to the classical theory, concepts should be definable in terms of set of necessary and sufficient conditions. Nowadays, it is widely agreed that most ordinary concepts do not conform to this assumption; rather, they exhibit prototypical effects and can be characterized in terms of prototypical information. As an alternative to the classical theory, different positions and theories on the nature of concepts have been proposed within the field of cognitive psychology. All of them are assumed to account for (some aspects of) prototypical effects in conceptualization. Usually, they are grouped in three main classes, namely: prototype views, exemplar views and theory-theories (see e.g. [3], [4]). According to the prototype view, knowledge about categories is stored in terms of prototypes, i.e. in terms of some representation of the best instances of the category. For example, the concept CAT should coincide with a representation of a prototypical cat. In the simpler versions of this approach, prototypes are rep-resented as (possibly weighted) lists of features. According to the exemplar view, a given category is mentally represented as set of specific exemplars explicitly stored within memory: the mental

A. Chella et al. (Eds.): Biologically Inspired Cognitive Architectures 2012, AISC 196, pp. 131–136.
springerlink.com

representation of the concept CAT is the set of the representations of (some of) the cats we encountered during our lifetime. Theory-theory approaches adopt some form of holistic point of view about concepts. According to some versions of the theory-theories, concepts are analogous to theoretical terms in a scientific theory. For example, the concept CAT is individuated by the role it plays in our mental theory of zoology. In other version of the approach, concepts themselves are identified with micro-theories of some sort. For example, the concept CAT should be identified with a mentally represented micro-theory about cats. These approaches turned out to be not mutually exclusive. Rather, they seem to succeed in explaining different classes of cognitive phenomena, and many researchers hold that all of them are needed to explain psychological data (see again [3], [4]). Given this state of affairs, we propose to integrate some of them in computational representations of concepts. More precisely, we focus on prototypical and exemplar based approaches, and propose to combine them in a hybrid representation architecture in order to account for category representation and prototypical effects. In this phase, we do not take into consideration the theory-theory approach, since it is in some sense more vaguely defined if compared the other two points of view. As a consequence, its computational treatment seems at present to be less feasible.

2 Prototypes vs. Exemplars: Some Empirical Evidence from Psychology

According to the available experimental evidence, exemplar models are in many cases more successful than prototypes (for a more detailed review of these results, see [5]). It can happen for example that a less typical item is categorized more quickly and more accurately than a more typical category member if it is similar to previously encountered exemplars of the category ([6]). For example: a penguin is a rather atypical bird. However, let us suppose that some exemplar of penguin is already stored in my memory as an instance of the concept BIRD. In this case, it can happen that I classify new penguins as birds more quickly and more confidently than less atypical birds (such as, say, toucans or hummingbirds) that I never encountered before. Another important source of evidence for the exemplar model stems from the study of linear separable categories. According to the prototype approach, people should find it more difficult to form a concept of a non-linearly separable category. Subjects should be faster at learning two categories that are linearly separable. However, Medin and Schwanenflugel [7] experimentally proved that categories that are not linearly separable are not necessarily harder to learn. This is not a problem for exemplar based theories, which do not predict that subjects would be better at learning linearly separable categories. In the psychological literature, this result has been considered as a strong piece of evidence in favor of the exemplar models of concept learning. The above mentioned results seem to favor exemplars against prototypes. However, other data do not confirm this conclusion. An empirical research supporting the hypothesis of a multiple mental representation of categories is Malt [8]. This study was aimed to establish if people categorize and learn categories using exemplars or prototypes. The empirical data, consisting in behavioral measures such as categorization probability and reaction time, suggest that subjects use different strategies to categorize. Some use exemplars, a few rely on prototypes, and others appeal to both exemplars and prototypes. Summing up, prototype and exemplar representations present significant differences, and have different merits in explaining cognitive behavior. In addition, this distinction seems to have also neural plausibility, as it is witnessed by various empirical researches ([9], [10], [11]). For example, Squire and Knowlton [12] studied an amnesic patient, E.P., whose medial temporal lobes were severely injured in both hemispheres. E.P. was unable to recognize previously seen items. However, his categorization performances in some conditions were similar to those of normal subjects. It is worthy noting that, after training, E.P. was unable to recognize even training items.

3 Prototypes vs. Exemplars in Artificial Computational Systems

It is likely, in our opinion, that a dual, prototype and exemplar based representation of concepts could turn out to be useful also from a technological point of view, for the representation of non classical concepts in artificial systems. In the first place, there are kinds of concepts that seem to be more suited to be represented in terms of exemplars, and concepts that seem to be more suited to be represented in terms of prototypes. For example, in the case of concepts with a small number of instances, which are very different from one another, a representation in terms of exemplars should be more convenient. An exemplar based representation could be more suitable also for non linearly separable concepts (see the previous section). On the other hand, for concepts with a large number of very similar instances, a representation based on prototypes seems to be more appropriate. Consider for example an artificial system that deals with apples (for example a fruit picking robot, or a system for the management of a fruit and vegetable market). Since it is no likely that a definition based on necessary/sufficient conditions is available or adequate for the concept APPLE, then the system must incorporate some form of representation that exhibits typicality effects. But probably an exemplar based representation is not convenient in this case: the system has to do with thousands of apples, which are all very similar one another. A prototype would be a much more natural solution. In many cases, the presence of both a prototype and an exemplar based representation seems to be appropriate. Let us consider the concept BIRD. And let us suppose that a certain number of individuals b1, ., bn are known by the systems to be instances of BIRD (i.e., the system knows for sure that b1, ., bn are birds). Let us suppose also that one of these bi's (say, bk) is a penguin. Then, a prototype PBIRD is extracted from exemplars b1, ., bn, and it is associated with the concept BIRD. Exemplar bk concurs to the extraction of the prototype, but, since penguins are rather atypical birds, it will result to be rather dissimilar from PBIRD. Let us suppose now that a new exemplar bh of penguin must be categorized. If the categorization process were based only on the comparison between the target and the prototype, then bh (which in its turn is rather dissimilar from PBIRD) would be categorized as a bird only with a low degree of confidence, in spite of the fact that penguins are birds in all respects. On the other hand, let us suppose that the process of categorization takes advantage also of a comparison with known exemplars. In this case, bh, due to its high degree of similarity to bk, will be categorized as a bird with full confidence. Therefore, even if a prototype for a given concept is available, knowledge of specific exemplars should be valuable in many tasks involving conceptual knowledge. On the other hand, the prototype should be useful in many other situations.

4 Conceptual Spaces

In the rest of this paper, we shall consider conceptual spaces (CSs) [13] as a possible framework to develop some aspects of the ideas presented in the above sections. CSs are geometrical representations of knowledge that consist of a number of quality dimensions. In some cases, such dimensions can be directly related to perceptual mechanisms; examples of this kind are temperature, weight, brightness, pitch. In other cases, dimensions can be more abstract in nature. To each quality dimension is associated a geometrical (topological or metrical) structure. The central idea beyond this approach is that the representation of knowledge can take advantage from the geometrical structure of CSs. For example, instances (or exemplars) are represented as points in a space, and their similarity can be calculated in the terms of their distance according to some suitable distance measure. Concepts correspond to regions, and regions with different geometrical properties correspond to different kinds of concepts. As an example, let us briefly consider a CS for color ([13], sect. 1.5). One possibility to describe colors consists in choosing three parameters: brightness, saturation and hue. Such parameters can be viewed as the dimensions of a chromatic CS: brightness varies from white to black, so it can be represented as a linear dimension with two end points; saturation (i.e., color intensity) ranges from grey to full intensity, therefore, it is isomorphic to an interval of the real line; hues can be arranged in a circle, on which complementary colors (e.g. red and green) lie opposite to each other. As a result, a possible CS for colors is a tridimensional space with the structure of the familiar color spindle. From our point of view (as we shall argue in the next session), CSs could offer a computational and representational framework do develop some aspects of our proposal of representing concepts in terms of both prototypes and exemplars.

5 Conceptual Spaces for Representing Prototypes and Exemplars

Conceptual spaces are well suited to represent concepts in non classical terms. For example, in a CS the region corresponding to a concept must not have necessarily crisp boundaries. Moreover, in many cases, within CSs typicality effects emerge in a natural way, and an explicit description of typical traits is not re-quired. In his writings, Gardenfors emphasizes these facts ([13], sect. 3.8). However, Gardenfors concentrates almost exclusively on representations based on prototypes. In our opinion, CSs are well suited also for the exemplar based modeling of concepts, and for developing hybrid, prototype and exemplar based, solutions. Let us consider prototypes first. According to the theory of CSs, convex regions play a special role. Given a CS, concepts that correspond to convex regions are called properties ([13], ch. 3). Properties enjoy a special status. In particular, in the case of properties, prototypes have a natural geometrical interpretation: they correspond to the geometrical center of the region itself. Given a certain property, a degree of centrality can be associated to each point that falls within the corresponding region. This degree of centrality can be interpreted as a measure of its typicality. Conversely, given a set of n prototypes represented as points in a CS, a tessellation of the space in n convex regions can be determined in the terms of the so-called Voronoi diagrams. In our opinion, CSs are well suited also for representing concepts according to the exemplar based approach. As said before (sect. 4), exemplars are represented as points in a CS. Therefore, if the prototypical representation of some concept C in a CS corresponds to a convex region (with the prototype of C corresponding to the center of the region), then it is easy to keep within the same representation also the information concerning (some) known exemplars of C. This can facilitate many forms of conceptual reasoning, which are psychological plausible, and which can turn useful in many application contexts. This is due to the fact that CSs are not specifically designed to represent prototypes; rather, they are a general framework for knowledge representation in which, at certain conditions, prototypes emerge as a consequence of the global geometric properties of the model. In this respect, CSs deeply differ from traditional approaches in which prototypes are explicitly represented as local data structure - for example, as frames, or as (possibly weighted) lists of features. An approach that, in certain respects, is akin to our proposal has been developed by Mastrogiovanni et al. [14] in the context of a robotic reasoning and planning architecture. They adopt a metric space based on Kohonen self-organizing neural maps, which, in many respects, is similar to a conceptual space. Exemplar and classes (represented in terms of prototypes) are both coded as regions of this space.

Moreover, the theory of CSs is compatible with the possibility of representing concepts that do not correspond to properties, i.e. to convex regions in the space. Non-linearly separable categories ([7] - see sect. 2 above) are exactly "non convex" concepts of this sort. Exemplar based representation are probably an adequate choice for representing of a non linearly separable categories, and this can be achieved in the framework of CSs. Also in this case, however, it is likely that the geometrical structure of CSs allows many relevant forms of reasoning (based for example on the metric associated to the CS itself).

6 Conclusion

In conclusion, data from empirical psychology and cognitive neuroscience favor the hypothesis that concepts in human memory are represented as both prototypes and as sets of exemplars. We argue that a similar, prototype and exemplar based hybrid solution can be useful also for the development of computational intelligent artifacts. We individuated conceptual spaces as a possible theoretical framework to investigate some aspects of this hypothesis. CSs have the advantage of resting on a well understood formal basis. Moreover, they enjoy also of a good biological plausibility.

References

1. Wittgenstein, L.: Philosophische Untersuchungen. Blackwell, Oxford (1953)
2. Rosch, E.: Cognitive representation of semantic categories. Journal of Experimental Psychology 104, 573–605 (1975)
3. Murphy, G.L.: The Big Book of Concepts. The MIT Press, Cambridge (2002)
4. Machery, E.: Doing without Concepts. Oxford University Press, Oxford (2009)
5. Frixione, M., Lieto, A.: Prototypes Vs Exemplars in Concept Representation. In: Proc. KEOD, Barcelona (2012)
6. Medin, D.L., Schaffer, M.M.: Context theory of classification learning. Psyc. Rev. 85(3), 207–238 (1978)
7. Medin, D.L., Schwanenflugel, P.J.: Linear separability in classification learning. J. of Exp. Psyc.: Human Learning and Memory 7, 355–368 (1981)
8. Malt, B.C.: An on-line investigation of prototype and exemplar strategies in classification. Journal of Experimental Psychology: Learning, Memory, and Cognition 15(4), 539–555 (1989)
9. Asby, F.G., Maddox, W.T.: Human Category Learning. Annu. Rev. Psychol. 56, 149–178 (2005)
10. Kri, S.: The cognitive neuroscience of category learning. Brain Research Reviews 43(1), 85–109 (2003)
11. Smith, E.E., Jonides, J.: The Cognitive Neuroscience of Categorization. In: Gazzaniga, M.S. (ed.) The New Cognitive Neurosciences, 2nd edn. MIT Press (2000)
12. Squire, L., Knowlton, B.: Learning about categories in the absence of memory. Proc. Nat. Acad. Sc. USA 92(26), 12470–12474 (1995)
13. Gardenfors, P.: Conceptual Spaces: The Geometry of Thought. The MIT Press/Bradford Books, Cambridge, MA (2000)
14. Mastrogiovanni, F., Scalmato, A., Sgorbissa, A., Zaccaria, R.: Problem Awareness for Skilled Humanoid Robots. The Int. J. of Machine Consciousness 3(1), 91–114 (2011)

The Small Loop Problem: A Challenge for Artificial Emergent Cognition

Olivier L. Georgeon[1,2] and James B. Marshall[3]

[1] Université de Lyon, CNRS, Lyon, France
[2] Université Lyon 1, LIRIS, UMR5205, F-69622, Villeurbanne Cedex, France
[3] Sarah Lawrence College, Bronxville, NY 10708, USA

Abstract. We propose the Small Loop Problem as a challenge for biologically inspired cognitive architectures. This challenge consists of designing an agent that would autonomously organize its behavior through interaction with an initially unknown environment that offers basic sequential and spatial regularities. The Small Loop Problem demonstrates four principles that we consider crucial to the implementation of emergent cognition: environment-agnosticism, self-motivation, sequential regularity learning, and spatial regularity learning. While this problem is still unsolved, we report partial solutions that suggest that its resolution is realistic.

Keywords: Self-motivation, decision process, early-stage cognition.

1 Introduction

We introduce an idealized learning problem for an artificial agent that serves as a benchmark to demonstrate four principles of emergent cognition. We named this problem the *Small Loop Problem* (SLP). In a review of benchmarks for artificial intelligence, Rohrer [10] argued that a benchmark can constitute a formal statement of one's research goals. Accordingly, in parallel to presenting the SLP, this paper presents a statement and an argumentation in favor of four principles that we consider fundamental to emergent cognition: (a) environment-agnosticism, (b) self-motivation, (c) sequential regularity learning, and (d) spatial regularity learning.

The principle of environment-agnosticism (a) was proposed to account for the idea that the agent should not implement ontological presuppositions about the environment [8]. The SLP requires that the designer of the agent must not include predefined knowledge of the environment in the agent. Classical ways of including such knowledge would be in the form of logical rules that would exploit predefined semantics associated with sensory input, or in the form of a set of predefined states of the world that would be made desirable to the agent by the implementation of a reward function. Instead, the SLP requires environment-agnostic agents to learn the semantics of sensorimotor information and the ontological structure of their world by themselves.

The SLP approach to self-motivation (b) relates to the problem of implementing a *Discrete Time Decision Process* that learns a *policy function* P(t) to maximize a *value*

A. Chella et al. (Eds.): Biologically Inspired Cognitive Architectures 2012, AISC 196, pp. 137–144.
springerlink.com

function V(t) over time. Such a process implements motivation because the agent learns behaviors that fulfill an innate value system. This view assumes that such an innate value system was selected through phylogenetic evolution in the case of natural organisms to favor the survival of the organism and of its species. An agent that solves the SLP must implement a mechanism that learns such a policy function without relying on ontological presuppositions about the world. Note that traditional algorithms of reinforcement learning [12] do not fulfill this requirement because they require the designer to associate a reward value to predefined states of the world. Even Partially Observable Markov Decision Processes (POMDPs) require prior knowledge of a *state evaluation function* to assess *believed states* from observations [1].

In the SLP, the agent has a set of 6 possible actions $A=\{a_1, \ldots a_6\}$ and a set of two possible observations $O=\{o_1, o_2\}$ (binary feedback). We define the set of possible interactions $I = A \times O$ as the set of the 12 tuples $i = [a_j,o_k]$ that associate a possible action with the possible observation resulting from that action. Each interaction i has a predefined numerical value v_i. The *value function* V(t) equals the value v_i of the interaction i enacted on step t. The policy function must learn to choose the action a_j at each time step t that would maximize the value function in an infinite horizon. Note the formal difference from reinforcement learning, which requires a reward function as a function of the state of the world and of the action, whereas the SLP formalism does not involve a formalization of the environment in terms of a set of states. When adapted to the SLP formalism, traditional reinforcement learning algorithms can learn short-term dependencies between observations and actions, but they fail to learn long-term temporal and spatial regularities that they need to learn to fully solve the SLP.

The principle of sequential regularity learning (c) follows from the fact that the agent must discover, learn, and exploit temporal regularities in its interaction with the environment to maximize the value function V(t). Doing so without prior assumptions on the environment remains an open challenge, which the SLP is designed to address.

Finally, the agent must discover, learn, and exploit spatial regularities that exist in its "body" structure and in the structure of the environment (d). Many neuroanatomists who study the evolution of animal brains argue that organization of behavior in space is a primordial purpose of cognition [e.g., 4]. Natural organisms generally have inborn brain structures that prepare them to deal with space (the tectum or superior colliculus). These observations suggest that spatial regularity learning is a key feature of emergent cognition. We designed the SLP to investigate how this feature could be integrated into a biologically inspired cognitive architecture with the other principles presented above.

Section 1 describes the SLP in detail. Section 2 reports our partial solution that implements environment-agnosticism, self-motivation, and sequence learning. Section 3 presents how we envision coupling our current solution with spatial regularity learning to move toward the full solution. The conclusion recapitulates the challenges raised by the SLP. While this problem may seem simplistic, it is still unsolved, and, we argue, it is important for the study of emergent cognition.

2 The Small Loop Problem (SLP)

For an artificial agent, the SLP consists of "smartly" organizing its behavior through autonomous interaction with the *Small Loop Environment*. The Small Loop Environment is the loop of white squares surrounded by green walls shown in Figure 1. Note that the SLP differs from benchmarks traditionally used in AI [e.g., 10] by the fact that the environment does not provide a final goal to reach.

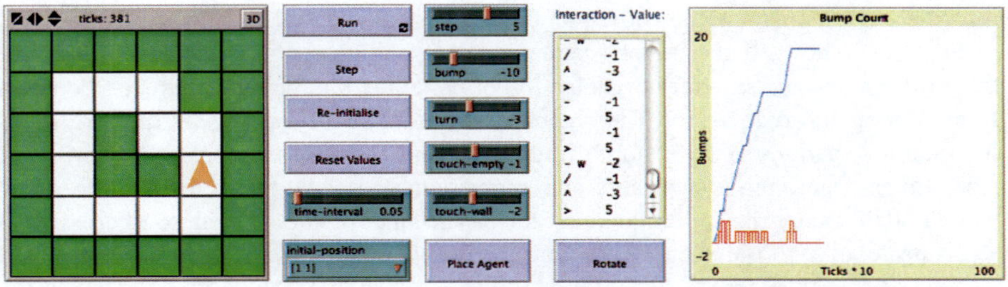

Fig. 1. The Small Loop Platform in NetLogo

The set of possible actions (A) contains the 6 following actions: try to move one square forward (a_1), turn 90° left (a_2), turn 90° right (a_3), touch front square (a_4), touch left square (a_5), touch right square (a_6). Each action returns a single bit observation as feedback ($O = \{o_1, o_2\}$). The 10 possible interactions are then: *step* ($[a_1,o_1]$), *bump* ($[a_1,o_2]$), *turn left* ($[a_2,o_1]$), *turn right* ($[a_3,o_1]$), *touch front/left/right empty* ($[a_4,o_1]$/$[a_5,o_1]$/$[a_6,o_1]$), *touch front/left/right wall* ($[a_4,o_2]$/$[a_5,o_2]$/$[a_6,o_2]$). Note that turn actions always return feedback o_1, which makes 10 interactions rather than 12.

The principle of environment agnosticism implies that the agent has no initial knowledge of the meaning of interactions: an interaction's label is meaningless to the agent. To demonstrate this, the SLP requires that swapping any label $[a_j,o_k]$ with any other label $[a_m, o_n]$ would still give rise to the same behavior when the agent is rerun.

The experimenter presets the values of interactions before running the agent (using the controls shown in Figure 1). We specify the following reference values: *step*: 5; *bump*: -10; *turn*: -3; *touch (empty or wall)*: -1. With these values, we expect the agent to learn to maximize moving forward, and avoid bumping and turning. To do so, we expect the agent to learn to use *touch* to perceive its environment and only turn in appropriate categories of situations (because *touch* has a less negative value than *bump* or *turn*). Additionally, if there is a wall ahead, the agent should touch on the side and turn towards that direction if the square is empty, so as to subsequently move forward.

Note that on each decision cycle, the best action to choose does not depend only on a single previous interaction but may depend on a sequence of several previous interactions and on the possibility of enacting several next interactions. This makes the SLP suitable to demonstrate sequential regularity learning. The highest-level most satisfying sequence consists of making a full tour of the loop, which can be repeated indefinitely. The value of this sequence is equal to 5 x 12 (move forward) -3 x 6 (turn) = 42, corresponding to 2.33 points/step.

The principles of self-motivation together with environment-agnosticism imply that the agent must adapt to any set of values. Trivial examples are those in which positive values are associated with *turn* or *bump* or *touch*: the agent would learn to keep spinning in place, or bumping, or touching indefinitely. The SLP thus consists of implementing a mechanism that tends to enact interactions with high values and to avoid interactions with negative values without any presupposition of what these interactions mean in the environment. To solve this problem, the agent must learn hierarchies of sequential regularities so it can use certain interactions to gain information to anticipate the consequences of later interactions.

Section 2 shows that a purely sequential learning mechanism can partially solve this problem. For a "smarter" organization of behavior, we, however, expect the agent to exploit spatial regularities. The agent should construct a *self model* that organizes interactions spatially. For example: touch left and turn left concern the agent's left side, touch front, move forward, and bump concern the agent's front. The agent should also categorize situations in the environment with regard to their spatial structure relative to the agent's position, for instance the categories: *left corner, right corner,* and *long edge of the loop.* This need for spatial categorization makes the Small Loop Problem suitable to demonstrate spatial regularity learning.

3 The Sequential Solution

We have reported an algorithm that brings a partial solution to a similar problem [7]. We now offer a NetLogo simulation online[1] to demonstrate the behavior of this algorithm on the Small Loop Environment. This demonstration shows that the agent usually learns to avoid bumping after approximately 300 steps and reaches a stable satisfying behavior that consists of circling the loop after approximately 600 steps. This demonstration also shows that the agent has difficulties in the upper right area of the loop because of the inverted corner.

Figure 2 shows the trace of an example run. This trace shows that the interactions were unorganized and poorly satisfying in the agent's terms until step 150. From step 190 on, the agent learned to touch ahead before trying to move forward, but it still got puzzled in the upper right area around step 220 and 270. In this particular instance, it learned to characterize *left corners* by the sequence that leads to them when circling the loop counterclockwise: "touch left empty, turn left, move forward, touch front wall". In this left corner context, the agent learned to chose *turn right* (steps 318, 354, 390), which allowed it to engage in full tours of the loop.

In the trace in Figure 2, Tape 1 represents the interactions: touch empty (white squares), touch wall (green squares), turn (half-circles), move forward (white triangles), bump (red triangles); the upper part represents interactions *to the left*, the lower part interactions *to the right*. Tape 1 shows that the agent learned to avoid bumping after step 276 by always touching ahead before moving forward. Tape 2

[1] `http://liris.cnrs.fr/ideal/demo/small-loop/`

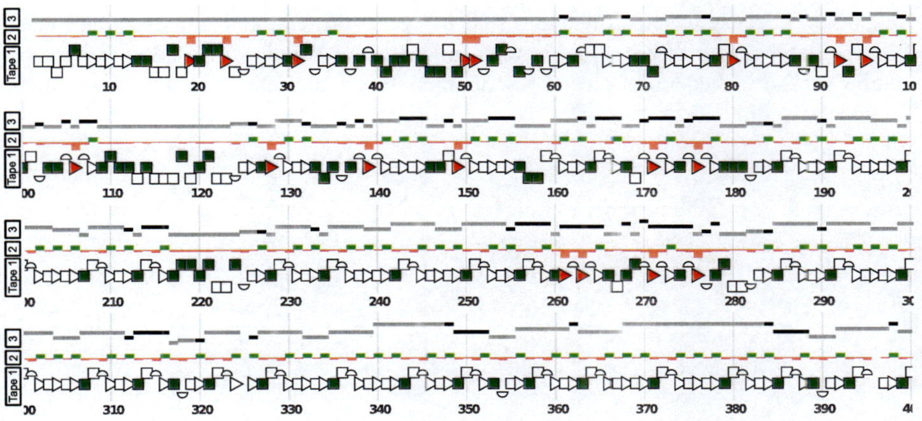

Fig. 2. Example activity trace of an agent that learns sequential regularities

represents the interactions' values as a bar-graph (green when positive, red when negative); it shows that the agent got more consistently positive interactions from step 290 on. Tape 3 represents the level of the enacted sequence in the hierarchy of learned sequences; it shows that the agent gradually exploited higher-level sequences. The value obtained when the behavior was stabilized was of 5x12 (move forward) – 3x6 (turn) – 1x17 (touch), corresponding to 0.71 points/step.

4 Towards the Spatio-sequential Solution

To give an idea of what would constitute a full solution, we started to implement a mechanism of spatial awareness. To do so, we endowed the agent with a *spatial memory* that kept track of the spatial location where interactions were enacted. The algorithm updates the spatial memory by translating its content when the agent moves forward and rotating it when the agent turns (a basic form of Simultaneous Localization And Mapping, SLAM). This solution, however, violates the principle of agnosticism because it assumes the relation between interactions and transformations in spatial memory. We also hard-wired the agent's "self model" to the spatial memory. For example, we hard coded the spatial position of the different *touch* interactions relative to the agent (left side, front, right side).

In essence, the algorithm learns *bundles of interactions* that represent observable *phenomena* in the environment. We define a bundle as the set of interactions afforded by a phenomenon [6]. The SLP provides two kinds of observable phenomena: *empty squares* and *walls*. The agent constructs bundles gradually as it explores the environment. Over time, the agent recognizes the phenomena that surround it and represents them by bundles localized in the agent's spatial memory. In turn, bundles in spatial memory generate weighted propositions (positive or negative) to enact the interactions that they contain. This mechanism increases the speed of the agent's adaptation because it helps the agent select interactions adapted to its spatial context. Figure 3 shows the effects of this mechanism in an example. A video is available online to show the entire run, the sequential trace, and the content of the spatial memory dynamically[2].

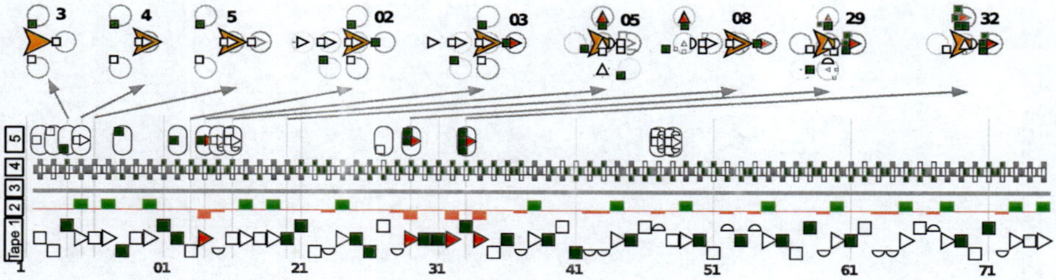

Fig. 3. Example activity trace of an agent that learns spatial and sequential regularities

In Figure 3, tapes 1 to 3 correspond to the same as in Figure 2. Tape 1 shows that the agent bumped only four times (steps 13, 28, 31, and 33). Tape 4 represents the four surrounding squares in the agent's spatial memory (squares whose content is unknown are gray). Tape 5 represents the construction of bundles over time (gray rounded rectangles that contain interactions). The upper part of Figure 3 shows snapshots of the agent's spatial memory at different steps. Gray circles represent bundles localized in spatial memory. These circles are fading to represent decay in spatial memory.

On steps 1 to 3, the agent successively touched the three squares surrounding it. On step 4, it moved forward. Because the touching forward made on step 2 and the moving forward on step 4 concerned the same spatial location, these interactions were bundled together to represent an *empty square phenomenon*. On step 5, a new touching forward activated an *empty square bundle* in front of the agent. The interaction *move forward*, now belonging to this bundle, generated additional positive support to move forward (in the agent's decisional mechanism). In a similar way, the interaction *touching front wall* and *bumping* were bundled together on step 13. On step 19, the *touching front wall* activated the newly-created *wall bundle* in front of the agent. The bump interaction, then belonging to the wall bundle, generated negative support for trying to move forward, preventing the agent from bumping into the wall. On step 96 (not in Figure 3), the learned sequence *turn right – move forward* was added to the *empty square bundle*, which led the agent to subsequently enact this sequence when an empty square was again recognized on the right. This experiment shows that this mechanism significantly improves the agent's management of the upper right area. The agent started to circle the loop on step 70 (clockwise). The value obtained after stabilization was again of 0.71 points/step.

This example illustrates two limitations that we expect the full solution to solve. (a) The average value per step obtained after the learning phase should tend to the highest value made possible by the initial settings (2.33 with the settings proposed in

[2] `http://e-ernest.blogspot.fr/2012/04/ernest-112.html`

Section 1). This requires implementing more elaborated learning mechanisms that make the agent appropriately renounce touching when the structure of the environment is better known. (b) The agent's "self-model" (i.e., the effects and positions of interactions in space) should not be assumed but rather learned. We, nonetheless, consider it valid to implement a spatial memory that assumes the two-dimensional grid structure of the environment. This assumption is justified by the fact that natural organisms have inborn brain structures that prepare them to deal with the spatial nature of their environment (e.g., the superior colliculus). The purpose of the SLP is, however, not to have the agent learn a full map of its environment but to adapt its behavior to local temporal and spatial context. Therefore, we posit that the area covered by the spatial memory should be smaller than the full environment space.

5 Conclusion

We propose the Small Loop Problem as a benchmark to evaluate agents that implement four principles of emergent cognition: environment agnosticism, self-motivation, sequential regularity learning, and spatial regularity learning. This benchmark contrasts with most existing benchmarks for unsupervised learning agents [e.g., 5, 11] in that it does not involve a final goal to reach. Instead, the agent's self-motivation comes from the fact that primitive interactions have different values.

We present two partial solutions. The first partial solution is based on an original sequential decision process. It demonstrates that the agent can organize its behavior by learning and exploiting sequential regularities of interactions without presuppositions on the meaning of interactions. The second partial solution illustrates an architecture that associates the sequential decision process with a spatial regularity learning mechanism. In its current version, this solution, however, conflicts with the agnosticism principle because it requires assumptions on the agent's self model. The Small Loop Challenge requires eliminating these assumptions.

Different solutions exist to learn spatial structures from uninterpreted sensors [e.g., 9], to learn self models [e.g., 3], and to categorize situations on the basis of self-motivation [e.g., 2]. Yet, the question of integrating these solutions together remains unsolved. Studies of natural organisms such as insects and archaic vertebrates suggest that these organisms manage to solve these problems through fundamental mechanisms of cognition whose replication in an artificial cognitive architecture remains a challenge. The Small Loop Problem offers a simple formalization of this challenge, which hopefully makes its resolution realistic. Solving this challenge will open the way to developing self-motivated agents capable of dealing with more complex spatiotemporal regularities in their interactions with the environment.

Acknowledgement. This work was supported by the French *Agence Nationale de la Recherche* (ANR) contract ANR-10-PDOC-007-01 and by a research fellowship from the *Collegium de Lyon*. We gratefully thank Frank Ritter and Christian Wolf for their comments on this report.

References

[1] Aström, K.: Optimal control of Markov processes with incomplete state information. Journal of Mathematical Analysis and Applications (10), 174–205 (1965)
[2] Blank, D.S., Kumar, D., Meeden, L., Marshall, J.: Bringing up robot: Fundamental mechanisms for creating a self-motivated, self-organizing architecture. Cybernetics and Systems 32(2), 125–150 (2005)
[3] Bongard, J., Zykov, V., Lipson: Resilient machines through continuous self-modeling. Science 314, 1118–1121 (2006)
[4] Cotterill, R.: Cooperation of basal ganglia, cerebellum, sensory cerebrum and hippocampus: Possible implications for cognition, consciousness, intelligence and creativity. Progress in Neurobiology 64, 1–33 (2001)
[5] Dietterich, T.G.: An Overview of MAXQ Hierarchical Reinforcement Learning. In: Choueiry, B.Y., Walsh, T. (eds.) SARA 2000. LNCS (LNAI), vol. 1864, pp. 26–44. Springer, Heidelberg (2000)
[6] Gay, S., Georgeon, O.L., Kim, J.W.: Implementing spatial awareness in an environment-agnostic agent. In: Proceedings of BRIMS 2012, 21st Annual Conference on Behavior Representation in Modeling and Simulation, Amelia Island, Florida, pp. 62–69 (2012)
[7] Georgeon, O.L., Ritter, F.E.: An intrinsically-motivated schema mechanism to model and simulate emergent cognition. Cognitive Systems Research 15-16, 73–92 (2012)
[8] Georgeon, O.L., Sakellariou, I.: Designing environment-agnostic agents. In: Proceedings of ALA 2012, Adaptive Learning Agents Workshop at AAMAS 2012, 11th International Conference on Autonomous Agents and Multiagent Systems, Valencia, Spain, pp. 25–32 (2012)
[9] Pierce, D., Kuipers, B.: Map learning with uninterpreted sensors and effectors. Artificial Intelligence 92, 169–227 (1997)
[10] Rohrer: Accelerating progress in Artificial General Intelligence: Choosing a benchmark for natural world interaction. Journal of Artificial General Intelligence 2, 1–28 (2010)
[11] Sun, R., Sessions, C.: Automatic Segmentation of Sequences through Hierarchical Reinforcement Learning. In: Sun, R., Giles, C.L. (eds.) Sequence Learning. LNCS (LNAI), vol. 1828, pp. 241–263. Springer, Heidelberg (2001)
[12] Sutton, R.S., Barto, A.G.: Reinforcement learning: An introduction. MIT Press, Cambridge (1998)

Crowd Detection Based on Co-occurrence Matrix

Stefano Ghidoni, Arrigo Guizzo, and Emanuele Menegatti

Intelligent Autonomous Systems Laboratory (IAS-Lab),
Department of Information Engineering,
The University of Padova,
via Gradenigo, 6/B – 35131 Padova – Italy

Abstract. This paper describes a new approach for crowd detection based on the analysis of the gray level dependency matrix (GLDM), a technique already exploited for measuring image texture. New features for characterizing the GLDM have been proposed, and both Adaboost and Bayesian classifiers have been applied to the new feature introduced, and the system has been tested on a real-case scenario inside a stadium.

1 Introduction

Intelligent video surveillance systems have recently started tackling tasks at a sensibly higher level with respect to what happened in the past. At the same time, new challenges have also been faced: behavior analysis, application of social force models and face recognition are just few examples of recently addressed tasks that show a complexity level sensibly higher with respect to classic people tracking or patrolling activities.

Among such new tasks, crowd detection represents a strong challenge, since crowd is difficult to model, and does not have a precise shape; on the other hand, this is an important feature for applications concerning security and safety.

Since crowd is often spread over large areas, video surveillance of such scenarios would benefit from being built following a multi-agent approach: each agent should keep under control a limited portion of the crowded environment, and communicate with other agents. Eventually, such surveillance agents will need to interact with human operators, that will still be responsible for final decisions regarding crowd safety. Interaction will involve agents reporting about salient events that happened, in order to let security operators take action in short time. At the same time, interaction will happen also in the opposite direction, as software agents will also be able to learn from humans which are the salient events to be kept under control.

Even though agents in this scenario are evolving, they need to rely on lower level algorithms that let them deal with crowded scenes: this is the case of the crowd detection and localization system described in this paper, which represents a capability that agents should possess and on which they can build higher level behaviors and interfaces with humans. The capability of detecting and measuring

A. Chella et al. (Eds.): Biologically Inspired Cognitive Architectures 2012, AISC 196, pp. 145–152.
springerlink.com

crowd has the same importance, in this context, as the ability of detecting people in a number of other applications of video surveillance systems.

2 Related Work

The task of crowd detection has been tackled following a number of different approaches. The three main methods rely on texture analysis, motion analysis, and features.

Texture analysis for crowd detection has been described in [1]: in this work, authors exploited a technique originally proposed in [2] and applied it to approach the problem of crowd detection. Such method is based on the Gray-Level Dependency Matrix (GLDM), which is a histogram-like function describing color transitions between adjacent pixels. The analysis of the GLDM is a complex task: for this reason, it is performed by means of a set of indicators, each one defined in order to gather a specific characteristic of the GLDM. A large set of features has been originally proposed in [2], and other have been added for adapting GLDM-based approaches to new applications [3]. Texture and its orientation are concepts exploited also in biological systems, and are part of the mid-level visual processes of the brain [4]; such processes are employed for extracting higher level information, as the extraction of 3D shape information [5].

Features represent another possible way for detecting crowd. Several approaches have been investigated, including density of SIFT features [6]. Such approaches represent indirect methods for measuring texture: for instance, the density of SIFT features is related to the amount of texture in the image, and the same can be said for edge density.

3 GLDM-Based Texture Analysis

The system presented here aims at detecting a specific type of crowd, constituted by the public in a football stadium. The detection algorithm has been developed following a texture-based approach, which turned out to be more reliable for the specific application addressed. The crowd detector described here is based on the system presented in [7], from which the evaluation of the GLDM and the multilayer shape analysis, as well as the crowd localization method based on the pyramidal image subdivision, are inherited.

In this paper, several advances over the system described in [7] are presented. A novel set of features is in fact proposed here, that improves the performance over the set previously presented.

3.1 GLDM Shape Analysis

The GLDM is a rather complex histogram defined over a two-dimensional domain bearing a high amount of information, that is only partially expressed by means of basic indicators proposed in the literature. Instead of exploiting indicators proposed in [2], a deeper shape analysis is proposed: the GLDM, which is basically a 3D function, is analyzed considering contours obtained by intersecting it with horizontal planes at several heights. A new set of features for describing shapes obtained in this way has then been created. Since multiple shapes are considered for each GLDM, this is called a multilayer approach.

Elliptic Approximation. Each cut of the GLDM may contain one or more contours, among which one is clearly larger than the others, and is called the principal component, which has an ellipse-like shape. To measure the regularity of the principal component, an ellipse is fitted on it, then deviations from this approximation are measured, similarly to what a convex hull does, but adapted to the concept of histogram. Two indicators are evaluated in this way: Low In (LI), and High Out (HO). To describe how they are evaluated, let $G^h_{\theta,d}$ be the region occupied by the GLDM cut of the principal component at height h, and E^h be the fitting ellipse. Then, LI is evaluated as:

$$\mathrm{LI}^h = \| (x,y) : (x,y) \in G^h_{\theta,d}, (x,y) \notin E^h \| , \tag{1}$$

where $\| \cdot \|$ indicates the cardinality of a set. The indicator LI therefore is the number of bins inside the fitting ellipse that do not belong to the GLDM contour cut. The complementary concept is expressed by HO:

$$\mathrm{HO}^h = |(x,y) : (x,y) \in E^h, (x,y) \notin G^h_{\theta,d}| \, . \tag{2}$$

The LI and HO indicators may be considered as a measure of the regularity of the principal component.

Symmetry. Symmetry is another key aspect for GLDM classification. It is measured by considering both the GLDM as a whole, and single layers separately. In the first group can be found the Volume Difference (VD) and Equal bins (EB) features. To evaluate the first one, the GLDM is divided into two halves by the principal diagonal; volumes of the two parts are then subtracted:

$$\mathrm{VD} = \sum_{(x,y):x<y} G_{\theta,d}(x,y) - \sum_{(x,y):x>y} G_{\theta,d}(x,y) \, . \tag{3}$$

The EB indicator is evaluated as the number of symmetrical bin pairs having the same height:

$$\mathrm{EB} = \| (x,y) : x > y, G_{\theta,d}(x,y) = G_{\theta,d}(y,x) \| \, . \tag{4}$$

Symmetry is evaluated also on single layers: for each point P_i of the diagonal inside the principal component, intersection points of the contour with a line perpendicular to the diagonal, and crossing it on P_i are found; then the distances D_i^+ and D_i^- between each of them and P_i itself are evaluated, together with their differences $\Delta_i = |D_i^+ - D_i^-|$. The Layer Symmetry (LS) feature of the i-th layer is evaluated as the variance of the normalized values for Δ_i:

$$\mathrm{LS}_i = \mathrm{var}\left\{ \frac{\Delta_i}{\max_i \Delta_i} \right\} . \tag{5}$$

The LS feature of the whole GLDM is finally evaluated as the average of all LS_i.

Another features exploited in shape classification, Width Variance (WV) is evaluated in a similar way, but instead of considering the differences Δ_i, the sum $S_i = D_i^+ - D_i^-$ is exploited. The same procedure is then applied, leading to:

$$WV_i = \text{var}\left\{\frac{S_i}{\max_i S_i}\right\}, \tag{6}$$

which are used to obtain the final value of WV by averaging WV_i.

Number of Components. Previously described features refer to the principal component of the GLDM. However, information about the whole function cannot neglect other components, that are considered by other features. The number of such components is evaluated in an approximate way, based on the assumption that all of them will lie around the diagonal: for this reason, instead of dealing with the whole GLDM function domain, only the diagonal is considered, and only those components intersecting it at a certain height will be counted.

Data about the number of components for every height, C_i, are then exploited to evaluate the maximum value (maximum number of components, MC), its average (AC) and variance (VC). Differences between layers are also considered, leading to a set of features. First of all, their variance (DV) is evaluated; moreover, to measure how smooth is the increase or decrease of the number of components, the Component Evenness (CE) is calculated as:

$$CE = \sqrt{\frac{1}{N}\sum_{i=1}^{N-1} i\left[c_i - c_{i+1} - \overline{\Delta c}\right]^2}. \tag{7}$$

4 Classification

Machine learning has been exploited to analyze feature vectors in order to detect crowd. Two methods have been tried and evaluated: AdaBoost and Naive Bayes. The first one has shown good performance in computer vision literature [8], while the second one has been chosen because of its simplicity and because it solves natively multi-class problems.

4.1 AdaBoost

In this work, the Real variant [9] of the algorithm has been used, which provides a confidence measure for each prediction instead of a simple class label. Decision trees have been selected as weak classifiers. Each decision tree is built-up recursively based on the training examples. Every internal node is created representing a test on values of a different feature; based on tests results, every training example walks a path to a specific leaf node. A leaf node represents a subset of training examples and is labelled with the most frequent class in its subset.

Using a set of decision trees combined with AdaBoost provided feature characterization: internal nodes indicated which features were used in class prediction in each AdaBoost training iteration, acting like feature selectors. Moreover, this setup does not need data normalization, which is often recommended for other classifiers to get better performance.

For what concerns multi-class classification problems, they have been broken down into binary classification problems using the One-Vs-All approach.

4.2 Naive Bayes Classifier

This is the simplest probabilistic classifier. It is based on the strong assumption that the features are independent of each other. Predictions are made by computing the probability of the class C given the features $F_1, F_2, \ldots, F_n$.

In this work, feature values have been approximated to a Gaussian distribution. Therefore, in the training phase, mean and variance have been computed for every feature on all the training examples; predictions have been done selecting the class with the highest probability:

$$\mathrm{BC}\left(f_1, \ldots, f_n\right) = \mathrm{argmax}_c\, p\left(C = c\right) \prod_{i=1}^{n} p\left(F_i = f_i | C = c\right).$$

This classifier has been chosen because it supports natively multi-class problems; moreover, it does not need parameters nor data normalization or other preprocessing operations.

5 Results

The crowd detection system has been tested on a set of 22 sequences, for a total of 901 frames at the resolution of 640×480.

Recalling the pyramidal structure described in [7], a training session has been run for each subdivision, that is, for the whole image, and for divisions into 4 and 16 sub-images. For each training session, features have been evaluated on two GLDMs, with $d = 1$ and $\theta \in \{\pi/4, 3\pi/4\}$, since these two diagonal orientations provided best results.

Crowd has been divided into classes, based on the number of people that are in the scene; classes are: no crowd (0-1 people), low crowd (2-4), mid (5-9) and high (10 or more). This classification scheme is meaningful when the image is subdivided into 4 or 16 sub-images; when considering the whole image, the two central classes should be neglected. In table 1 the number of samples of each class available in the dataset are detailed: it can be seen that for 'low' and 'mid' classes very few cases can be observed.

For training the AdaBoost classifier, 50 weak classifiers have been used, consisting in decision trees with at most 3 decision stumps. For both AdaBoost and Bayes classifiers, cross-validations have been run, applying the leave-one-out method to the 22 sequences present in the dataset.

Table 1. Number of samples available for each class, for the whole image and for 4 and 16 subdivisions

Regions	No crowd	Low	Mid	High	Total
1	328	41	41	498	901
4	1843	93	508	1160	3604
16	9609	2020	1541	1246	14416

5.1 Preliminary Feature Analysis

Since all features derived in section 3.1 are related to geometrical aspects of the GLDM, a certain level of correlation will be present among them. To characterize such dependency, Principal Component Analysis (PCA) has been employed.

In our work, dimensions are represented by features. PCA analysis reported that our features can be represented with a number of principal components between 4 and 5 (depending on regions number). According to [10], the number of principal components is also the number of the original features which are really needed to represent most of the information.

5.2 Exhaustive Search

An exhaustive search has been performed to point out the impact of features set on performance. Features by Marana have been combined with a non-empty subset of novel features; all possible non-empty combinations have been tried during simulations, which meant $2^{11} - 1$ runs for every classes-regions configuration. The best results are summarized in two tables; they show the performance increase obtained exploiting the new features presented, in two configurations: image divided into 4 regions (table 2, and into 16 regions (3). Both tables also show the performance increase determined by the new features.

The introduction of the new features determines a performance increase in every classes-regions configuration: as it is clear looking at the tables, performance increase is between 5.6% and 14.3% using AdaBoost, while it is even higher using

Table 2. Performance results using 4 classes

(a) AdaBoost classifier performance

Regions	Features	MAR + New	MAR	Var.
1	LS, WV, AC	0.670	0.586	+14.3%
4	EB, WV, DV, AC	0.646	0.574	+12.5%
16	VD, LI, AC	0.679	0.630	+ 7.8%

(b) Bayesian classifier performance

Regions	Features	MAR + New	MAR	Var.
1	VD, WV	0.609	0.585	+ 4.1%
4	VD, EB, LI, MC, VC	0.686	0.624	+ 9.9%
16	VD, EB, LS, LI, WV, DV, AC, VC	0.581	0.454	+28.0%

Table 3. Performance results using 16 regions

(a) AdaBoost classifier performance

Classes	Features	MAR + New	MAR	Var.
2	VD, LI, AC	0.800	0.746	+7.2%
3	VD, LI, AC	0.719	0.681	+5.6%
4	VD, LI, AC	0.679	0.630	+7.8%

(b) Bayesian classifier performance

Classes	Features	MAR + New	MAR	Var.
2	VD, EB, LS, LI, WV, DV, AC, VC	0.744	0.633	+17.5%
3	VD, EB, LS, LI, WV, DV, AC, VC	0.617	0.522	+18.2%
4	VD, EB, LS, LI, WV, DV, AC, VC	0.581	0.454	+28.0%

(a) (b)

Fig. 1. Example of classification output, for image divided into 4 (a) and 16 (b) regions. The color rectangles are superimposed on crowd detection output; colors indicate density: red stands for 'high' class, green for 'mid' class.

Bayesian classifier. Comparing both classifiers on the 16 regions configuration (table 3), AdaBoost provided better absolute performance; this configuration is considered more interesting because smaller regions have more uniform texture and they provide better crowd localization.

Concerning features subsets, the PCA esteems are consistent with the results obtained by AdaBoost: at most 5 features are needed to get the highest performance. The Bayes classifier requires more features, especially for high number of regions; this is caused by the probabilistic approach, which need a higher number of variables in order to provide more accurate results.

The best performance is achieved in the 16 regions and 2 classes configuration using AdaBoost. The 3 and 4 classes configurations provide more information about crowd density at the cost of a performance decrease (10.1% and 15.1% respectively).

In figure 1, two examples of classification output are presented for image divided into 4 and 16 regions; density information is expressed by superimposition of color rectangles. As it can be seen, crowd is correctly detected, and localization information is also provided.

5.3 Conclusions

In this paper, a novel crowd detection technique based on the GLDM analysis has been proposed. New features have been introduced in order to better characterize the 3D shape, and a validation phase exploiting machine learning techniques has been performed, demonstrating that the proposed technique outperforms the state of the art based on GLDM. Redundancy among developed features have been studied and the smallest feature set providing the best performance has been identified running an exhaustive search driven by results of the PCA.

References

1. Marana, A.N., Velastin, S.A., Costa, L.F., Lotufo, R.A.: Estimation of crowd density using image processing. In: IEE Colloquium on Image Processing for Security Applications (Digest No.: 1997/074), pp. 11/1–11/8 (March 1997)
2. Haralick, R.M.: Statistical and structural approaches to texture. Proceedings of the IEEE 67(5), 786–804 (1979)
3. Rahmalan, H., Nixon, M.S., Carter, J.N.: On crowd density estimation for surveillance. In: The Institution of Engineering and Technology Conference on Crime and Security, pp. 540–545 (June 2006)
4. Aspell, J., Wattam-Bell, J., Atkinson, J., Braddick, O.: Differential human brain activation by vertical and horizontal global visual textures. Experimental Brain Research 202, 669–679 (2010)
5. Orban, G.A.: The extraction of 3d shape in the visual system of human and non-human primates. Annual Review of Neuroscience 34(1), 361–388 (2011)
6. Arandjelović, O.: Crowd detection from still images. In: Proc. British Machine Vision Conference, BMVC (September 2008)
7. Ghidoni, S., Cielniak, G., Menegatti, E.: Texture-based Crowd Detection and Localisation. In: International Conference on Intelligent Autonomous Systems, IAS 2012 (in press, June 2012)
8. Viola, P., Jones, M.J., Snow, D.: Detecting pedestrians using patterns of motion and appearance. In: Proceedings of the Ninth IEEE International Conference on Computer Vision, vol. 2, pp. 734–741 (October 2003)
9. Friedman, J., Hastie, T., Tibshirani, R.: Additive logistic regression: a statistical view of boosting. Annals of Statistics 28, 2000 (1998)
10. Jolliffe, I.T.: Principal Component Analysis. Springer Series in Statistics. Springer (1986)

Development of a Framework for Measuring Cognitive Process Performance

Wael Hafez

WHA Research Inc.,
Alexandria, VA, USA
w.hafez@wha-research.com

Abstract. Cognitive architectures are concerned with the design of intelligent systems that should be able to perform cognitive tasks. The current paper introduces a theoretical approach to develop a general measurement framework for intelligent systems performance. The framework is based on concepts from communication theory and assumes the activity of an intelligent system to be the result of the communication between the system and its environment. Using the different entropies involved in this communication, the framework defines the system communication state and it is argued that this state captures the technical performance of the system.

Keywords: communication, entropy, intelligent systems, measurement.

1 Introduction

The performance of an intelligent system is a result of the interaction between the system and its environment. This interaction can be seen as a process in which the system receives input from the environment and responds in form of output back to it. In general, this process is captured in a cognitive model. As an intelligent system performs a cognitive process (and regardless of its nature and purpose) the system deploys some type of resources.

From a communication perspective, this interaction process can be seen as a communication process. In this communication, the system performs the cognitive task by receiving some messages from the environment and responds to them in form of messages sent back to it. From this perspective, the resources used to complete the cognitive task are defined in terms of the exchanged messages.

The current research is trying to validate the hypothesis that certain cognitive processes are associated with specific system-environment communication patterns. That is, a cognitive process such as learning would involve a specific communication pattern between the system and its environment. Such patterns can then be used to identify and quantify the associated cognitive processes.

An analogy is the correlation between specific higher cognitive processes and brain activation patterns as detected by brain scans. Although such correlations do not

A. Chella et al. (Eds.): Biologically Inspired Cognitive Architectures 2012, AISC 196, pp. 153–159.
springerlink.com

directly indicate the content of the cognitive task, they indicate the category of the task such as learning, attention, perception, etc. [1]. In the case of brain activation, the underlying patterns are spatial (or temporal). In the current approach to intelligent systems the underlying patterns are thought to be "communicational".

2 Approach Basic Idea

The correlation between the system cognitive performance and the underlying resources is established by considering the cognitive process from a communication perspective [2]. An intelligent system is assumed to be made up of a limited number of agents that interact together, and with the environment, through events (Fig. 1).

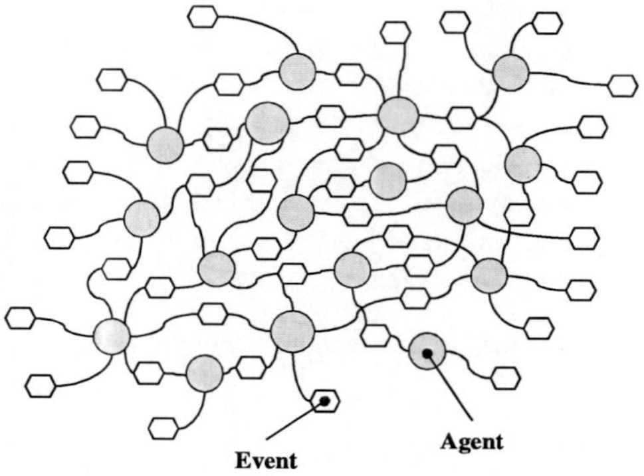

Fig. 1. Intelligent system general representation. To perform a task, the different agents interact with one another by initiating and responding to events.

To apply the communication approach to a cognitive process, the process is assumed to involve a limited number of agents and events as shown in Fig. 2. The process defines a boundary that separates the agents involved in this process from other agents in the system. By defining a boundary, the process events are separated into three different categories:

- Input events, the set of events received or collected by the agents involved in the process from the environment and/or other agents,
- Output events, the set of events send by the process agents to the environment, and
- Input-Output (I/O) dependent events, a set of the joint input and output events used to define how the output of the process depends on its input.

A cognitive process has then a limited number of input events and a limited number of output events. The various events are used to define the communication resources involved in the cognitive process (Fig. 3). The objective of the current research is to

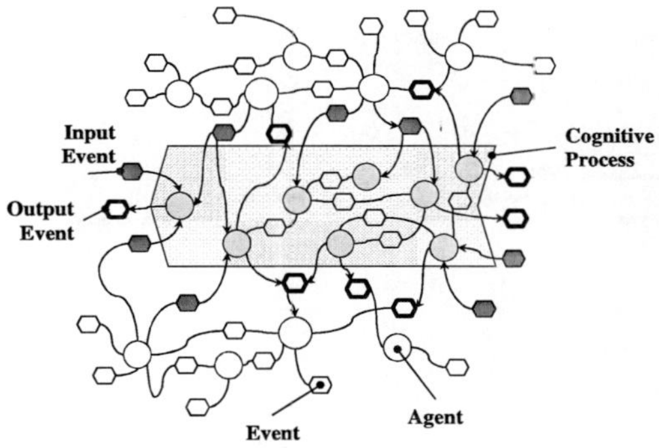

Fig. 2. Event categories involved in a cognitive process

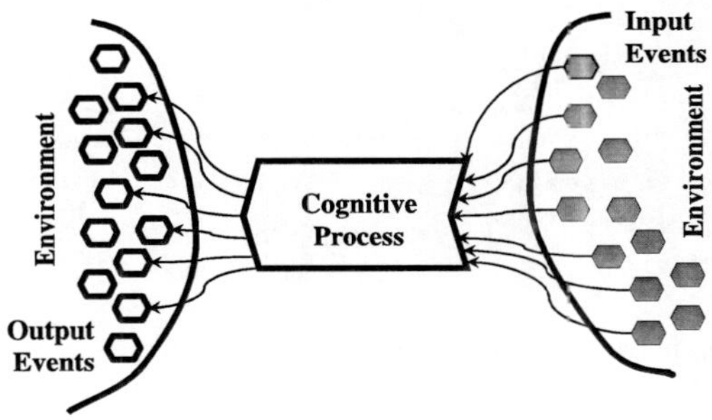

Fig. 3. Cognitive process as a communication process. A cognitive process can be performed by one or more agents. However, from a communication perspective, what is relevant to the process are the communication resources, i.e. the used events.

investigate the relationships among the events involved in the cognitive process. This is achieved by capturing the statistical characteristics of the different events sets involved in the process.

3 Cognitive Process Communication Entropies

As defined by communication theory [4], the statistical characteristics of a communication process are captured using the entropy of the involved sets. Accordingly, the resources used during a cognitive process are captured by determining the entropies of the involved sets.

Entropy is a function in the number of elements in a set and their corresponding probabilities. As a cognitive process (as defined here) involves three event sets, there are three entropies relevant to the performance of a cognitive process. Fig. 4 indicates how the communication model as defined by communication theory is used to identify the different cognitive process entropies.

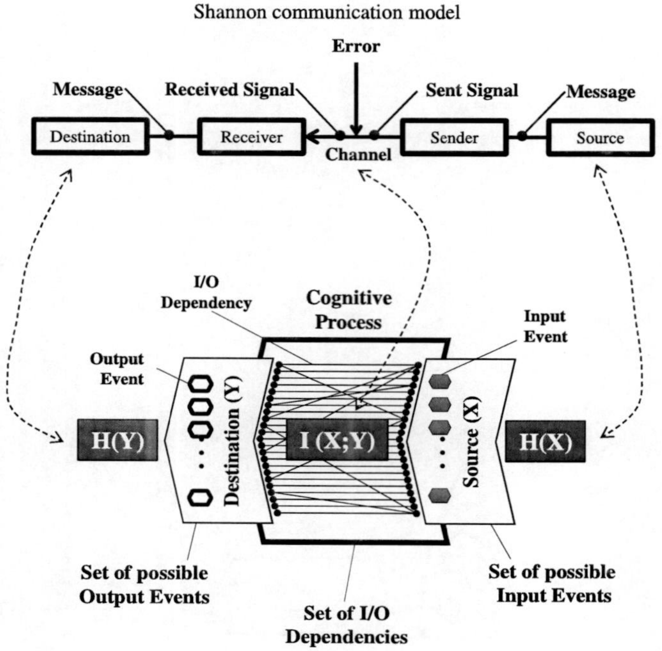

Fig. 4. From a communication perspective, the input events are considered to represent the communication source (X), the output events are considered to represent the communication destination (Y). The set or I/O dependent events represent the communication channel. H(X), H(Y) and I(X;Y) are the entropies used to define the three sets.

4 Why Entropy Can be Used to Capture the Performance of a Cognitive Process?

Entropy is a function in the number of elements in a set and their corresponding probabilities. Any change in the number of elements in the set, or in the probability of the elements, changes the entropy of the set.

Executing a cognitive process causes a change in one or all of the process related communication entropies. This is because executing the process can involve:

- Changes to the number of events used to perform the process, and/or
- Changes to the probability of using the events during the execution of the process

The approach is based on this observation and assumes that cognitive processes are associated with specific entropy change patterns.

5 Measuring Process Performance

The three communication entropies (input event set entropy, output event set entropy and I/O dependent events set entropy as shown in Fig. 4) are related to one another by what is called the system communication state (C-state) [3]. The C-state (Fig. 5) is a function in:

- the level of dependency between the input and output event sets (in terms of communication theory, channel capacity (I(X;Y)), and
- the efficiency with which the process relates the input and output events (relation between the channel capacity and input and output event sets joint entropy H(X,Y))

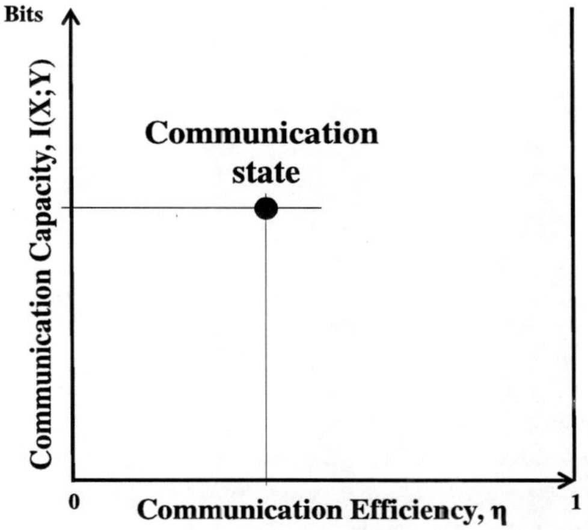

Fig. 5. Process communication state

A cognitive process might involve one or all of the following:

- use existing input and output events and existing I/O events dependencies, or
- include or define new input and output events, or
- define new I/O events dependencies.

Any of the above activities involve a change in either the number of involved events or their usage probability. This means that any process activity results in a change in at least one or all of the three communication entropies. This in return leads to a change in the process communication state. The current research tries thus to use C-state changes to indicate the outcome of the cognitive process (Fig. 6).

The correlation between a C-state (the technical aspect of a process) and the outcome of the process (functional aspect of a process) is achieved by successfully identifying C-state change patterns specific to each outcome of the cognitive process.

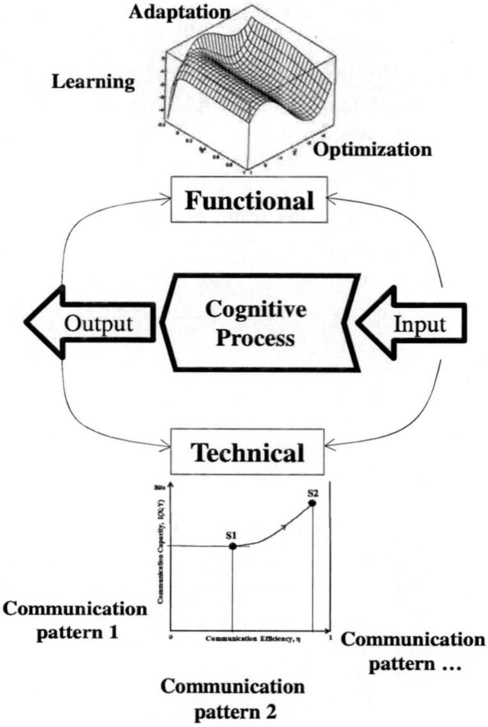

Fig. 6. Technical and functional views of a process. A cognitive process is executed for a purpose and to enable some functionality like learning. Changes in the associated C-state are assumed to be specific to each functional aspect of a process. That is each process (or process outcome) is assumed to be associated with a specific C-state change pattern.

6 Discussion

The paper introduced a theoretical approach for measuring the technical performance of a cognitive process based on the resources used during this process. Cognitive processes can involve one or more agents that can be different in nature and structure and each agent can operate according to a different model. The system C-state (or the communication dimension as represented by the communication state) is neutral to the differences among the various agents and their underlying models. That is, the communication dimension can be used to capture the behavior of the different agents and how they perform a cognitive task regardless of the differences in the used models. As such, the approach can be used to evaluate the overall performance of a cognitive architecture and to compare different models and cognitive designs options.

The approach can also be used to investigate the correlation between certain cognitive processes and the underlying C-state change patterns. If a cognitive process (e.g. learning) is associated with a specific communication pattern, then this pattern can be used as a reference to improve the configuration of this process and accelerate its implementation in other intelligent systems.

Intelligent systems are expected to operate in a changing environment. Some changes might be beyond the system ability to cope with. Identifying C-state change patterns associated with critical changes in the environment can provide insight into how to design the system for coping with such changes.

7 Future Work

The current approach addressed the theoretical aspect of the framework which is based largely on results from communication theory. The next steps in establishing the framework is to apply it to investigate the performance of actual cognitive architectures and cognitive processes.

References

1. Cabeza, R., Nyberg, L.: Imaging cognition. II. An empirical review of 275 PET and fMRI studies. Journal of Cognitive Neuroscience 12(1), 1–47 (2000)
2. Hafez, W.: Intelligent System-Environment Interaction as a Process of Communication. In: Proceedings of the IEEE International Conference on Systems, Man and Cybernetics, Istanbul, Turkey, October 10-13, pp. 3393–3396. IEEE (2010), doi:http://ieeexplore.ieee.org/10.1109/ICSMC.2010.5642335
3. Hafez, W.: Approach for Defining Intelligent Systems Technical Performance Metrics. In: PerMIS 2012, College Park, MD, USA, March 20-22 (in press, 2012)
4. Shannon, C.E., Weaver, W.: The Mathematical Theory of Communication. Univ. of Illinois Press, Illinois (1949)

I Feel Blue: Robots and Humans Sharing Color Representation for Emotional Cognitive Interaction

Ignazio Infantino, Giovanni Pilato, Riccardo Rizzo, and Filippo Vella

National Research Council of Italy (CNR), Institute of High Performance Computing and Networking (ICAR), V.le delle Scienze edif. 11, 90128, Palermo (PA), Italy

Abstract. The paper presents a representation of colors integrated in a cognitive architecture inspired by the Psi model. In the architecture designed for a humanoid robot, the observation and recognition of humans and objects influence the emotional state of the robot. The representation of color is an additional feature that allows the robot to be "*in tune*" with the humans and share with them a physical space and interactions. This representation takes into account the current hypothesis about how the human brain allows sophisticated process and manage the colors, considering both universals and linguistic approaches. The paper describes in detail the problems of color representation, the potential of a cognitive architecture able to associate them with emotions, and how they can influence the interactions with the human.

1 Introduction

The color is an important feature detected by the human visual system that allows you to have relevant information for the recognition of objects, to discriminate the components of the scene, to get information about light sources. The perception of color, that involves all the cognitive processes of the brain, has been studied by neurobiologists, psychologists, philosophers, engineers, trying to understand the workings of the brain, to create more detailed models and, in some case, how to reproduce artificially its functionality.

The color has raised much interest since it provides a limited context to address the problem of the so-called semantic gap, i.e. of how to tie the concepts and the meaning: "the apple is red", "that spoon is light green", "to feel blue ".

A humanoid with a cognitive infrastructure inspired by the human biology provides an ideal experimental platform to investigate these aspects: (1) a visual perception system based on standard cameras is roughly comparable to human visual system (most of the processing carried out by hardware and firmware acquisition device is designed to highlight what the human perceptual system determines as relevant), (2) the humanoid may interact in the real environment to have its own color experience (including associations with objects or parts of the scene), (3) it can physically or verbally interact with a human in order find the color label for a given perceived stimulus.

The purpose of our work is to integrate into an architecture with cognitive capabilities, mechanisms of perception, conceptualization and creating of the language dealing with colors according to the so-called universalist approach. This model is

A. Chella et al. (Eds.): Biologically Inspired Cognitive Architectures 2012, AISC 196, pp. 161–166.
springerlink.com

opposed to the linguistic relativism approach asserting that the categories of perceived colors are primarily influenced by cultural processes. The universalist approach instead identifies eleven possible basic colors: white, black, red, green, yellow, blue, brown, purple, pink, orange, and gray. Berlin and Kay found that languages with less than eleven color categories follow a specific evolutionary pattern [14].

Both the universalist approach and the linguistic relativism fail to capture the richness and complexity of human color categorization. In order to improve the identification of universal categories using automatic classification algorithms additional factors have been proposed[2]: the choice of the appropriate color space representation; the consideration that in reality the colors are not evenly distributed but have a specific statistical distribution; the influence of various biological components of the perceptual system [8]. If you want a complete model then you must also consider the influence of cultural and linguistic diversity, for example by introducing negotiation mechanism (considering agents who perceive the same scene), sharing of linguistic terms at the social scale [7].

On that basis we try to ensure that the humanoid, interacting with humans, may share a similar and consistent color representation of the current knowledge of the human visual system [5].

In previous work we presented an implementation of a Psi architecture that manages emotional based human-humanoid interaction[10]. The recognition of human and objects influenced the emotional state of the robot by means of an index of pleasure/distress, and it was made manifest turning on the LEDs in the head of the robot with an appropriate color. This paper introduces another aspect of the environment that might influence the emotional state of the robot: the color. We introduce this specific perceived feature because it is a key aspect in the understanding of the human cognitive processes and the links between perception and language.

2 Colors and Emotions in Cognitive Architecture

The architecture proposed in[10] was based on the Psi model, and the motivation of the robot was driven by an index influenced by the emotional associations learned from the recognition of people and objects. Detected human expression and the identification of the object and the person act positively or negatively on the emotional index of the robot. For example, a positive feedback occurs when a person and an object are recognized and the human perceived the object as pleasing.

In this first implementation of our system the predominant color of the object, which influence the mood of the robot according to an associative patterns of color-emotion psychology, is detected. The robot builds its own experience and knowledge of color using his perception. Of course, the experience of color is limited by the environment where it acts and percepts that will provide a limited statistical distribution of color[3]. Figure 1 shows all entities involved in calculating the emotional index of the humanoid. Some components are implemented using the APIs of the NAO, while color and emotion are processed through our new modules written in python. The mood of the robot is shown to an extern observer by changing the color of the LEDs of robot head.

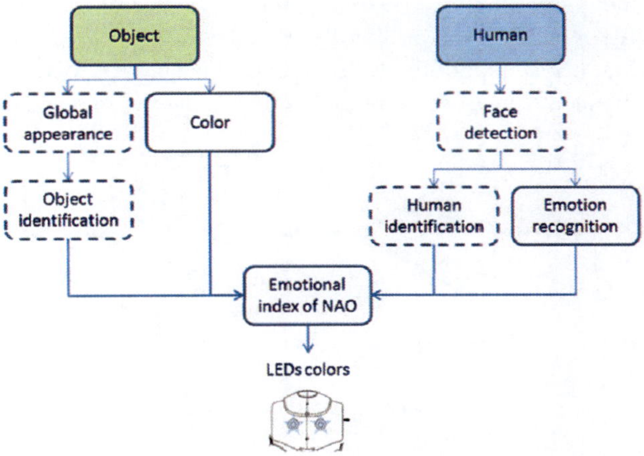

Fig. 1. Entities involved in the computation of the emotional index of the humanoid robot. Dashed components are implemented by standars APIs of the NAO software.

3 Color Representation in Nao Robot

According to [3] we assume that clustering of colors and lexicalization are separated processes that can be executed by different, although strongly connected, subsystems. Color categories are presented and memorized as points in the color space; these points work as prototypes of a color, so that the system can categorize a color input stimulus comparing it to such prototypes.

In the present work our goal is to verify if a similar mechanism can be implemented without the explicit prototypes used in [3], by using a distributed representation of the color stimuli. Such representation can be implemented by a self organizing artificial neural networks: these networks are inspired by some structures in the human brain; these neurons activate in a way that reflects the similarity of the input signals i.e. similar inputs activate neighborhood neurons.

In our approach the color category came from the training process of a neural gas network(NG) [12] and it can be modified only using the training process. The NG distributes the neural units in the color space according to the input color distribution, so that a color that is not present in the training input data will not be represented in the network. The color categories came from a labeling process on these neural units. In this case many neural units corresponding to similar colors will be labeled using the same name: e.g. different variants of yellow will have the same label, i.e there is not a single color prototype. In our experimental set-up the visual system in Nao robot acquires the images and represents color in RGB space.

In this work we decided to implement a similar approach to what illustrated in [3], but we used a neural gas network in order to compress the input color signal and represent the colors as activation patterns on a neural network [11,1]. The neural gas units u_i, $i = 1,2,\ldots,N$ are not constrained on a lattice but are free to move in the input space. The training algorithm can be found in[12]. The neural network acts as a compression subsystem because it is a binary activation pattern that is propagated in the color representation system, not the RGB signal. This neural network is trained using a set of color obtained from a set of images of the robot environment.

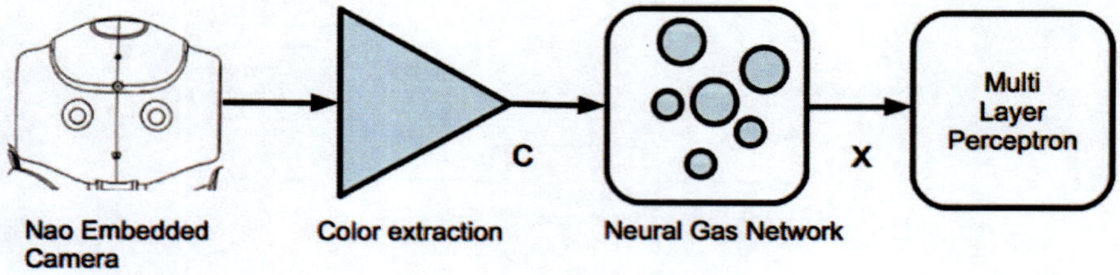

Fig. 2. The Nao Color Subsystem

3.1 Labeling the Color Space

Using a distributed color representation, the lexicalization must be obtained by recognizing the activity pattern generated by the input from the Nao cameras. The labeling process in our case is much more complicated than the one considered in [3]. We considered a multilayer perceptron MLP network which creates a mapping from the neural activity patterns corresponding to the color input to the color labels used in lexicalization. In order to create this mapping it is necessary to collect a suitable set of examples for the training phase of the MLP.

The advantage of this method is that the mapping can be clearly defined and it guarantees that some colors will be clearly recognized, moreover the MLP allows creating very complex mappings. In Figure 3 there is an example: the set of neurons is represented as a set of white circles inside a gray manifold that represent the area were the input colors of the training set were distributed. The labels Color 1, Color 2, ... and so on represent the labels associated by the mapping after the training of the MLP. A new color, indicated with as Input Color in figure 3 will activate the nearest neurons, their degree of activation is indicated as scale of gray. Due to the fact that the activated neurons are all inside the area labeled Color 1 the new color will be labeled as Color 1 either.

We make use of a Multi Layer Perceptron (MLP) classifier to realize the lexicalization phase. In particular the activation pattern of the neural gas units constitutes the input pattern for the MLP, while output units are associated to linguistic labels.

To test the validity of the proposed approach, three MLPs architectures have been considered: a traditional MLP with sigmoids as activation function of the hidden units, a MLP with sinusoids as activation function of the hidden units and an alphaNet feedforward network[9]. In our experiments we have used a 300-input pattern associated with the 300 units of the neural gas. Outputs are constituted by 13 units representing the linguistic labels used to train the MLPs.

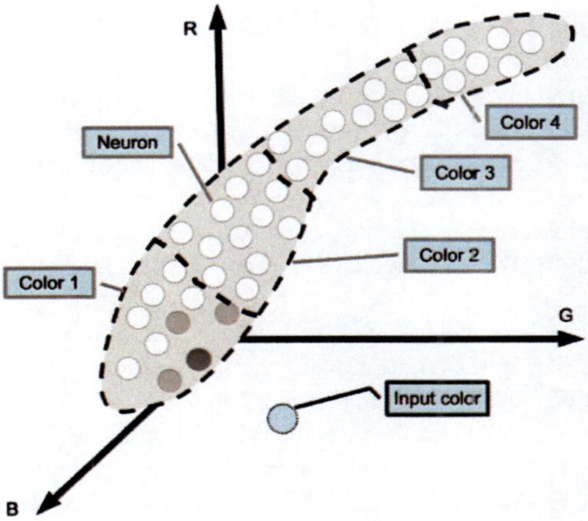

Fig. 3. Activation patten and labeling of a new color

Experimental results show that all three architectures learned the 300 patterns by using only three hidden neural units. The best performing architecture during the training phase is that one which makes use of the sinusoidal activation function (2 seconds for training), followed by the alphaNet (3 seconds for training) and the traditional architecture, which uses sigmoids as activation functions of its hidden units (4 seconds for training).

4 Conclusions

In this paper an empathetic communication system about color representation has been presented. We have integrated mechanisms of perception, conceptualization and creation of linguistic representation of colors according to the universalist approach into an architecture with cognitive capabilities. A neural system, which makes a joint use of neural gas and a MLP architecture has been employed. The prototype that implements the proposed methodology shows that a humanoid robot is capable to learn and share a cognitive representation of colors with a human user.

References

1. Ardizzone, E., Chella, A., Rizzo, R.: Color image segmentation based on a neural gas network. In: Proc. ICANN 1994, Int. Conf. on Artificial Neural Networks, vol. II, pp. 1161–1164 (1994)
2. Baronchelli, A., Gong, T., Puglisi, A., Loreto, V.: Modeling the emergence of universality in color naming patterns. PNAS (2010)
3. Belpaeme, T., Bleys, J.: Explaining Universal Color Categories Through a Constrained Acquisition Process. Adaptive Behavior 13, 293–310 (December 13, 2005), doi:10.1177/105971230501300404

4. Berlin, B., Kay, P.: Basic Color Terms: Their Universality and Evolution. University of California Press, Berkeley (1969)
5. Cangelosi, A.: Grounding language in action and perception: From cognitive agents to humanoid robots. Physics of Life Reviews 7(2), 139–151 (2010)
6. Cook, R., Kay, P., Regier, T.: Dataset of the World Color Survey, `http://www.icsi.berkeley.edu/wcs/data.html`
7. De Greeff, J., Belpaeme, T.: The development of shared meaning within different embodiments. In: 2011 IEEE Intl. Conf. on Development and Learning (ICDL), August 24-27, vol. 2, pp. 1–6 (2011)
8. Jäger, G.: Natural Color Categories Are Convex Sets. In: Aloni, M., Bastiaanse, H., de Jager, T., Schulz, K. (eds.) Logic, Language and Meaning. LNCS, vol. 6042, pp. 11–20. Springer, Heidelberg (2010) (revised selected papers)
9. Gaglio, S., Pilato, G., Sorbello, F., Vassallo, G.: Using the Hermite Regression Formula to Design a Neural Architecture with Automatic Learning of the 'Hidden' Activation Functions. In: Lamma, E., Mello, P. (eds.) AI*IA 1999. LNCS (LNAI), vol. 1792, pp. 226–237. Springer, Heidelberg (2000)
10. Gaglio, S., Infantino, I., Pilato, G., Rizzo, R., Vella, F.: Vision and emotional flow in a cognitive architecture for human-machine interaction. In: Proceedings of the Meeting of the BICA, Frontiers in Artificial Intelligence and Applications, vol. 233 (2011)
11. Huang, H.Y., Chen, Y.S., Hsu, W.H.: Color image segmentation using a self-organization map algorithm. Journal of Electronic Imaging 11(2), 136–148 (2002)
12. Martinetz, T., Schulten, K.: A "neural gas" network learns topologies. In: Artificial Neural Networks, pp. 397–402 (1991)
13. Schauerte, B., Fink, G.A.: Web-based Learning of Naturalized Color Models for Human-Machine Interaction. In: Proc. of Intl. Conf. on Digital Image Computing, DICTA (2010)
14. Steels, L., Belpaeme, T.: Coordinating perceptually grounded categories through language. A case study for colour. Behavioral and Brain Sciences 24(8), 469–529 (2005)

Investigating Perceptual Features for a Natural Human - Humanoid Robot Interaction Inside a Spontaneous Setting

Hiroshi Ishiguro[1,2], Shuichi Nishio[2], Antonio Chella[3], Rosario Sorbello[3], Giuseppe Balistreri[3], Marcello Giardina[3], and Carmelo Calí[4]

[1] Graduate School of Engineering Science, Osaka University, 1-3 Machikaneyama Toyonaka Osaka, Japan
[2] ATR Intelligent Robotics and Communication Laboratory, 2-2-2 Hikaridai Seikacho Sourakugun Kyoto, Japan
[3] DICGIM, RoboticsLab, Universitá di Palermo, V. delle Scienze, Palermo, Italy
[4] Dipartimento FIERI-AGLAIA Universitá di Palermo, Palermo, Italy
rosario.sorbello@unipa.it

Abstract. The present paper aims to validate our research on human-humanoid interaction (HHI) using the minimalistic humanoid robot Telenoid. We have conducted human-robot interactions test with 100 young people with no prior interaction experience with this robot. The main goal is the analysis of the two social dimension (perception and believability) useful for increasing the natural behavior between users and Telenoid. We administrated our custom questionnaire to these subjects after a well defined experimental setting (ordinary and goal-guided task). After the analysis of the questionnaires, we obtained the proof that perceptual and believability conditions are necessary social dimensions for a successfully and efficiency HHI interaction in every daylife activities.

Keywords: Telenoid, Geminoid, Social Robot, Human-Humanoid Robot Interaction.

1 Introduction

Since humanoid robots are going to be part of the lives of human beings, specific studies are oriented to investigating collaborative and social features related to human-humanoid interaction (HHI)[1][2][3]. The HHI oriented toward a cohabitation environment where human and humanoid will share common tasks and goals is one of main direction where the Biologically Inspired Cognitive Architecture (BICA) community wants to invest their efforts [4].

In particular Kanda et al.[5] focused their attention to the concept of communication humanoid robot thinking as partner to help the human activities. Oztop et al.[6] put their attention to understand the perceptive relation between human and humanoid robots. The iCat, developed by Poel et al.[7], is a user-interface robot able to exhibit a range of emotions through its facial features

A. Chella et al. (Eds.): Biologically Inspired Cognitive Architectures 2012, AISC 196, pp. 167–174.
springerlink.com

and it is generally controlled by predefined animations. Metta et al.[8], describes ICub, a child humanoid robot used in embodied cognition research.

In contrast to these typical humanoid robots, Geminoid HI-1 is a humanoid robot with the external appearance of its inventor, Prof. Hiroshi Ishiguro and it is thought of being indistinguishable from real humans at first sight [9][10][11]. In particular, much relevant literature appeared on the features of natural character of agents interaction [12] [13][14].

The minimal agency used to solve BICA challenge include a key aspect that is defined as "sense of co-presence" [15][16].We oriented our research in the HHI field in the direction of "sense of being together with other people in a shared virtual environment" [17]. In particular we are interested in the study related to the sense of a person to be present in a remote environment with a robot ("Telepresence") and to the sense of a person to be present in a common environment with a robot where humans and humanoid are "accessible, available and subject to one another" [18].

To conduct our research studies,in this direction, in order to understanding areas of human cognition, which have not been tested or clarified until now, we used Telenoid as shown in the figure (Fig. 1) where it is represented the experimental setting used for our tests. This is a special humanoid robot because with its minimalistic human appearance it is designed to appear and behave like a human. The goal of our experimental paradigm is devoted to understand how some features of perceptual behavior work during the HHI. We want to assess the degree of Believability of interaction along dimensions that can be reasonably taken as meaningful indicators of social interaction, both in free and task directed conditions.

We have conducted a human-humanoid interaction test with 100 young people who did not have prior interaction experience with humanoid robots. Given the data analysis performed, we may claim only to have individuated two interaction dimensions, that is the Perceptual behavior and Believability that can serve to increase the natural looking- like of interaction behavior in human-humanoid interaction.

2 The Proposed Approach

From the relevant literature, we derived perceptual and cognitive features of overt behaviour used as parameters for the evaluation of natural HRI. We used Telenoid robot, to study such perceptually accessible features as meaningful clues for social interaction. Hence, the interaction setting and the research methodology were modeled as reported in [2].

Students of the Faculty of Architecture and Engineering (University of Palermo) were recruited for the tests and they did not have prior interaction experience with humanoid robots. All participants (100 total, 62 male and 38 female with average age 22) were introduced to the Telenoid, to the interaction setting structure that required a two stage interaction with the robot and to

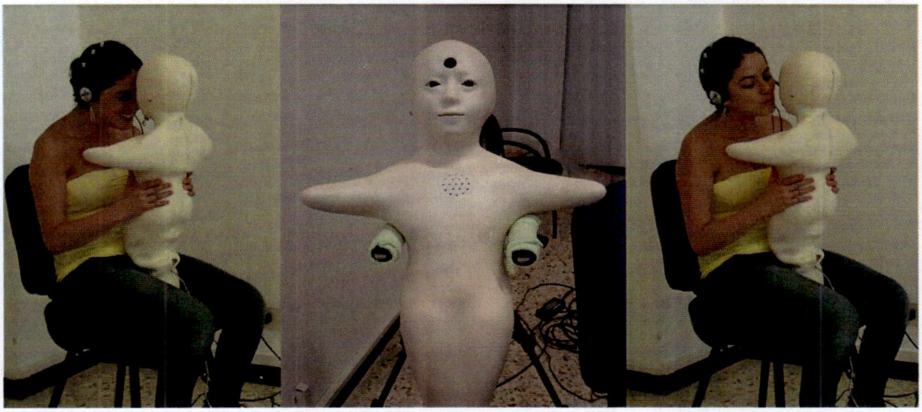

Fig. 1. The Telenoid and the experimental setting

fill up a questionnaire. All interactions were videotaped. A first free interaction stage, meant to allow subjects to adapt either to interact with the humanoid robot or to acquire as early as possible the skills for operating the robot through the control box.

A second interaction stage was instead task driven. Participants were allowed to choose an interactions scenario among a proposed range that spanned booking a hotel reservation, making a phone call to a mobile company to get a contract or services information, to matriculate or to enter his/her name or one of his/her fellows ones for a course examination by talking directly with the robot. The questionnaire was meant to recover some perceptual and social aspects of natural conditions of agents interaction. These emerged aspects mirrored some salient ordinary cognitive abilities, which agents could specify in such cases to improve the efficacy of interaction. The first construct is intended to cover perceptual features of overt interactive behavior. the perceptual awareness of sharing a common environment with the robot is assumed to be of momentous importance for the interacting subjects to ascribe intentions and actions to the robot itself. This aspect of interaction can prove to be the perceptual link with the second construct: believability. The concept is defined in Dautenhahn (1998)[19], and Poel et al. (2009)[7] operationalise it designing a construct whose aspects are represented by items grouped according to the indicators of personality, emotion, responsiveness, and self motivation. We assumed that meaningful descriptive units of behaviour must be picked up at the meso-scale level of what it is looking like to agents, we treated the perceptual features of interaction as sensory qualities that appear to have a meaning in themselves only when appropriately fit to one another rather than appearing as mere proxies of intentions or meanings to be retrieved beyond them.

3 Experimental Results

Standard item analysis has been performed [2] on the codified data using Split-half Spearmann-Brown coefficient, Pearson Coefficient,and Cronbach alphas were calculated to test the reliability of the scales and the correlation among the multiple items of each single construct of the questionnarie. We surveyed with these group of 100 young people the degree of acceptability of these artificial robots in every-day life.

Asking to the users, that had either none or mixed previous experience and knowledge of robotics, their reaction to the possibility to have the robot in the next 5 years inside the working and home space we surveyed the high degree of the acceptability of this artificial robotic agent in every-day life. Instead of using Bogardus scale we arranged the items of the questionnaires in a Likert scale because we wanted to avoid to narrow down the measurements of willingness of the users to have friendship interaction.

We had concrete measures of the attitudes towards non-human agents with the following dimensions:

1. the varying degree of acceptance in first person or on the ground of others choices;
2. the degree of diffusion weighted by the artificial tasks of the robotic device either in household or at work;
3. the range of attitudes from favorable reception to interest, distrust, annoyance.

For a first descriptive survey of our data, it emerged that subjects consider robots as potential useful device for many daylife. The percentage of skeptical subjects is quite low with a value of 20-25% across all contexts.

The question reported in table 1 is representative of one of the dimensions used for the perception construct, and also provides a conceptual link with the believability construct. The items of this question deal with the perceptual basis of the source of behavioral consistency. The Telenoid is then alleged to allow the human agent to see on the ground of the overt behavior given its degrees of freedom and voice-motion coupling. We conceived of observable behavior consistency as a fundamental underpinning of believability, along with coherence, that characterizes peoples construal during the interaction with other agents.

We reasoned that a consistent agent need to be able to assess the information available in the context of interaction, and this ability would require a value system composed of traits, and behaviour inclinations. Human agents must judge this system as the main guide for the robot when it has to recognize the users requests or it has to decide how to act toward them. Accordingly to the items of the question, the source of the robots value system appear to depend, whether on an inherent collection of traits, emotion-motivation system or on a set of inclinations necessary to meet the human agent requests. This value system might be either endowed from the outside (i.e. the operator) or ascribed by human interacting agents themselves. Subjects had to judge the fit to observable behaviors of sentences, which describe the Telenoid as assessing information and meeting requests on the ground of an inherent system of actual feelings and drives.

These emotional feelings are respectively defined in the items of the question as: simulated, imitated, played, replicated, and projected on behavioral screen. For the time being, our descriptive analysis showed that only a small fraction (5.77%) of subjects consider the Telenoid as equipped by an actual inherent value system that allows it to make appraisals of the context. Almost one third (30.70%) consider this description slightly unfit. We also expected a high overall unfit judgement (i.e. slightly unfit and unfit) given the Investigating Perceptual Features Human-Humanoid Interaction 5 conditions of our controlled interaction set up (44.23%). Even though it is worth noticing that as much as 23.38% subjects find this sentence as slightly fit to the overt behaviour of the humanoid robot. The imitation description scores the highest judgment of fit and the least one of unfit, and it is the one that subjects are least inclined to judge as neither fit nor unfit. That which fares best is the sentence describing the Telenoid as simulating the source of consistent behavior that is judged as slightly fit by 40.38% subjects and scores 48.07% for the overall judgment of fit.

The description of the Telenoid as a playing agent is judged as the most slightly unfit, and it records the highest unfit judgment (43.13%) after the sentence describing it as proving what he actually displays. The as-if projection screen description fares comparatively worse. It scores as the least fit and slightly fit, the most unfit, and records the highest percentage of subjects that find neither fit nor unfit. Finally, the sentence that describes the Telenoid as reproducing a consistent behaviour source records the third higher percentage of judgmnent of being slightly fit (35.29%). The results of this first survey are not so easily interpretable as they stand. These experimental findings may be consistent with the hypothesis that the Telenoid appears to acting on the ground of a sort of embedded value system also if the robots knowledge is tele-operated.

The questions reported on table 2 and 3 are related to the two dimensions that are at the basis of believability construct building: motivation and communication.

Regarding the question showed on table 2 the cumulative judgment of subjects fares rather well. Users consider that the Telenoid looks like as a believable agent. In particular the robot appears to pay attention to them, to acknowledge their requests, to show to be interested in them, and to display the expertise necessary to meet their demands in the guided task. A high rate of respondents expressed neither agreement nor disagreement, and this means that the overall performance of the Telenoid as believable interacting agent is efficient in defined tasks. Finally, the overall judgements of disagreement are at very low rates across all variables of this dimension. Regarding the question showed on table 3, the main focus is concerned the emotional expression of the Telenoid as important cognitive clues. In particular the question try to evaluate the capability of the robot to manage the variety of circumstances of interaction, especially where specific task driven requests are formulated by human agents. Evaluating the above mentioned experimental findings, subjects rated high the chance that the

Table 1. Perceptions about Telenoids behavior during interaction

	unfit	slightly unfit	Neither fit Nor unfit	slightly fit	fit
It Shows emotions	1,10%	15,38%	20,88%	49,45%	13,19%
It Simulates emotions	2,20%	21,98%	18,68%	42,86%	14,29%
It Imitates emotions	6,59%	17,58%	40,66%	30,77%	4,40%
It play emotions	6,67%	15,56%	37,78%	33,33%	6,67%
It reproduce emotions	18,68%	46,15%	18,68%	16,48%	0%
It is an interface for showing emotions	35,96%	38,20%	20,22%	4,49%	1,12%

Table 2. Credibilitys Evaluation of Telenoid as Interacting Agent

	Strongly Disagreement	Disagreement	Neither Agreement Nor Disagreement	Agreement	Strongly Agreement
It pays attention	0%	3,45%	27,59%	63,79%	5,17%
It meets users needs	0%	22,41%	55,17%	22,41%	0%
It shows involvement	1,75%	12,28%	31,58%	47,37%	7,02%
It shows expertise	1,72%	13,79%	31,03%	48,28%	5,17%

Table 3. Credibilitys Evaluation of Telenoid as Interacting Agent

	Strongly Disagreement	Disagreement	Neither Agreement Nor Disagreement	Agreement	Strongly Agreement
It shows spontaneous and uncontrolled emotions	10,34%	36,21%	29,31%	18,97%	5,17%
It shows personal and users emotions	3,51%	24,56%	43,86%	24,56%	3,51%
It shows reactions and emotions based on users needs	0%	22,81%	26,32%	49,12%	1,75%
It shows reactions and emotions that change during interaction	3,45%	12,07%	22,41%	55,17%	6,90%

Telenoids copying behavior ability is centered on the agents requests, which are to be met (49.12%), and on the course of emotions-requests cycle humanoid and human agents are engaged in (55.17%).

The view that the Telenoid copying behavior might be tuned to available information that proves relevant for itself and the human agents records a high response of neither agreement nor disagreement (43.86%), that again can attest the difficulty of the item of this question, or its sounding as somewhat ill-defined. But for the time being, the fit of this finding describe the lack of a sense of shared environment but it shows a high rate of approval when the Telenoid well simulates or imitates the source of its behavior consistency.

4 Conclusion and Future Works

The sense of "togetherness" between persons and Humanoid considered as BICA Robot Agent is "inherently social" [20] and is highly connected with the concepts of particular behavior defined "sensible" because [4] capable to express cognitive functionality.

The present research aims at a descriptive analysis of the main perceptual and social and cognitive features of natural conditions of agent interaction, which can be specified by agent in human-humanoid robot interaction.

Based on the data we already collected in this first stage of experiments, in the next future we will try to discern the impact of cognitive factors emerged from the questionnaires: easy of use, satisfaction, efficiency, personality, common sense, motivation and naturalness (thanks to the movements of the arms and the head of the humanoid robot). These parameters were obtained aggregating all users answers along the perception of the Telenoids features and the evaluation of the roles of these parameters in the interaction. A factor analysis need to be performed on subjects responses to item clusters within and across constructs. A regression on the data is also required to verify whether subjects knowledge of robotics, or their implicit evaluation of the widespread diffusion and use of artificial robotic devices will push in a correct direction the high value of their ratings. For some items, it is interesting to evaluate the reason of the high percentage of neutral ratings in the users responses. Some cases will require also a preliminary differential semantics analysis, before using in the next stage of the research. For other issues such as *the correct meaning of perceptual position and distances in the space of interaction between user and robot*, the analysis might be complemented by video-taped material. It will be important to link subjects responses to successfully or unsuccessfully task achievements. The rates of positive and negative achievements will be used as dependent variables of an appropriate set of features of the Telenoid. We will then be able to measure the efficiency of the interaction and success of the task assigned to the user using these indicators as independent proxies.

References

1. Balistreri, G., Nishio, S., Sorbello, R., Ishiguro, H.: Integrating Built-in Sensors of an Android with Sensors Embedded in the Environment for Studying a More Natural Human-Robot Interaction. In: Pirrone, R., Sorbello, F. (eds.) AI*IA 2011. LNCS, vol. 6934, pp. 432–437. Springer, Heidelberg (2011)
2. Ishiguro, H., Nishio, S., Chella, A., Sorbello, R., Balistreri, G., Giardina, M., Calí, C.: Perceptual Social Dimensions of Human - Humanoid Robot Interaction. In: Lee, S., Cho, H., Yoon, K.-J., Lee, J. (eds.) Intelligent Autonomous Systems 12. AISC, vol. 194, pp. 409–421. Springer, Heidelberg (2012)
3. Balistreri, G., Nishio, S., Sorbello, R., Chella, A., Ishiguro, H.: Natural human robot meta-communication through the integration of android's sensors with environment embedded sensors. Frontiers in Artificial Intelligence and Applications, vol. 233, pp. 26–37. IOS Press (2011)

4. Chella, A., Lebiere, C., Noelle, D., Samsonovich, A.: On a roadmap to biologically inspired cognitive agents. Frontiers in Artificial Intelligence and Applications, vol. 233, pp. 453–460. IOS Press (2011)
5. Kanda, T., Miyashita, T., Osada, T., Haikawa, Y., Ishiguro, H.: Analysis of humanoid appearances in human–robot interaction. IEEE Transactions on Robotics 24(3), 725–735 (2008)
6. Oztop, E., Franklin, D., Chaminade, T., Cheng, G.: Human-humanoid interaction: is a humanoid robot perceived as a human? International Journal of Humanoid Robotics 2(4), 537 (2005)
7. Poel, M., Heylen, D., Nijholt, A., Meulemans, M., Van Breemen, A.: Gaze behaviour, believability, likability and the icat. AI & Society 24(1), 61–73 (2009)
8. Metta, G., Sandini, G., Vernon, D., Natale, L., Nori, F.: The icub humanoid robot: an open platform for research in embodied cognition. In: Proceedings of the 8th Workshop on Performance Metrics for Intelligent Systems, pp. 50–56. ACM (2008)
9. Ishiguro, H.: Android science–toward a new cross-interdisciplinary framework. Robotics Research 28, 118–127 (2007)
10. Kanda, T., Ishiguro, H., Ono, T., Imai, M., Nakatsu, R.: Development and evaluation of an interactive humanoid robot robovie. In: Proceedings of IEEE International Conference on Robotics and Automation, ICRA 2002, vol. 2, pp. 1848–1855. IEEE (2002)
11. Shimada, M., Minato, T., Itakura, S., Ishiguro, H.: Evaluation of android using unconscious recognition. In: 2006 6th IEEE-RAS International Conference on Humanoid Robots, pp. 157–162. IEEE (2006)
12. Argyle, M., Dean, J.: Eye-contact, distance and affiliation. Sociometry, 289–304 (1965)
13. Argyle, M., Cook, M.: Gaze and mutual gaze (1976)
14. Argyle, M., Ingham, R., Alkema, F., McCallin, M.: The different functions of gaze. Semiotica 7(1), 19–32 (1973)
15. Durlach, N., Slater, M.: Presence in shared virtual environments and virtual togetherness. Presence: Teleoperators & Virtual Environments 9(2), 214–217 (2000)
16. Zhao, S.: Toward a taxonomy of copresence. Presence: Teleoperators & Virtual Environments 12(5), 445–455 (2003)
17. Slater, M., Sadagic, A., Usoh, M., Schroeder, R.: Small-group behavior in a virtual and real environment: A comparative study. Presence: Teleoperators & Virtual Environments 9(1), 37–51 (2000)
18. Goffman, E.: Behavior in public places: Notes on the social organization of gatherings. Free Pr. (1966)
19. Dautenhahn, K.: The art of designing socially intelligent agents: Science, fiction, and the human in the loop. Applied Artificial Intelligence 12(7-8), 573–617 (1998)
20. Biocca, F.: Communication within virtual reality: Creating a space for research. Journal of Communication 42(4), 5–22 (1992)

Internal Simulation of an Agent's Intentions

Magnus Johnsson[1] and Miriam Buonamente[2]

[1] Lund University Cognitive Science, Kungshuset, Lundagård, 22222 Lund, Sweden
{magnus}magnusjohnsson.se
[2] Department of Chemical, Management, Computer,
Mechanical Engineering of the University of Palermo, Palermo, Italy
{miriambuonamente}@gmail.com

Abstract. We present the Associative Self-Organizing Map (A-SOM) and propose that it could be used to predict an agent's intentions by internally simulating the behaviour likely to follow initial movements. The A-SOM is a neural network that develops a representation of its input space without supervision, while simultaneously learning to associate its activity with an arbitrary number of additional (possibly delayed) inputs. We argue that the A-SOM would be suitable for the prediction of the likely continuation of the perceived behaviour of an agent by learning to associate activity patterns over time, and thus a way to read its intentions.

To interact means to understand what others want and to foresee their intentions. Verbal communication is not enough, but intentions must be inferred from behaviour. Humans are able to do this, but an interesting question is how this ability could be implemented in artificial agents. One way to approach this problem would be to look closely at how this ability could arise in humans.

It has been argued that humans and animals can simulate perceptions, different actions and evaluate their likely consequences [1][2] by eliciting similar activity in perceptual and motor parts of the brain as if the actions and their likely perceptual consequences actually happened [3]. Among other things, internal simulation could also explain the appearance of an inner world [4]. We propose that humans understand others intentions by simulating the likely continuation of the perceived behaviour, or by simulating what they would have done in a similar situation.

To provide an artificial agent with an ability for internal simulation we propose the Associative Self-Organizing Map (A-SOM) [5][7], which is related to the Self-Organizing Map (SOM) [9]. The A-SOM is able to learn to associate its activity with the, possibly time delayed, activity of other neural networks.

By associating the activity of the A-SOM with its own earlier activity the A-SOM becomes able to remember perceptual sequences. This provides the ability to receive some initial sensory input and to continue to elicit the activity likely to follow in the nearest future even though no further input is received. This could be considered as sequence completion of perceptual activity over time. This has been tested in several simulations [6][7][8].

A. Chella et al. (Eds.): Biologically Inspired Cognitive Architectures 2012, AISC 196, pp. 175–176.
springerlink.com

We propose that the A-SOM could be used in a cognitive architecture for an agent to predict the intentions of an observed agent as well as its own intentions.

This could be done because the A-SOM should be able to learn to associate its own activity over time elicited by either the observed agents behaviour or its own. For example, the behaviour of the observed agent could be its particular way of moving towards certain goals. After enough observation of the agent, the A-SOM should have learnt the typical perceptual sequences associated with the agent's typical behaviours when it is moving towards different goals.

Then, due to the ability of the A-SOM for sequence completion of perceptual activity over time, it should be possible for it to internally simulate the sequence of activity likely to follow the activity elicited by the agent's initial behaviour. In this way the A-SOM should be able to reach the activity corresponding to the likely goal intended by the observed agent by internal simulation alone.

References

1. Barsalou, L.W.: Perceptual symbol systems. Behavioral and Brain Sciences, 577–609 (1999)
2. Grush, R.: The emulator theory of representation. Behavioral and Brain Sciences, 377–442 (2004)
3. Hesslow, G.: Conscious thought as simulation of behaviour and perception. Trends Cogn. Sci. 6, 242–247 (2002)
4. Hesslow, G., Jirenhed, D.-A.: The inner world of a simple robot. J. Consc. Stud. 14, 85–96 (2007)
5. Johnsson, M., Balkenius, C., Hesslow, G.: Associative Self-Organizing Map. In: Proceedings of IJCCI 2009, Funchal, Madeira, Portugal, pp. 363–370 (2009)
6. Johnsson, M., Gil, D., Balkenius, C., Hesslow, G.: Supervised Architectures for Internal Simulation of Perceptions and Actions. In: Proceedings of BICS 2010, Madrid, Spain (2010)
7. Johnsson, M., Martinsson, M., Gil, D., Hesslow, G.: Associative Self-Organizing Map. In: Mwasiagi, J.I. (ed.) Self Organising Maps - Applications and Novel Algorithm Design, pp. 603–626. INTECH (2011)
8. Johnsson, M., Gil, D., Hesslow, G., Balkenius, C.: Internal Simulation in a Bimodal System. In: Proceedings of SCAI 2011, Trondheim, Norway, pp. 173–182 (2011)
9. Kohonen, T.: The self-organizing map. Proceedings of the IEEE 78(9), 1464–1480 (1990)

A Model of Primitive Consciousness Based on System-Level Learning Activity in Autonomous Adaptation

Yasuo Kinouchi[1] and Yoshihiro Kato

[1] Tokyo University of Information Sciences, Chiba, Japan
kinouchi@rsch.tuis.ac.jp

Abstract. Although many models of consciousness have been proposed from various viewpoints, these models have not been based on learning activities in a whole system with capability of autonomous adaptation. Through investigating a learning process as the whole system, consciousness is basically modeled as system level learning activity to modify both own configuration and states in autonomous adaptation. The model not only explains the time delay of Libet's experiment, but also is positioned as an improved model of Global Workspace Theory.

Keywords: model of consciousness, autonomous adaptation, phenomenal consciousness, global workspace theory, and reinforcement learning.

1 Introduction

Although many models of consciousness have been proposed from various viewpoints, these models have not been based on learning activities in a whole system with capability of autonomous adaptation.[1,2] To clarify the functions and configuration needed for learning in a system that autonomously adapts to the environment Kinouchi et al have been investigating a simplified system using artificial neural nodes. On the system they have showed that phenomenal consciousness is explained using a "virtualization" method in the information system and that learning activities in a whole system adaptation are related to consciousness. However, they have not sufficiently clarified the learning activities of such a system.[3-5]

Through investigating a learning process as the whole system, consciousness is basically modeled as system level learning activity to modify both own configuration and states in autonomous adaptation. The model not only explains the time delay of Libet's experiment [6], but also is positioned as an improved model of Global Workspace Theory (GWT).[7-10]

2 Basic Conditions and Modeling Method

2.1 Basic Conditions of System

We assumed the basic conditions of autonomous adaptation as shown below.

A. Chella et al. (Eds.): Biologically Inspired Cognitive Architectures 2012, AISC 196, pp. 177–185.
springerlink.com

(i) The system must autonomously adapt to a complex environment without a teacher. This means that the system must learn from its own experiences that consist of the sequential perceptions, actions, and rewards in the environment as quickly and effectively as possible.

(ii) To adapt autonomously to the environment, the system has self-action-decision functions to adapt to certain circumstances and a learning-control that varies the system configuration itself based on a value-evaluation mechanism, such as reward and punishment.

(iii) Artificial neural nodes with stochastic characteristics, in which information is represented by random pulse frequency of activated nodes, are implemented. We assumed that signal processing speed of the nodes is at a comparable level to humans. Although these neural nodes operate slowly, the system must decide and execute as quickly as possible upon an appropriate or suitable action in accordance with the system's current situation.

(iv) The system must reduce as much of its resources, such as the nodes, and energy used in the system itself as possible.

2.2 Basic Configuration

To grasp complex functions of consciousness clearly and to realize a speedy response in spite of using slow speed neural nodes, the system configuration is simplified as much as possible. The basic configuration is depicted in Fig.1. The main characteristics are shown below.

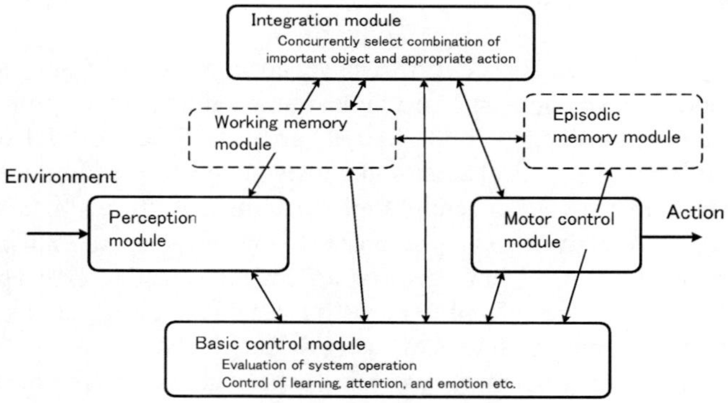

Fig. 1. Basic configuration of system

(a) Many processing flows are executed in parallel, and each flow is executed as quickly as possible.

(b) Selection of important objects and decision for appropriate action are executed by integration module concurrently in one operation.

(c) The motor control module immediately drives motors on the basis of the decision from the integration module as an action for environment. At the same time, in many cases, learning activities as a system are carried out by using multiple modules in the system.

(d) Above activities are control of basic control module.

2.3 Classification of Processing Level

The processing of the system is classified into two flows, system and non-system level. In system level processing, the activities are decided on the basis of information from the whole system; for example, the decision of approach or avoidance of the system. To execute these flows, various pieces of perceived information need to be gathered from a wide area within the system and interact. On the other hand, in non-system level processing, various activities are directly controlled by the corresponding local or restricted information. System level processing needs more time and resources than non-system level processing. We assume that system level processing is activated only for situations where the system has to operate as a whole and for this processing an integration module is required.

2.4 Basic Learning Method and Levels of Learning Activities

System level learning activities are executed by the actor-critic method under the framework of reinforcement learning.[11] The integration module functions as an actor, and evaluation mechanism in basic control module functions as a critic.

Learning that modify the system itself are classified into two activities, system level learning activities and non-system level learning activities. System level learning activities are executed under the influence of the unique result of evaluation mechanism in the basic control module as a whole system. However, non-system level learning activities are executed based on local activities. These two activities are not exclusive. In many cases these activities operate concurrently. We assume that system level learning activities can be executed by only a part of the system level processing.

2.5 Phenomenal Consciousness and Functional Consciousness

In modeling of consciousness, it is important to note the difference between phenomenal consciousness and functional consciousness.[12] To clarify the difference, a model with two layers, a physical layer and a logical layer, is proposed. The physical layer is composed of neural nodes. In contrast, the logical layer is composed of only the information selected and mapped from the physical layer.

Selected information is classified to three groups corresponding to our daily feeling as below.

Group-a: Real-time information of real space, situation, and our body as the actual phenomena experienced outside our brains.

Group-b: Thinking or recollected images.

Group-c: States of mind as a mental or emotional phenomena inside the brain.

3 Physical Configuration

Physical configuration is shown in Figure 2.

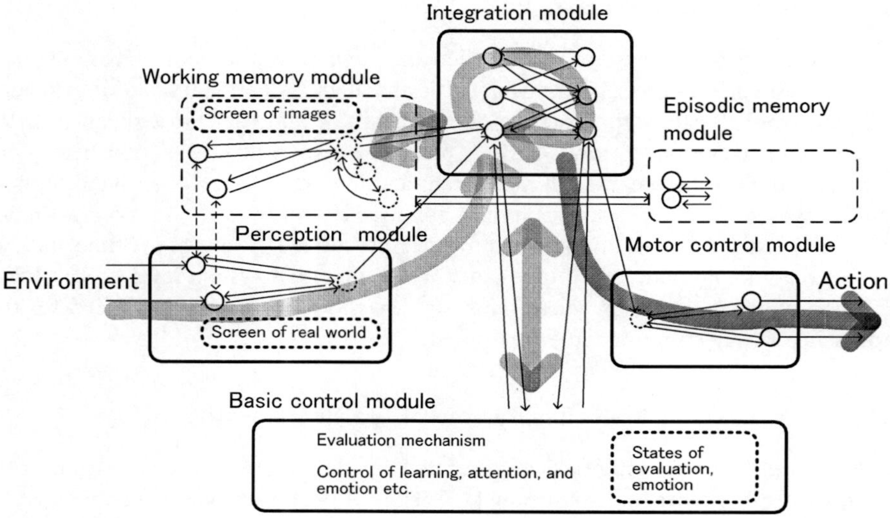

Fig. 2. Physical configuration

(1) The perception module consists of micro-feature nodes, concept perception nodes and the screen of the real world. The micro-feature nodes represent corresponding sensor signals mainly from the environment. The concept perception nodes recognize concepts based on related micro-features. The screen of the real world is depicted by micro-features that are active at that time. (Here, the information needed to depict the screen is investigated, but how the screen is depicted is not referred to).

(2) The integration module composed of interconnected massive neural nodes selects a dominant situation represented by nodes related to concepts or objects in its environment, and as shown below calculates approximately desirable candidates for an action concurrently and speedily by the steepest descent method based on iterative interactions between looped nodes.

$$\frac{\Delta D}{\Delta x_i} = a_i + \sum_j b_{i,j} x_j \tag{1}$$

$$D = \sum_i a_i x_i + \sum_{i,j} b_{i,j} x_i x_j \tag{2}$$

D : System desirability
x_i : Necessity of object or action i
a_i : Expected reward corresponding to x_i
$b_{i,j}$: Expected reward corresponding to a pair of x_i and x_j

The selected dominant situation is transferred to the working memory module, and the candidate for an action is transferred to the motor control module. The integrated module operates as an “actor” in reinforcement leaning. Processing flows executed in this module belong to system level processing.

(3) The working memory module has image micro-feature nodes, image concept nodes and the screen of images. Each image micro-feature node corresponds to the micro-feature node with the same attribute in the perception module. Each image concept node represents the concept node that corresponds to it in the perception module, and has functions of delayed memory that memorize multiple states at time t, t-1,...,t-n. First, output signals from the integration module stimulate the image concept node. The node stimulates image micro-feature nodes that belong to the image concept node. When these image micro-feature and concept nodes have sufficient support signals from the basic control module, nodes activation are maintained by mutual stimulation between nodes for a short time, similar to the perception module. Images on the screen are depicted by these activated image micro-feature nodes, and the states of image concept nodes are transmitted to the integration module.

(4) The motor control module transmits orders to the motor for action.

(5) The episodic memory module sequentially memorizes a group of information mainly buffered in working memory that corresponds to a screen of real world, images, and emotional states including the results of the evaluation.

(6) The basic control module has automatic or semi-automatic control functions of evaluation mechanism, system level learning, processing path, resource, and emotional states.

4 Processing Steps in System Level Learning

Processing steps are shown in Fig. 3.

Step 1: Recognition and estimation of influence on the system

Recognitions by micro-feature and concept nodes are executed. At the same time, the evaluation mechanism estimates the degree of various influences that an individual node gives the system.

Step 2: Calculation of nearly optimum plan

The integration module outputs a nearly optimal plan $P(t)$ that is in accordance with the measurements of the system desirability $D(t)$ at time t. Here, $D(t)$ is determined by the degree of influences evaluated at the previous step, and system states corresponding to control functions of emotion and homeostasis. Motor signals are transmitted to motor modules, and signals of selected dominant objects are transmitted to the working memory module.

Step 3-1: Calculation of a total reward $R_{total}(t)$

The evaluation mechanism estimates $E(p(t))$, that is expected reward of plan $P(t)$, and evaluate real reward $R_{real}(t)$. Total reward $R_{total}(t)$ is then calculated.

$$R_{total}(t) = E(P(t-1)) - E(P(t)) + R_{real}(t) \tag{3}$$

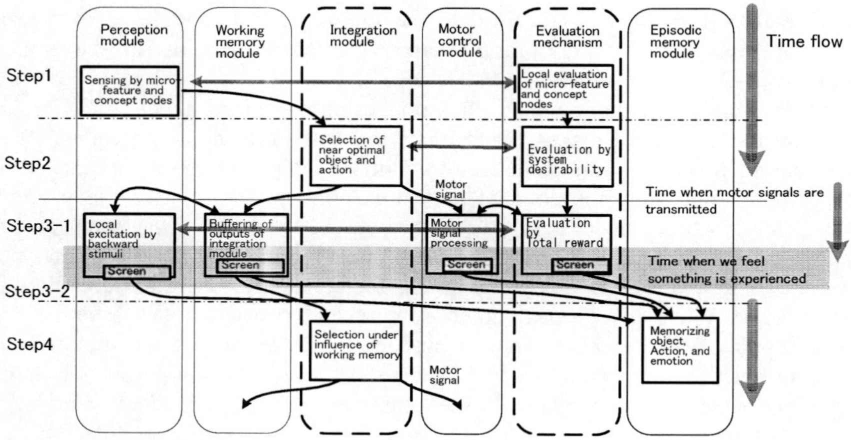

Fig. 3. Processing steps in system level processing

Step3-2: Execution of system level learning in accordance with total reward

When $R_{total}(t)$ is sufficiently high, and working memory and micro feature nodes belonging to selected objects did not engage in high priority other work, the basic control module executes system level learning.

In executing system level learning activities, nodes states that correspond to situation selected by the integration module are fixed or maintained temporarily and total reward $R_{total}(t)$ functions as a teacher signal. To fix or maintain nodes states, micro-feature and concepts nodes excite concurrently based on mutual stimulation between them under the control of the basic control module. We assume that, at this moment, we feel the selected situation as a scene recognized in environment, and reward values as emotion in daily life. Differences from the time when a motor signal is transmitted to the time when we feel corresponds to a delay such as that described by Libet. [6] To cope with Stability-Plasticity Dilemma [13], the basic control module memorizes the information, combination of fixed states and a teacher signal, into episodic memory sequentially. The data are recalled when the system is not in a busy state, such as sleeping. Recalled information slowly and repeatedly modifies the actor's main configuration, such as connection weights.

Step 4: System level processing based on contents of working memory

In the same way as output of the perception module, contents memorized on the working memory are input to the integration module and are able to execute system level processing. We assume that our "conscious" action when we act purposefully is based on this processing, additionally repeated processing between the integration and working memory module without using stimuli from environment are corresponding to our daily "thinking".

5 Configuration and Function of Logical Layer

The phenomenal consciousness is modeled in upper layers as shown in Fig. 4.

Information for the system level learning, group-a, b, and c, are selected from the physical layer, and mapped to the logical layer using a "virtualization" method in the information system. [14] The operations in the logical layer are represented by interactions between only the selected information. The information not selected from the physical layer is invisible in the logical layer.[15]

The concept of self is composed of many types of self as advocated by Damasio [16] and Franklin [12]. In autonomous adaptation, one of the most important functions is an evaluation of rewards, then the value evaluation mechanism represents the core part of the self, and we assume that it corresponds to Damasio's core self. Additionally, the self that includes screen of images and declarative memory corresponds to Damasio's autobiographical self. This self not only feels autobiographical experiences, but also feels a reflection of self that the system feels or thinks the system's states.

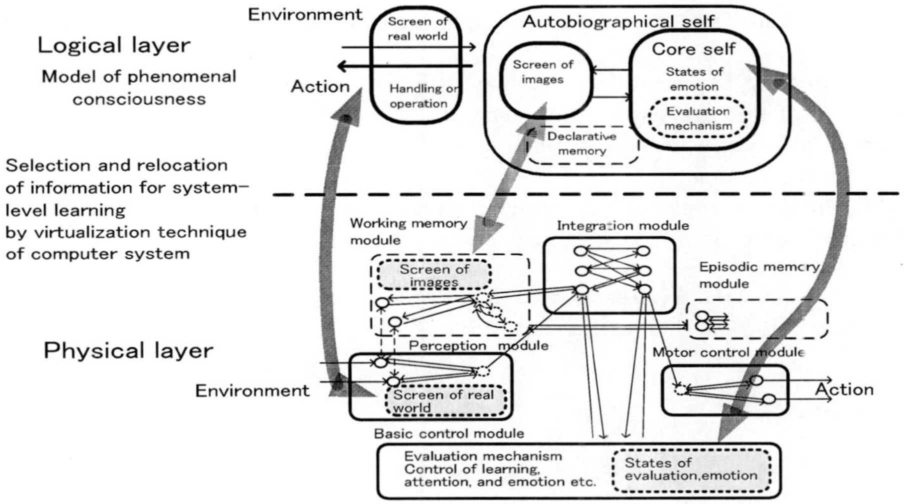

Fig. 4. Configuration of logical layer (model of phenomenal consciousness)

6 Summary and Discussion

A model of primitive consciousness based on "system level" or "global" learning activities in autonomous adaptation is proposed. In the model, consciousness is indispensable and essential functions in autonomous adaptation and dominant phenomenon in the learning activities correspond to our daily feelings.

Given that the integration module functions as a kind of Global Workspace, our model and GWT have a similarity. However, GWT does not distinguish system-level learning from system-level processing, and assumes broadcast of output of the system-level processing to be phenomenal consciousness.[7,8,10] In our model, execution of system-level learning, corresponding to main conscious feeling, becomes feasible after system-level processing was performed. To execute system-level processing in sequence hastily or rapidly, system-level learning is not necessary. Moreover, in many situations, system-level learning interrupts the flow of action and slows down performance. Learning that modify internal state and configuration is preparing activity for the next time or future, but action related to the environment is necessary at that time. In autonomous adaptation, as action has a higher priority than learning, transmission of action signals have to precede conscious feeling. Through grasping a primary purpose of the consciousness as system-level learning and that phenomenal consciousness is a kind of phenomena resulting from activity using information to be used in learning, our model explains the time delay in Libet's experiment.[6]

Although consciousness is essentially based on learning functions, contents memorized on the working memory felt as a conscious experience are able to influence the next action through the integration module or perception module. We think this case corresponds to "broadcast for recruitment" in GWT and "executive summary" of Koch. [17]

Additionally, as system level learning activities are executed under the control of the value of one evaluation mechanism, the processing stream inevitably becomes single. This explains the reason behind the assertion that "human consciousness usually displays a striking unity." [18]

We are now investigating the system configuration and programming a whole system simulator to clarify its operational characteristics in more detail.

Acknowledgments. We thank Mr. Shoji Inabayashi and Prof. Igor Aleksander of Imperial College for their useful suggestions.

References

1. Seth, A.: Models of Consciousness. Scholarpedia 2(1), 1328 (2007)
2. Taylor, J., Freeman, W., Cleeremans, A.(eds.): Brain and Consciousness, special issue of Neural networks. Neural Networks 20(9) (2007)
3. Kinouchi, Y.: A logical model of consciousness on an autonomously adaptive system. IJMC 1(2) (2009)
4. Kinouchi, Y., Kato, Y., Hayashi, H., Katsumata, Y., Kitakaze, K., Inabayashi, S.: A model of primitive consciousness in autonomously adaptive system under a framework of reinforcement learning. In: AISB 2011 Machine Consciousness (2011)
5. Kinouchi, Y., Kato, Y., Inabayashi, S.: Model of primitive consciousness based on learning activity in autonomously adaptive system. In: ASSC, vol. 16 (2012)
6. Libet, B.: Mind time: The temporal Factor in Consciousness. Harvard University Press (2004)

7. Baars, B.J.: A cognitive theory of consciousness. Cambridge University Press (1988)
8. Baars, B., Franklin, S.: How conscious experience and working memory interact. Trend in Cognitive Sciences 7(4) (2003)
9. Changeux, J.P., Dehaene, S.: The Neural Workspace Model: Conscious Processing and Learning. In: Learning Theory and Behavior, vol. 1. Elsevier (2008)
10. Franklin, S., Strain, S., Snaider, J., McCall, R., Faghihi, U.: Global workspace theory, its LIDA model and the underlying neuroscience. Biologically Inspired Cognitive Architectures 1(1) (2012)
11. Sutton, R., Barto, A.: Reinforcement Learning: An Introduction. MIT Press, Cambridge (1998)
12. Franklin, S., D'Mello, S., Baars, B., Ramamurthy, U.: Evolutionary Pressures for Perceptual Stability and Self as Guides to Machine Consciousness. International Journal of Machine Consciousness 1(1), 99–110 (2009)
13. Carpenter, G.A., Grossberg, S.: Pattern recognition by self-organizing neural networks. The MIT Press (1991)
14. Patterson, D., Hennessy, J.: Computer Organization and Design: The hardware/software interface. Morgan Kaufman Publishers, San Francisco (1994)
15. Norman, D.: Invisible Computer. MIT Press (1998)
16. Damasio, A.: Descartes' error-Emotion, Reason, and the Human Brain. Putnam's Sons (1994)
17. Koch, C.: The Quest for Consciousness: A Neurobiological Approach. Roberts and Company Publishers, Colorado (2004)
18. Brook, A., Raymont, P.: The Unity of Consciousness: The Stanford Encyclopedia of Philosophy (Fall 2010 Edition), http://plato.stanford.edu/entries/consciousness-unity/

Decision-Making and Action Selection in Two Minds

Muneo Kitajima[1] and Makoto Toyota[2]

[1] Nagaoka University of Technology,
16031-1, Kami-Tomioka Nagaoka Niigata 940-2188, Japan,
mkitajima@kjs.nagaokaut.ac.jp
http://oberon.nagaokaut.ac.jp/ktjm/

[2] T-Method, 3-4-9-202 Mitsuwadai, Wakaba-ku, Chiba-city, Chiba, Japan

Abstract. This paper discusses the differences between decision-making and action selection. Human behavior can be viewed as the integration of output of System 1, *i.e.*, unconscious automatic processes, and System 2, *i.e.*, conscious deliberate processes. System 1 activates a sequence of automatic actions. System 2 monitors System 1's performance according to the plan it has created and, at the same time, it activates future possible courses of actions. Decision-making narrowly refers to System 2's slow functions for planning for the future and related deliberate activities, *e.g.*, monitoring, for future planning. On the other hand, action selection refers to integrated activities including not only System 1's fast activities but also System 2's slow activities, not separately but integrally. This paper discusses the relationships between decision-making and action selection based on the architecture model the authors have developed for simulating human beings' *in situ* action selection, Model Human Processor with Real time Constraints (MHP/RT) [3] by extending the argument we have done in the argument we have made in previous work [5].

Keywords: decision-making, action planning, Two Minds.

1 Decision-Making and Action Selection

Decision-making is the act or process of choosing a preferred option or course of action from a set of alternatives. It precedes and underpins almost all deliberate or voluntary behavior. *Action selection* is the process for selecting "what to do next" in dynamic and unpredictable environments in real time. The outcome of decision-making is regarded as part of resources that are available when selecting actions [9]. As dual-processing theories suggest (*e.g.*, [2]), two qualitatively different mechanisms of information processing operate in forming decisions. The first is a quick and easy processing mode based on effort-conserving heuristics. The second is a slow and more difficult rule-based processing mode based on effort-consuming systematic reasoning. The first type of process is often unconscious and tends to automatic processing, whereas the second is invariably conscious and usually involves controlled processing.

A. Chella et al. (Eds.): Biologically Inspired Cognitive Architectures 2012, AISC 196, pp. 187–192.
springerlink.com

Kahneman, winner of the Nobel Prize in economics in 2002, introduced behavioral economics, which stems from the claim that decision-making is governed by the so-called "Two Minds" [2], a version of dual processing theory, consisting of System 1 and System 2. System 1, the first type of process, is a fast feed-forward control process driven by the cerebellum and oriented toward immediate action. Experiential processing is experienced passively, outside of conscious awareness (one is seized by one's emotions). In contrast, System 2, the second type of process, is a slow feedback control process driven by the cerebrum and oriented toward future action. It is experienced actively and consciously (one intentionally follows the rules of inductive and deductive reasoning).

This paper discusses the relationships between decision-making and action selection based on the architecture model the authors have developed for simulating human beings' *in situ* action selection, Model Human Processor with Real time Constraints (MHP/RT) [3] by extending the argument we have done in the argument we have made in previous work [5]. MHP/RT defines how System 1 and System 2 work together to generate coherent behavior being synchronized with ever-changing environment.

2 MHP/RT: Model Human Processor with Realtime Constraints

We proposed Model Human Processor with Realtime Constraints (MHP/RT) as a simulation model of human behavior selection. It stems from the successful simulation model of human information processing, Model Human Processor (MHP) [1], and extends it by incorporating three theories we have published in the cognitive sciences community. The Maximum Satisfaction Architecture (MSA) deals with coordination of behavioral goals [7], the Structured Meme Theory (SMT) involves utilization of long-term memory, which works as an autonomous system [10], and Brain Information Hydrodynamics (BIH) involves a mechanism for synchronizing the individual with the environment [6].

MHP/RT includes a mechanism for synchronizing autonomous systems (square-like shapes with rounded corners in Figure 1), working in the "Synchronous Band." MHP/RT was created by combining MHP and Two Minds by applying our conceptual framework of Organic Self-Consistent Field Theory [4].

MHP/RT works as follows:

1. Inputting information from the environment and the individual,
2. Building a cognitive frame in working memory (not depicted in the figure but it resides between the conscious process and the unconscious process to interface them),
3. Resonating the cognitive frame with autonomous long-term memory to make available the relevant information stored in long-term memory; cognitive frames are updated at a certain rate and the contents in the cognitive frames are frames are a continuous input to long-term memory to make pieces of information in long-term memory accessible to System 1 and System 2,

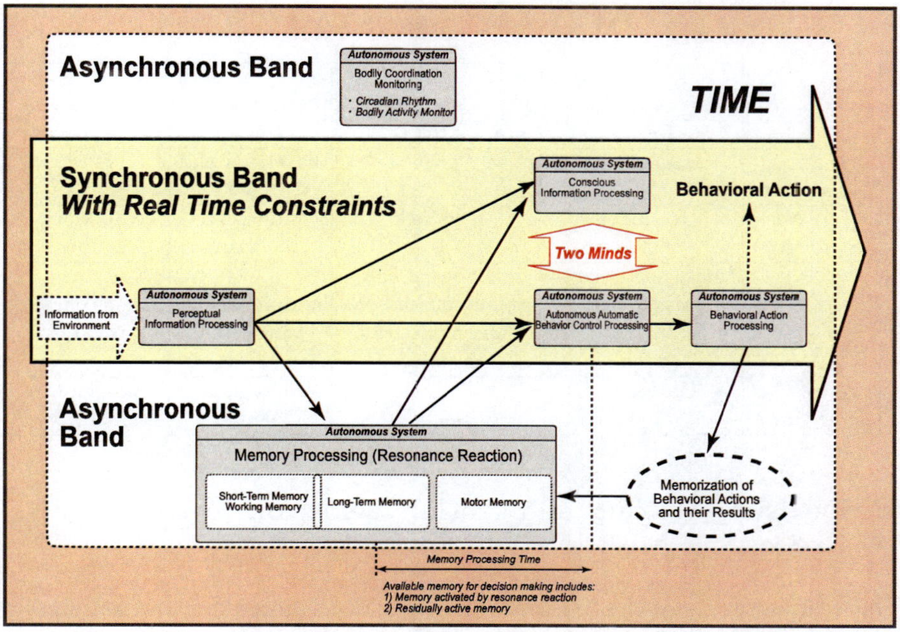

Fig. 1. MHP/RT (adapted from [3])

4. Mapping the results of resonance on consciousness to form a reduced representation of the input information, and
5. Predicting future cognitive frames to coordinate input and working memory.

3 Four Processing Modes of Human Behavior

In [5], the authors introduced Four Processing Modes of *in situ* human behavior that are derived by augmenting the theory of decision-making, Two Minds [2], by taking into account the different nature of decision-making before the boundary event and after the boundary event, that is captured by Newell's time scale of human action [8]. Table 1 shows the resultant Four Processing Modes of *in situ* human behavior; at each moment along the time dimension human behaves in one of the four modes and he/she switches among them depending on the internal and external states.

Decision-making processes before the boundary event and those after the boundary event are significantly different in terms of the impact of real time constraints on the decision-making processes. Considering that decision-making is the result of the workings of System 1 and System 2, there are four distinctive behavior modes, 1) conscious (System 2) behavior before the boundary event, 2) conscious (System 2) behavior after the boundary event, 3) unconscious (System 1) behavior before the boundary event, and 4) unconscious (System 1) behavior after the boundary event.

Table 1. Four Processing Modes [5]

	System 2 Conscious Processes		**System 1 Unconscious Processes**	
	Before	*After*	*Before*	*After*
Time Constraints	none or weak	exist	none or weak	exist
Network Structure	feedback	feedback	feedforward + feedback	feedforward + feedback
Processing	main serial conscious process + subsidiary parallel process	main serial conscious process + subsidiary parallel process	simple parallel process	simple parallel process
Newell's Time Scale	Rational / Social	Rational / Social	Biological / Cognitive	Biological / Cognitive

4 Decision-Making and Action Selection in the Four Processing Modes

This section discusses the differences between decision-making and action selection using Figure 2, adapted from [5], that illustrates the Four Processing Modes along the time dimension expanding before and after the boundary event.

4.1 Creation and Utilization of Event Memory

The four processing modes are defined by referring to a single event (boundary event). Therefore, it is useful to consider how each of the four processing modes works when one encounters an event for the first time, and it encounters the same event in the future.

When one encounters an event for the first time, "System 1 After" processing and/or "System 2 After" processing will work to create encodings of the event as an experiential memory frame. "System 2 After" processing will elaborate on the outcome of "System 1 After" processing. Usually, several times of repetition of encountering the same event will be necessary to establish a cohesive memory frame.

The experiential memory frame thus created may be activated before the event happens through "System 1 Before" processing and/or "System 2 Before" processing. This paper suggests that action selection corresponds to "System 1 Before" processing and decision-making corresponds to "System 2 Before" processing. Since characteristic times of System 1 and those of System 2 are significantly different, they have different meanings for the behavior to be taken for the event. As shown in Figure 3, "System 2 Before" processing, or decision-making, for the future event will work long before the event happens when there is time available for collecting possible actions through deliberate

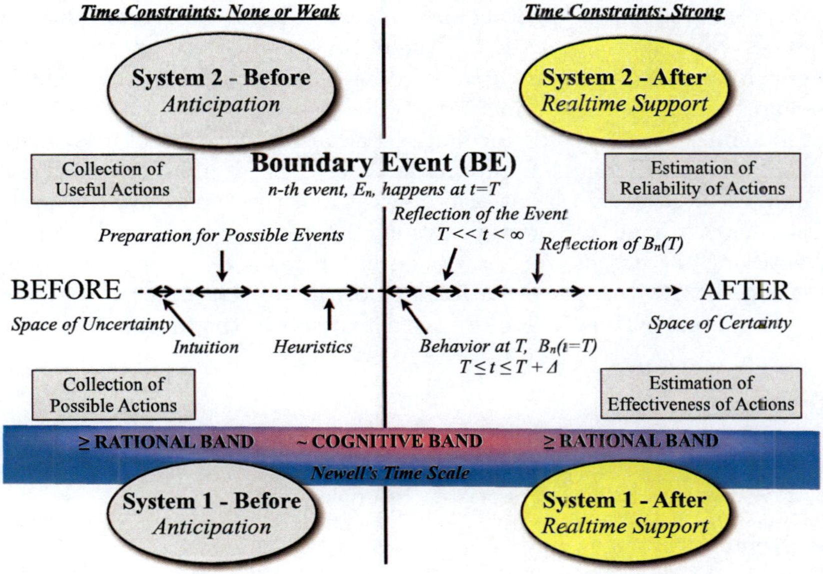

Fig. 2. How the Four Processing Modes work (adapted from [5])

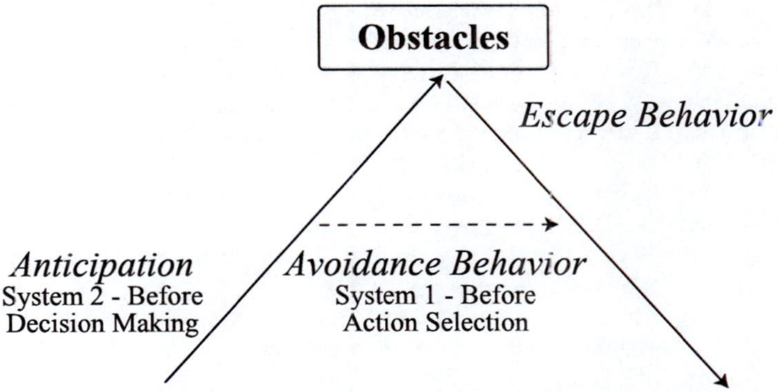

Fig. 3. Decision-making carried out by "System 2 Before" processing mode and action selection by "System 1 Before" processing mode of MHP/RT.

thinking, whereas "System 1 Before" processing, or action selection, for the immediate future anticipatory event will happen; one will be able to select action to behave appropriately, not only experiencing the event but also avoiding the event (not experiencing the event but an alternative event).

4.2 Transition from Experiential Memory to Prospective Memory

An experiential memory frame that "System 1 After" processing has created will be converted into an prospective memory frame, that can be used by "System 1

Before" processing for anticipating and preparing for future events. This conversion process can be automatic when "System 1 After" processing is able to identify the perceptual objects that are associated with the encoding of the event stored in the experiential memory frame.

For example, when one has encountered a harmful insect and been stung, he or she would immediately and automatically establish a link between the visual and auditory perceptual signals of the insect and the action to drive away the insect by his/her hand. Otherwise, "System 2 After" processing will be required for identifying the objects that might be useful for anticipating the event and associating them with perceptual features of the objects that can be detected by the perceptual system before the event happens in the future.

References

1. Card, S.K., Moran, T.P., Newell, A.: The Psychology of Human-Computer Interaction. Lawrence Erlbaum Associates, Hillsdale (1983)
2. Kahneman, D.: A perspective on judgment and choice. American Psychologist 58(9), 697–720 (2003)
3. Kitajima, M., Toyota, M.: Simulating navigation behaviour based on the architecture model Model Human Processor with Real-Time Constraints (MHP/RT). Behaviour & Information Technology 31(1), 41–58 (2012)
4. Kitajima, M.: Organic Self-Consistent Field Theory (2012), `http://oberon.nagaokaut.ac.jp/ktjm/organic-self-consistent-field-theory/`
5. Kitajima, M., Toyota, M.: Four Processing Modes of in situ Human Behavior. In: Samsonovich, A.V., Jóhannsdóttir, K.R. (eds.) Biologically Inspired Cognitive Architectures 2011 - Proceedings of the Second Annual Meeting of the BICA Society, pp. 194–199. IOS Press, Amsterdam (2011)
6. Kitajima, M., Toyota, M., Shimada, H.: The Model Brain: Brain Information Hydro- dynamics (BIH). In: Love, B.C., McRae, K., Sloutsky, V.M. (eds.) Proceedings of the 30th Annual Conference of the Cognitive Science Society, Austin, TX, p. 1453. Cognitive Science Society (2008)
7. Kitajima, M., Shimada, H., Toyota, M.: MSA: Maximum Satisfaction Architecture – a basis for designing intelligent autonomous agents on web 2.0. In: McNamara, D.S., Trafton, J.G. (eds.) Proceedings of the 29th Annual Conference of the Cognitive Science Society, Austin, TX, p. 1790. Cognitive Science Society (2007)
8. Newell, A.: Unified Theories of Cognition (The William James Lectures, 1987). Harvard University Press, Cambridge (1990)
9. Suchman, L.: Plans and situated actions: the problem of communication. Cambridge University Press, Cambridge (1987)
10. Toyota, M., Kitajima, M., Shimada, H.: Structured Meme Theory: How is informational inheritance maintained? In: Love, B.C., McRae, K., Sloutsky, V.M. (eds.) Proceedings of the 30th Annual Conference of the Cognitive Science Society, Austin, TX, p. 2288. Cognitive Science Society (2008)

Cognitive Chrono-Ethnography: A Methodology for Understanding Users for Designing Interactions Based on User Simulation with Cognitive Architectures

Muneo Kitajima[1] and Makoto Toyota[2]

[1] Nagaoka University of Technology,
16031-1, Kami-Tomioka Nagaoka Niigata 940-2188, Japan
mkitajima@kjs.nagaokaut.ac.jp
http://oberon.nagaokaut.ac.jp/ktjm/

[2] T-Method, 3-4-9-202 Mitsuwadai, Wakaba-ku, Chiba-city, Chiba, Japan

Abstract. A handful cognitive architectures have been proposed in the BICA society that are capable of simulating human beings' behavior selections. The purpose of this paper is to discuss the importance of designing interactions between users and the information provided to users via PC displays, traffic road signs, or any other information devices, and to suggest biologically-inspired cognitive architectures, BICA, are useful for designing interactions that should satisfy users by taking into account the variety of interactions that would happen and defining requirements that should satisfy user needs through user simulation using a cognitive architecture in BICA. A new methodology of defining requirements based on user simulations using a cognitive architecture, Cognitive-Chrono Ethnography (CCE), is introduced. A CCE study is briefly described.

Keywords: human beings' behavior selection simulation, Two Minds, human–computer interaction, interface design.

1 Designing Satisfactory Interactions between Users and Information Devices

1.1 "Know the Users" in Human-Computer Interaction

"Know the users" is the key principle for designing satisfactory interactions. Users interact with information devices in order to achieve the states where they want to be. During the course of interactions, users expect to have satisfactory experience. From design side, this can be accomplished by applying the principle, "know the users", and by designing interactions accordingly to provide as much satisfaction as possible to the users through their experience of using the information devices. However, it is often hard to practice this principle due to the diversity of users. Each user has his/her own experience in using interaction devices, and his/her past experience should affect significantly how he/she

A. Chella et al. (Eds.): Biologically Inspired Cognitive Architectures 2012, AISC 196, pp. 193–198.
springerlink.com

would interact with the devices at a particular situation. Since no one has the same experience, it seems no systematic way to practice the "know the users" principle.

1.2 "Know the Users" in Behavioral Economics

"Know the users" is an important study issue in the other domains such as behavioral economics. How does a user decide to purchase a new tablet PC for daily use? He or she selects one from a number of candidates in order to realize the states where he/she wants to be. This situation is very similar to the one described above. In the field of economics, the user's decision-making process has been studied extensively. Recently, Kahneman [2] revealed that the core process of human beings' decision-making is an integral process of so-called Two Minds [1,3]. We suggest that human brains would work similarly when people interact with information devices as when they engage in economic activities. If Two Minds is also working in human-computer interaction processes, we would be able to systematically apply the principle "know the user" for designing satisfactory interactions.

Two Minds refers to the following two systems; System 1, the automatic and fast unconscious decision-making process, oriented toward immediate action, and System 2, the deliberate and slow conscious decision-making process, oriented toward future action. We can easily imagine how Two Minds would work when users interact with information devices. In human-computer interaction, users deliberately consider what to do next and perform a series of actions on the device automatically. At the same time, they pay attention to the device's feedback and plan future actions accordingly. What we need to understand is how users switch between the slow and the fast processes of Two Minds, and explain and predict the behaviors we observe. The users' behaviors change depending on how the interaction is designed. The smoother the switching, the more the users would feel satisfaction. By taking into account explicitly the interaction between the slow deliberate processes and the fast automatic processes, we will be able to design interactions that surely satisfy the users' interaction experience.

1.3 "Know the Users" in BICA

Another domain that is relevant to "know the users" is a branch of robotics which studies "biologically inspired cognitive architectures (BICA)." Its purpose is to design an autonomous system that is capable of working in the ever-changing environment. We can simulate people's behaviors by using an appropriate cognitive architecture. We showed above that the smoothness of switching between the slow and the fast processes is important for satisfactory interaction design. This is a time-critical dynamic aspect of human-computer interaction, which behavioral economics cannot address but any models working on appropriate cognitive architectures would be able to deal with.

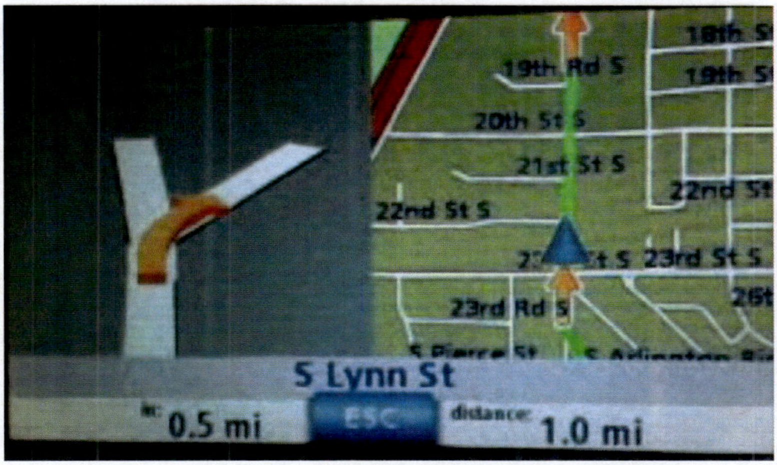

Fig. 1. Screenshot from a car-navigation system

1.4 Two Minds and Cognitive Architecture

What does it mean to the interaction design activities that Two Minds resides behind people's behaviors? We'd like to suggest that Interaction design is about designing time for the user in terms of a series of events that the user will be provided at a specific time T, by taking into account the fact that the user's process is controlled by Two Minds. This is because interactions happen at the interface of a system and a user, and the only and unique dimension that the system and the user's Two Minds can share is the time dimension. The user decides what to do next by using his/her Two Minds at time $T - \alpha$, carries it out at time T, the system responds to it at $T + \beta$, and this cycle continues. The system's response at $T + \beta$ needs to take into account how the user's Two Minds would process it. He/she may expect the system's response for consciously confirming or unconsciously matching whether he/she did right or not, or he/she may expect it for consciously planning or unconsciously triggering the next action. The user's expectations can become diverse but interactions designers need to take into account them appropriately in order for the designed system should satisfy the users' expectations.

Here is an example to illustrate the point. When you hear the car-navigation system start speaking in synthesized voice, you switch your attention to listening to what it says and try to comprehend it for planning your driving for the near future. The navigation system is designed to speak, for example, "Slight right turn in point five miles on South Lynn Street" with the screen shown in Figure 1 at some specific moment. The driver, who is not familiar with the route, is supposed to listen to the instruction and read the screen consciously and carefully, and integrate the provided information from the car-navigation system with the current driving situation for imagining and planning the immediate-future driving and creating a sequence of actions for the maneuver; when to start reducing speed, when to start braking, and so forth. When the navigation system starts speaking at time T "Slight right turn ...", it should intervene the driver's on-going processes and initiates a new interactive process stream on the part of the driver.

This interaction must be designed well by taking into account whatever Two Minds processes the driver engages in so that the newly initiated process does not interfere with the other on-going processes; some processes must be suspended and resumed at a proper timing with little cost, and the other processes should continue with no interference from the car-navigation system (*e.g.*, keep conversing with a person in the passenger seat). We can simulate switching between the unconscious automatic fast processes and the conscious deliberate slow processes by using an appropriate cognitive architecture developed in BICA such as proposed by us [4,5].

2 Cognitive Chrono-Ethnography: CCE

As described above, "know the users" can be practiced systematically by designing user studies based on a BICA simulation of users' mental operations controlled by Two Minds. By operationalizing this idea, we have developed a new methodology to study users, Cognitive Chrono-Ethnography.

2.1 CCE Steps

Figure 2 shows a typical flow of a CCE study. The purpose of a CCE study is to answer the study question in the form "what such-and-such people would do in such-and-such way in such-and-such circumstance?"

We start with an ethnographical filed observation. We visit the field where our target users engage in the activities we are to support by using information devices. Then we look into the results of observation to identify parameters that are related to the fast processes, *e.g.*, slamming on the brakes, and the slow processes, *e.g.*, planning detour by using a navigation system. Then we take an appropriate cognitive architecture and map the parameters on the cognitive architecture. We then run simulations to see how the parameters could be related with users' behaviors, *e.g.*, planning detour would slow down the initiation of slamming on the brakes by approximately 100msec. Through the simulations, we identify significant parameters that need to explore in detail by conducting a field study by a number of selected participants. For instance, we may recruit six participants who use a car navigation system daily for planning detour and six who don't. In addition, a half of the six participants are good at recognizing hazardous situations and the other half are not. These are the steps for preparation (shown in the purple area in the figure).

Then we record the participants' behavior in the field of the activities. We collect the following data in a CCE study: behavior observation records created by investigators, behavior measurement records (*e.g.*, a pin microphone to record their vocalization, a small ear-mounted camera to record the scene they are viewing, and an electrocardiograph to record their physiological responses

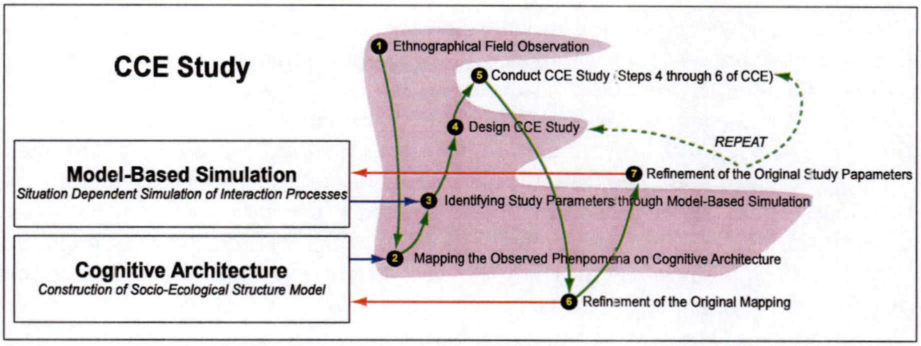

Fig. 2. CCE steps and brain models

to the events), on-site self-reports (participants themselves take photos, brief notes, and voice recording concerning their activities while their memories of the events remain fresh). After the recording, we conduct retrospective interviews. Using behavioral observation records, behavioral measurement records, and on-site self-reports, we have the participants describe their recorded behavior as detail as possible. We then compile the results of retrospective interviews to identify commonality among the participants who have the same behavioral features.

We may design a next CCE study to extend the result of the initial CCE study. We need to redo the steps 2 and 3 shown in the figure to design a new CCE study.

2.2 An Example of CCE Study

We'd like to illustrate the train station navigation study [4] to show how an actual CCE study was conducted. We were interested in how elderly passengers use guide signs at railway stations when they wanted to use facilities, *e.g.*, toilets, coin-operated lockers, *etc.*, or they had to transfer to another line. We identified critical parameters including such cognitive functions as *planning* for searching something necessary at train stations, *attention* for selectively focussing on task relevant information from the environment, and *working memory* for keeping the task relevant information active for performing actions smoothly. We conducted simulations by mapping these cognitive functions on our cognitive architecture [4], and derived ideas for a field study; people who don't have any problem in these cognitive functions would perform navigation tasks at train stations smoothly, on the other hand, people who have any problems in these cognitive functions would show some problems. We wanted to understand what people who don't have sufficient level of attention, for example, would do to accomplish searching for a toilet task, for instance. In this study, we found that the participants with weak attention tended not to use complicated signboards because they had difficulty in coordinating the slow process to decide which direction to go by comprehending signs and the fast process to gather information from the environment that changes rapidly as they walked.

Since the mental processes for accomplishing train-station-navigation task are slower than those for the tasks to follow the directions of a car navigation system, the detailed workings of System 1 and System 2 would be different. However, since both share the time-critical features of interactions between human beings (passengers or drivers) and the environments, the case study suggests that more suitable interactions design will be possible for those with specific cognitive characteristics for performing the tasks the interactions design should support satisfactorily. For example, for those with information gathering problem, or attention problem, the critical indications, which are signboards in the case of train station and navigation directions in the case of car navigation systems, have to be placed where they expect to find them.

3 Conclusion

A handful of models concerning brain functioning have been proposed in such a community as BICA. From now on, while designing a CCE study, one can select an appropriate brain functioning model for the study interest from the pool of models there. CCE is proved effective for studying dynamic Two Minds activities through several case studies. In the future, wide use of it is expected, and we expect that people use well-designed interactions in their daily lives with great satisfaction and feel happiness being with them.

References

1. Evans, J.S.B.T., Frankish, K. (eds.): In Two Minds: Dual Processes and Beyond. Oxford University Press, Oxford (2009)
2. Kahneman, D.: A perspective on judgment and choice. American Psychologist 58(9), 697–720 (2003)
3. Kahneman, D.: Thinking, Fast and Slow. Farrar, Straus and Giroux, New York (2011)
4. Kitajima, M., Toyota, M.: Simulating navigation behaviour based on the architecture model Model Human Processor with Real-Time Constraints (MHP/RT). Behaviour & Information Technology 31(1), 41–58 (2012)
5. Kitajima, M., Toyota, M.: Four Processing Modes of in situ Human Behavior. In: Samsonovich, A.V., Jóhannsdóttir, K.R. (eds.) Biologically Inspired Cognitive Architectures 2011 - Proceedings of the Second Annual Meeting of the BICA Society, pp. 194–199. IOS Press, Amsterdam (2011)

Emotional Emergence in a Symbolic Dynamical Architecture

Othalia Larue, Pierre Poirier, and Roger Nkambou

GDAC Research Laboratory, Université du Québec à Montréal
Montréal, QC, Canada
larue.othalia@courrier.uqam.ca,{poirier.pierre,nkambou.roger}@uqam.ca

Abstract. We present a cognitive architecture based on a unified model of cognition, Stanovich's tripartite framework, which provides an explanation of how reflective and adaptive human behaviour emerges from the interaction of three distinct cognitive levels (minds). In this paper, we focus on the description of our emotional model which is deeply rooted in neuromodulatory phenomena. We illustrate the emergence of emotional state using a psychological task : the emotional Stroop task.

1 Introduction

We present a three-level neuromodulated cognitive architecture implemented in a multi-agent system. As a proof of concept of the architecture as a whole, we already simulated a number of psychological task [1]. In this paper, we extend our studies to the emergence of emotional behaviour. Our architecture is based on a unified model of cognition, Stanovichs tripartite framework [2], which explains how reactive and sequential human behaviour emerges from the interaction of three distinct cognitive levels. The Reactive level, responsible for fast context-sensitive behaviours, includes instinctive processes, over-learned processes, domain-specific knowledge, emotional regulation and implicit learning. The Algorithmic level, responsible for cognitive control, can suppress the operation of reactive processes and effect decoupling (simulation) as well as serial associative processes. The Reflective level, responsible for deliberative processing and rational behaviour, also can suppress reactive processes, as well as trigger or suppress decoupling and serial associative processes. To study emotional emergence in this framework, we enhanced the systems reactive level semantic knowledge with the NRC emotion lexicon [3] and activated two neuromodulator systems: a dopaminergic system (modulating the architectures goal-seeking processes) and a serotonin system (modulating the speed of its semantic attentional control mechanism). We illustrate the emergence of emotional behaviour by performing an emotional Stroop task.

2 Description of the Architecture

Our architecture is implemented in a multi-agent simulation platform (Madkit) that allows the creation of agents of different complexity and in which large

A. Chella et al. (Eds.): Biologically Inspired Cognitive Architectures 2012, AISC 196, pp. 199–204.
springerlink.com

numbers of agents can operate in parallel. Madkit implements the AGR [4] (Agent/Group/Role) model, which we found particularly suitable to implement various groups of agents and a diversity of agents in each group. Each level in the architecture is composed of groups of agents acting in parallel, where each agent has one or more role. The message passing activity is essential to support the behaviour of the system (see Fig. 1). Agents send messages at various frequencies; those frequencies define the dynamics of the system. The main roles assigned to agents at the Reactive level are"sensor","effector" and"knowledge". The network of *Knowledge* agents (agents with the"knowledge" role) is initialized with a conceptual map that makes up the systems declarative knowledge (see A in Fig. 1). Knowledge agents therefore have two roles:"knowledge" and a word from the knowledge base (e.g., "Red"). Knowledge agents are connected together according to the links in the conceptual map. Upon receiving a message from a Sensor agent (see B in Fig. 1) or from another Knowledge agent, Knowledge agents send a message to those Knowledge agents they are connected to, therefore spreading activation in the network (as happens in semantic memories [5]). At the algorithmic level, control over the system is achieved through morphology [6]. *RequestStatus* agents query Knowledge agents (C) at regular intervals about their status (number of messages sent during that interval). Based on this, *Status* agents (D) represent the systems activity at a given time in the form of a distance matrix describing the systems message passing activity at that time. The distance between two concepts in the conceptual map is determined by the number of messages sent between the Knowledge agents bearing them as their role. Status agents also send a reduced representation (E) of this activity the Reflective level (see below). The systems short-term goals are sent by the Reflective level to the Algorithmic level in the form of a simple directed graph of *Goal* agents (F). Each Goal agent contains as its role a distance matrix that specifies the distance needed between Knowledge agents if the system is to reach its current goal. *Delta* agents compute the difference (G) between the matrix provided by the Status agents and the one provided by the Goal agents and provide the resulting difference to *Control* agents (H), which in turn send regulation messages (I) to agents at the Reactive level to modify their activation

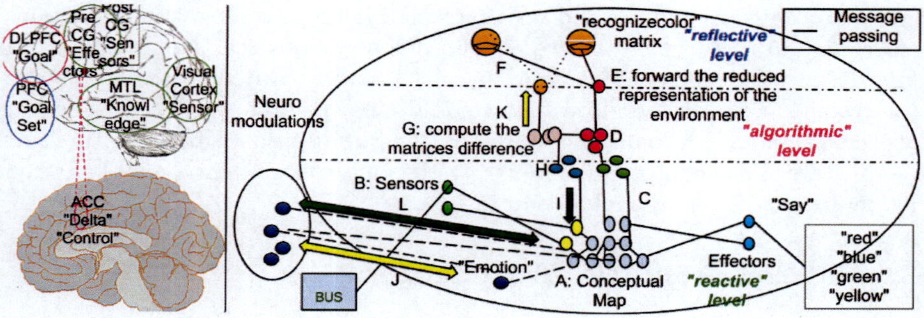

Fig. 1. Architecture and structural parallel with the brain

so that systems activity more closely matches the current short-term goal. At the reflective level, each agent has a shape (a distance matrix) as its role, which represents the state the system must be in to achieve a simple goal. Goal agents at this level are organized in a complex directed graph (J), in which each path represents a plan that can be applied to achieve a complex behaviour. A goal in the graph is activated if it best matches the reduced representation sent by the Status agents. The limited serial associative cognition of the Algorithmic level will execute this path step by step. Although the serial execution of each goal in the path provides the system with the sequentiality needed to achieve complex goals, the system does not lose its dynamic nature. Indeed, reduced representations (E) of the environment are sent on a regular basis by the Status agents so that serial cognitive association can be interrupted at any moment at the Reflective level by setting a new starting point or taking a new branch in the graph.

3 Intertwined Nature of Neuromodulators and Emotions

3.1 Emotions: From Neurophysiological States to Emotional States

In a dynamical system, a signal is called diffuse if it rules secondary variables in larger areas [7]. In our system, neuromodulations are diffuse signals that control the overall state of the system. They act on message passing activity between groups and within groups; more precisely, they modulate the frequencies upon which the different agents send messages. Each modulated agent is built with a transfer function which defined its sensitivity to the neuromodulation. For the purpose of this article we used simple lineary transfer functions. We integrated four neuromodulatory systems to our system: dopaminergic, we focus on dopaminergic and serotonergic neuromodulation[8].

Concerning the *Dopamine*, the mesolimbic pathway based in the midbrain projecting to the limbic system via the amygdala and hippocampus will be represented by an increased activation of the"Emotion agents" (in the system:"fear","anger","disgust","joy","sadness") to mirror its link with the saliency of emotional memory (J). Emotion agents belong to the reactive level. They are"knowledge" agents that bear an emotion word as one of their roles and have the ability to be influenced by and act on neuromodulators. The reward and motivation effect will occur thanks to the reinforcement (K) of goals that have led to a success in the past.

Concerning the *Serotonin*, we replicate the pathway projecting from the raphea nuclei to the thalamus, subthalamus, limbic system and cerebral cortex, especially its role in long-term memory and executive control. The influence on long-term memory will be implemented through the modulation of the messages exchanged between Knowledge agents. The influence on executive control (I) is reproduced by modulating messages between Control agents of the Algorithmic Group and agents of the reactive group. Low levels of serotonin are associated with negative emotional states (L), in our system there is a co-dependency between negative emotion agents (anger, fear, disgust) and serotonin.

3.2 Emotions: From Concepts to Emotional State

For Lindquist et al. [9], the emotional gestalt results from the combination of psychological processes. Lindquist et al.s work is an extension of the work of Barsalou [10] on conceptualization: concepts are represented through situated simulations of prior experience. The notion of situated conceptualization is extended to emotional concepts[9]. We are inspired by this constructivist approach for the implementation of emotions. The emotional concept is situated by different concepts in the knowledge base system. The NRC emotion lexicon [3] enabled us to add emotional values (links to emotion words in our system) to our initial database (based on the ConceptNet [11]). Activation of agents that constitute the conceptual map of the system is a result of the interaction at different levels (between groups and within groups), it will be the same in part for emotional concepts. They will be treated partly as other concepts: the emotional concept emerges from interactions between agents (message passing) which will determine its activation. However, unlike other concepts, emotions are directly linked to neuromodulations though the mutual influence between Emotional agents and Neuromodulation agents. We distance ourselves here a little from Lindquist's approach by letting these agents influence each other. The activation of an emotion will therefore depend on: the conceptual map of the system, reaction to the environment of the agents belonging to the reactive group, the influence of the neuromodulations on the various functions (agents) of the architecture that we described in the previous section, a bidirectional influence (neuromodulation on certain emotions/certain emotions on neuromodulations). To summarize, since the activation of links between Knowledge agents (including Emotion agents) depends on what has been extracted from the environment by the sensors and is regulated by agents of the algorithmic level (whose different activities are also regulated by different neuromodulators): emotional states emerge from the dynamics of it all. This state can be a composition of several "Semantically compatible" emotions: a complex emotion (i.e: "surprise" is linked to both "joy" and "fear").

4 Experiment and Results

In the classic Stroop task [12] is a task commonly used in cognitive psychology to evaluate a subjects attentional capacities. The subject is asked to respond to two opposing stimuli illustrating the existence of two competing pathways (one being more compelling than the other but opposed to the assigned goal). The emotional Stroop task is a variant of this task where the focus is put on the impact of emotional content on the reaction times of the subject. Where in the classical Stroop task the written color interfered with the naming of the color ink, in the emotional Stroop task the naming of the words color takes a significantly longer time for emotional words than for neutral words. We used a similar procedure to Gotlib and McCann [13]by presenting a block of mixed 50 trials of emotionally valued words ("death", "spider", "cancer",) and 50 trials of neutral words ("nature", "cinnamon", "bus",) to the system. We observed the systems behaviour in three conditions: "only concepts" system, "normal" neuromodulated system, "depressed" neuromodulated system. The only concept" system is a system in which emotional concept and links have been added at the reactive level but the links to neuromodulation agents are inactive. The "normal" and "depressed" systems are systems in which the neuromodulation model presented in section 3 has been activated. In the "depressed" system, serotonin and dopamine levels are significantly low (meaning that messages are sent by these two neuromodulators at intervals two times longer than in the "normal" condition). Results are presented in Fig. 2 (A). In the conceptual condition, we can observe that the response time (RT) for neutral words is slightly lower for neutral words than for emotional words. When adding emotional word to the database, we ascribed a high value to links between them and some other words in the database (as per the NRC database); however since no neuromodulations have been added to the system, little Stroop interference is shown. In the "normal" condition, since the neuromodulation have been activated, the system is more sensitive to emotional words. When emotional words are presented to the system, emotions related to them are strongly activated and are salient (dopamine effect) at the reactive level of the system. Control agents have therefore more work to do to enhance the activity of the agents in regards to their goal (color recognition). Furthermore, when the system is confronted to negative emotions, such as fear or sadness, serotonin is more activated which results in a change in the dynamics of the system: mainly system becomes highly emotional and its attentional mechanism (Control agents) is hypoactive, resulting in a higher response time for emotional words. Finally, in the "depressed" condition, as seen in Fig. 2(A) these factors are sharpened (latency of 159ms between the "normal" and "depressed" condition for emotional words). We lowered dopamine and serotonin parameters, meaning mainly that there was a decrease in the frequency upon which messages were sent to and therefore from Control agents (serotonin - inhibition) and Goal agents (dopamine - motivation), an increase of negative emotions activation (serotonin), a globally higher emotional saliency (dopamine). As a result, the system was highly "emotional". A reaction difference between the two conditions ("normal", "depressed") of the same word/agent "cancer" can be seen in Fig. 2 (B).

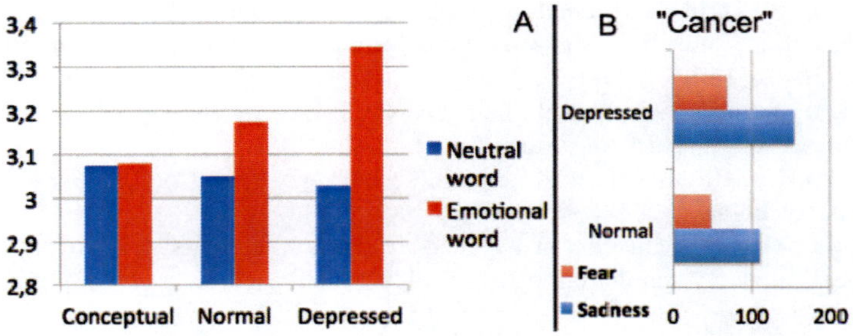

Fig. 2. A) Mean RT (s) for neutral and emotional word in "conceptual", "normal" and "depressed" conditions. B) "cancer" agent connections in "depressed" and "normal" condition.

5 Conclusion

With an emotional Stroop task, we illustrated emotional emergence in an architecture where emotional states are deeply rooted in neuromodulatory phenomena, focusing on the serotonin and dopamine neuromodulations. As the Stroop task mostly implicates the algorithmic level (cognitive control), we observed the impact of neuromodulations at this level; however, it would be interesting to focus on tasks that tap reflective level abilities (Wisconsin Card Sorting Task, rule discovery, goal maintenance). In this paper, we considered neuromodulations as they decrease the systems performance. In future works, we plan to work on the opposite direction in pushing the systems limits toward high performances by looking for the optimal neuromodulatory combination. An important feature of the neuromodulations we left out in this paper is their co-dependency [8]. Ideally, neuromodulations would form a dynamical system of their own within the global architecture and act upon each others activity.

References

1. Larue, O., Poirier, P., Nkambou, R.: A Three-Level Cognitive Architecture for the Simulation of Human Behaviour. In: Kosseim, L., Inkpen, D. (eds.) Canadian AI 2012. LNCS, vol. 7310, pp. 337–342. Springer, Heidelberg (2012)
2. Stanovich, K.E.: Rationality and the reflective mind. Oxford University Press (2010)
3. Mohammad, S., Turney, P.: Emotions Evoked by Common Words and Phrases. In: Proceedings of the NAACL-HLT 2010 (2010)
4. Ferber, J., Gutknecht, O., Michel, F.: From Agents to Organizations: An Organizational View of Multi-agent Systems. In: Giorgini, P., Müller, J.P., Odell, J.J. (eds.) AOSE 2003. LNCS, vol. 2935, pp. 214–230. Springer, Heidelberg (2004)
5. Anderson, J.R.: The architecture of cognition. Harvard Univ. Press (1983)
6. Cardon, A.: La complexit organise. Herms Science Publications, Paris (2005)
7. Gros, C.: Complex and Adaptive Dynamical Systems: A Primer. Springer (2010)
8. Cools, R., Nakamura, K., Daw, N.D.: Serotonin and Dopamine. Neuropsychopharmacology Rev. 36, 98–113 (2011)
9. Lindquist, K.A., Wager, T.D., Kober, H., Bliss-Moreau, E., Barrett, L.F.: The brain basis of emotion. Behavioural and Brain Sciences (2011)
10. Barsalou, L.: Simulation, situated conceptualization, and prediction. Phil. Trans. R. Soc. B 364(1521) (2009)
11. Havasi, C., Speer, R., Alonso, J.: ConceptNet 3. In: Proceedings of Recent Advances in Natural Languages Processing (2007)
12. Stroop, J.R.: Studies of interference in serial verbal reactions. Journal of Experimental Psychology 18, 643–662 (1935)
13. Gotlib, I., McCann, C.: Construct accessibility and depression. Journal of Personality and Social Psychology 47(2), 427–439 (1984)

An Integrated, Modular Framework for Computer Vision and Cognitive Robotics Research (icVision)

Jürgen Leitner[1], Simon Harding[2], Mikhail Frank[1],
Alexander Förster[1], and Jürgen Schmidhuber[1]

[1] Dalle Molle Institute for Artificial Intelligence (IDSIA), USI/SUPSI, Lugano, Switzerland
juxi@idsia.ch
[2] Machine Intelligence, Ltd UK
simon@machineintelligence.co.uk

Abstract. We present an easy-to-use, modular framework for performing computer vision related tasks in support of cognitive robotics research on the *iCub* humanoid robot. The aim of this biologically inspired, bottom-up architecture is to facilitate research towards visual perception and cognition processes, especially their influence on robotic object manipulation and environment interaction. The *icVision* framework described provides capabilities for detection of objects in the 2D image plane and locate those objects in 3D space to facilitate the creation of a world model.

1 Introduction

Vision and the visual system are the focus of much research in psychology, cognitive science, neuroscience and biology. A major issue in visual perception is that what individuals 'see' is not just a simple translation of input stimuli (compare *optical illusions*). The research of Marr in the 1970s led to a theory of vision using different levels of abstraction [10]. He described human vision as processing inputs, stemming from a two-dimensional visual array (on the retina), to build a three-dimensional description of the world as output. For this he defines three levels: a 2D (or primal) sketch of the scene (using feature extraction), a sketch of the scene using textures to provide more information, and finally a 3D model.

Visual perception is of critical importance, as the sensory feedback allows to make decisions, trigger certain behaviours, and adapt these to the current situation. This is not just the case for humans, but also for autonomous robots. The visual feedback enables robots to build up a cognitive mapping between sensory inputs and action outputs, therefore closing the sensorimotor loop. Thus being able to perform actions and adapt to dynamic environments. We are aiming to build a visual perception system for robots, based on human vision, that allows to provide this feedback leading to more autonomous and adaptive behaviours.

Our research platform is the open-system humanoid robot *iCub* [16] developed within the EC funded 'RobotCub' project. In our setup, as shown in Figure 1 (left), it consists of two anthropomorphic arms, a head and a torso and is roughly the size of a human child. The *iCub* was designed for object manipulation research. It also is an excellent

A. Chella et al. (Eds.): Biologically Inspired Cognitive Architectures 2012, AISC 196, pp. 205–210.
springerlink.com

experimental, high degree-of-freedom (DOF) platform for artificial (and human) cognition research and embodied artificial intelligence (AI) development [12]. To localise objects in the environment the *iCub* has to rely solely, similarly to human perception, on a visual system based on stereo vision. The two cameras are mounted in the head. Their pan and tilt can jointly be controlled, with vergence providing a third DOF. The neck provides 3 more DOF for gazing.

We describe a framework, named *icVision*, supporting the learning of hand-eye coordination and object manipulation, by solving visual perception issues in a biologically-inspired way.

2 The *icVision* Framework

Research on perception has been an active component of developing artificial vision (or computer vision) systems, in industry and robotics. Our humanoid robot should be able, like the human mind, learn to perceive objects and develop a representation that allows it to detect this object again. The goal is to enable adaptive, autonomous behaviours based on visual feedback by combining robot learning approaches (AI and machine learning (ML) techniques), with computer vision.

This framework was developed to build a biologically-inspired architecture (in-line with the description by Marr). It processes the visual inputs received by the cameras and builds (internal) representations of objects. It facilitates the 3D localisation of the detected objects in the 2D image plane and provides this information to other systems (e.g. motion planner). Figure 1 (right) sketches the *icVision* architecture. The system consists of distributed YARP modules[1] interacting with the *iCub* hardware and each other. The specialised modules can be connected and form pathways to perform, for example, object detection, similarly to the hierarchies in human perception in the visual cortex (V1, V2, V3, ...) [6].

The main module, the *icVision* **Core**, handles the connection with the hardware and provides housekeeping functionality (e.g., GUI, module start/stop). Implemented

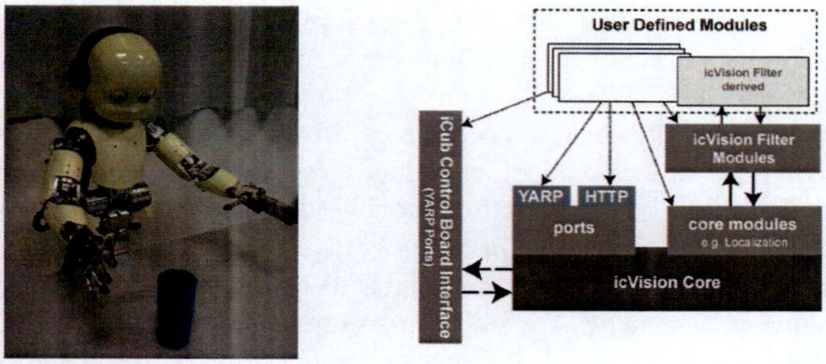

Fig. 1. Left: The *iCub* humanoid robot **Right:** Architecture of the *icVision* framework

[1] YARP [11] is a middleware that allows easy, distributed access to the sensors and actuators of the *iCub* humanoid robot, as well as, to other software modules.

modules include object detection (aka filters), 3D localisation and a gaze controller interface based on the position of the object in the image plane (as provided by the filters). These are reachable via standardised interfaces allowing for easy swapping and reuse of modules and extending functionality. For example, other brain-inspired modules, a saliency & a disparity map, have recently been added.

2.1 Detecting Objects (*icVision* Filter)

The first thing done by the human visual system, and investigated by us, is the segmentation (detection) in the visual space (the 2D images). There exists a vast body of work on all aspects of image processing [3], using both classical and machine learning approaches. Our framework, more particularly the *icVision* filter modules, which relate to Marr's first and second level, provide object detection and identification in the streamed images. The result of the filter is generally a binary segmentation of the camera image for a specific object. Figure 2 shows a tea box being tracked by the *iCub* in real-time using our framework with a learned filter. Also in Figure 3 the binary output can be seen.

Using machine learning, more complicated filters can be generated automatically instead of engineered. We apply Cartesian Genetic Programming (CGP) [13,14] to provide automatic generation of computer programs making use of the functionality integrated in the OpenCV image processing library [1], therefore incorporating domain knowledge. It provides an effective method to learn new object detection algorithms that are robust, if the training set is chosen correctly [4].

2.2 Locating Objects (*icVision* 3D)

To enable the robot to interact with the environment it is important to localise the object first. Developing an approach to perform robust localisation to be deployed on a real humanoid robot is necessary to provide the necessary inputs for on-line motion planning, reaching, and object manipulation.

Our framework provides a 3D localisation module, allowing for conversion between camera image coordinates and 3D coordinates in the robot reference frame. Using the objects location in the cameras (provided by an *icVision* Filter module) and pose information from the hardware, this module calculates where the object is in the world. This information is then used to update the world model. Figure 3 describes the full 3D

Fig. 2. The detection of a tea box in changing lighting condition performed by a learned filter. The binary output of the filter is used as red overlay.

location estimation process, starting with the camera images received from the robot and ending with the localised object being placed in our MoBeE world model [2].

For the *iCub* platform several approaches have previously been developed. Generally, stereo vision describes the extraction of 3D information out of digital images and is similar to the biological process of stereopsis in humans. Its basic principle is the comparison of images taken of the same scene from different viewpoints. To obtain a distance measure the relative displacement of a pixel between the two images is used [5]. While these approaches, based on projective geometry, have been proven effective under carefully controlled experimental circumstances, they are not easily transferred to robotics applications. The 'Cartesian controller module' [15], for example, provides basic 3D position estimation functionality and gaze control. This module works well on the simulated *iCub*, however it is not fully supported and functional on the hardware platform, and therefore does not perform well. The most accurate currently available localisation module for the *iCub* exists in the 'stereoVision' module. It provides accuracy in the range of a few centimetres.

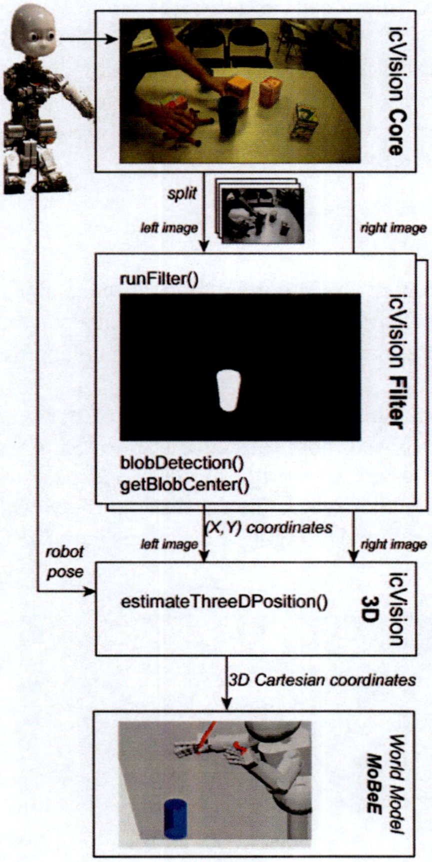

Fig. 3. The 3D location estimation works the following: At first the camera images are acquired from the hardware via YARP. The images are converted into grayscale, as well as, split into RGB/HSV channels and distributed to all active *icVision* filters. Each filter then processes the images received using OpenCV functions. (ending with a thresholding operation). The output of this is a binary image, segmenting the object to be localised. A blob detection algorithm is run on these binary images to find the (centre) location of the detected object in the image frame. The position of the object in both the right and left camera images is sent to the 3D localisation module, where together with the robots pose, i.e. the joint encoders, a 3D location estimation is generated. As the last step the localised object is then placed in the existing world model.

The *icVision* 3D localisation module provides an easy way of swapping between various localisation implementations, including the two mentioned here. We also provide an implementation estimating the location using machine learning [9].

3 Use Cases and Current Applications of the Framework

Here we give a short list of use cases, where the *icVision* framework, has already successfully been used in our research.

The learning of specific filters for certain objects was done using CGP (as mentioned above) [4,8]. The provided (human-readable) program allows for a unique identification of objects in the visual stream learned from a test set. For a more autonomous acquisition of object representations, *icVision* filters are learned from objects perceived by the cameras using a saliency map and standard feature detectors. This way we were able to provide the needed inputs to our CGP learner for building robust filters [7]. We are currently learning filters for the robot's fingers to perform research in how the humanoid can develop sensorimotor control.

The full system has been used together with a reactive controller to enable the *iCub* to autonomously re-plan a motion to avoid an object it sees [2]. The object is placed into the world model purely from vision, it is able to update the position of the object in real-time, even while the robot is moving.

To add our 3D localisation approach to the framework, we used a Katana robotic arm to teach the *iCub* how to perceive the location of the objects it sees. The Katana positions an object within the shared workspace, and informs the *iCub* about the location. The *iCub* then moves to observe the object from various angles and poses. Its pose and the 2D position outputs provided by an *icVision* filter are used to train simple, multi-layer artificial neural networks (ANN) to estimate the object's Cartesian location. We show that satisfactory results can be obtained for localisation [9]. Furthermore, we demonstrate that this task can be accomplished safely using collision avoidance software to prevent collisions between multiple robots in the same workspace [2].

4 Conclusions

We combine the current machine learning and computer vision research to build a biologically-inspired, cognitive framework for the *iCub* humanoid robot. The developed *icVision* framework facilitates the autonomous development of new robot controllers. Cognition and perception are seen as the foundation to developmental mechanisms, such as as sensorimotor coordination, intrinsic motivation and hierarchical learning, which are investigated on the robotic platform.

The reason for the focus on vision is twofold, firstly the limited sensing capabilities of the robotic platform and secondly, vision is the most important sense for humans. As we use a humanoid robot investigating how humans do this tasks of perception, detection and tracking of objects is of interest. These facilitate the building of a world model, which is used for tasks like motion planning and grasping. Realtime, incremental learning is applied to further improve perception and the model of the environment and the robot itself. Learning to grasp and basic hand-eye coordination are the areas of research this framework is currently applied.

References

1. Bradski, G.: The OpenCV Library. Dr. Dobb's Journal of Software Tools (2000)
2. Frank, M., et al.: The modular behavioral environment for humanoids and other robots (mobee). In: Int'l. Conference on Informatics in Control, Automation and Robotics (2012)
3. Gonzalez, R., Richard, E.W.: Digital image processing (2002)
4. Harding, S., Leitner, J., Schmidhuber, J.: Cartesian genetic programming for image processing. In: Genetic Programming Theory and Practice X. Springer (to appear, 2012)
5. Hartley, R., Zisserman, A.: Multiple view geometry in computer vision. Cambridge University Press (2000)
6. Hubel, D., Wensveen, J., Wick, B.: Eye, brain, and vision. Scientific American Library (1995)
7. Leitner, J., et al.: Autonomous learning of robust visual object detection on a humanoid (2012) (submitted to IEEE Int'l. Conference on Developmental Learning and Epigenetic Robotics)
8. Leitner, J., et al.: et al.: Mars terrain image classification using cartesian genetic programming. In: International Symposium on Artificial Intelligence (2012)
9. Leitner, J., et al.: Transferring spatial perception between robots operating in a shared workspace. In: IEEE/RSJ International Conference on Intelligent Robots and Systems (2012)
10. Marr, D.: Vision: A Computational Approach. Freeman & Co., San Francisco (1982)
11. Metta, G., Fitzpatrick, P., Natale, L.: YARP: Yet Another Robot Platform. International Journal of Advanced Robotics Systems 3(1) (2006)
12. Metta, G., et al.: The iCub humanoid robot: An open-systems platform for research in cognitive development. Neural Networks 23(8-9), 1125–1134 (2010)
13. Miller, J.: An empirical study of the efficiency of learning boolean functions using a cartesian genetic programming approach. In: Genetic and Evolutionary Computation Conference (1999)
14. Miller, J.: Cartesian genetic programming. Cartesian Genetic Programming, 17–34 (2011)
15. Pattacini, U.: Modular Cartesian Controllers for Humanoid Robots: Design and Implementation on the iCub. Ph.D. thesis, RBCS, Italian Institute of Technology, Genova (2011)
16. Tsagarakis, N.G., et al.: iCub: the design and realization of an open humanoid platform for cognitive and neuroscience research. Advanced Robotics 21, 1151–1175 (2007)

Insertion Cognitive Architecture

Alexander Letichevsky

Glushkov Institute of Cybernetics, Academy of Sciences of Ukraine
let@cyfra.net

Abstract. The main goal of the paper is to demonstrate that the Insertion Modeling System IMS, which is under development in Glushkov Institute of Cybernetics, can be used as an instrument for the development and analysis of cognitive (intellectual) agents that is agents that model human mind. Insertion cognitive architecture is real time insertion machine which realizes itself, has a center to evaluate the success of its behavior and is committed to achieving maximum success repeated. As an agent, this machine is inserted in its external environment and has the means to interact with it. The internal environment of cognitive agent creates and develops its model and the model of external environment. In order to achieve its goals it uses the basic techniques developed in the field of biologically inspired cognitive architectures as well as software development techniques.

Keywords: insertion modeling, process alebras, agents, environments, cognitive architectures.

1 Introduction

Insertion modeling is a trend that is developing over the last decade as an approach to a general theory of interaction of agents and environments in complex distributed multi-agent systems. The first works in this direction have been published in the middle of 90s [4,5,6]. In these studies, a model of interaction between agents and environments based on the notion of insertion function and the algebra of behaviors (similar to some kind of process algebra) has been proposed.

The main sources of insertion modeling are the model of interacting of control and operational automata, proposed by V. Glushkov back in the 60s [7,8] to describe the structure of computers, and the general theory of interacting information processes that has emerged in the 70s and is the basis for modern research in this area. It includs the CCS (Calculus of Communicated Processes) [17,18] and the π-calculus of R. Milner [19], CSP (Communicated Sequential Processes) of T. Hoare [10], ACP (Algebra of Communicated Processes) [2] and many other various branches of these basic theories. Fairly complete survey of the classical process theory is presented in the Handbook of Process Algebras [3], published in 2001.

In recent years, insertion modeling has been applied to the development of systems for the verification of requirements and specifications of distributed interacting systems [1,12,14,15,16]. The system VRS, based on insertion modeling,

A. Chella et al. (Eds.): Biologically Inspired Cognitive Architectures 2012, AISC 196, pp. 211–217.
springerlink.com

has been successfully applied to verify the requirements and specifications in the field of telecommunication systems, embedded systems, and real-time systems. A new insertion modeling system IMS [9], which is under development in the Glushkov Institute of Cybernetics of the National Academy of Sciences of Ukraine, is intended to extend the area of insertion modeling applications. We found many common features of the tools used in software development area based on formal methods and techniques used in biologically inspired cognitive architectures. This gives us hope to introduce some new ideas to the development of this subject domain.

Insertion modeling deals with the construction of models and study the interaction of agents and environments in complex distributed multi-agent systems. Informally, the basic principles of the paradigm of insertion modeling can be formulated as follows.

- The world is a hierarchy of environments and agents inserted into them.
- Environments and agents are entities evolving in time.
- Insertion of agent into environment changes the behavior of environment and produces new environment which is ready for insertion of new agents (if there is a room).
- Environments as agents can be inserted into higher level environment.
- Agents can be inserted from external environment as well as from internal agents (environments).
- Agents and environments can model another agents and environments on the different levels of abstraction.

All these principles can be formalized in terms of transition systems, behavior algebras, and insertion functions. This formalization can be used as high level abstractions of biological entities needed for computer modeling of human mind. Insertion modeling is consistent with M. Minsky's approach of the society of mind [20]. Mathematical foundation of insertion modeling is presented in [13].

2 The Main Notions

Transition system consists of states and transitions that connect states. Transitions are labeled by *actions* (signals, events, instructions, statements etc.). Transition systems are evolving in time changing their states, and actions are observable symbolic structures used for communication. We use the well-known notation $s \xrightarrow{a} s'$ to express the fact that transition system can evolve from the state s to s' performing action a. Usually transition systems are nondeterministic and there can be several transitions even labeled by the same action. If we abstract from the structure of states and concentrate only on (branching) sequences of observable actions we obtain the state equivalence called *bisimilarity* (originated from [21] and [17], exact definition can be found in [13]). Bisimilar states generate the same behaviors of transition systems.

To define the behaviors of transition systems we use equations of the form $u_i = F_i(u_1, u_2, \ldots), i = 1, 2, \ldots$ in *behavior algebra*. This algebra is defined by the set of actions and the set of behaviors (processes). It has two operations and termination constants. Operations are prefixing $a.u$ (a - action, u - behavior) and nondeterministic choice $u + v$ (u and v - behaviors). Termination constants are successful termination Δ, deadlock 0, and undefined behavior $\perp$. It has also approximation relation $\subseteq$, which is a partial order with minimal element $\perp$, and is used for constructing a complete algebra with fixed point theorem.

Transition systems with states considered up to bisimilarity (or up to behavior, which is the same) are called *agents*. The *type of an agent* is the set of actions it can perform (signals or messages to send, events in which an agent can participate etc.).

Environment by definition is an agent that possesses *insertion function*. Given the state of environment s and the state of agent u, insertion function computes the new state of environment which is denoted as $s[u]$. Note that we consider states up to bisimilarity and if we have some representation of behaviors, the behaviors of environment and agent can be used as states.

The state $s[u]$ is a state of environment and we can use insertion function to insert a new agent v into environment $s[u] : (s[u])[v] = s[u, v]$. Repeating this construction we can obtain the state of environment $s[u_1, u_2, \ldots]$ with several agents inserted into it. Environment is an agent with insertion function, so if we forget the insertion function, then environment can be inserted as agent into a higher level environment and we can obtain hierarchical structure like

$$s[s_1[u_{11}, u_{12}, \ldots]_{E_1}, s_2[u_{21}, u_{22}, \ldots]_{E_2}, \ldots]_E$$

Here notation $s[u_1, u_2, \ldots]_E$ explicitly shows the environment E to which the state s belongs (environment indexes can be omitted if they are known from the context).

A special type of environments is considered in IMS to have a sufficiently rich language for the description of environment states properties. These environments are called *attribute environments*. The state of attribute environment is the valuation of *attributes* - symbols that change their values while changing the state in time. Each attribute has type (numeric, symbolic, enumerated, agent and behavior types, functional types etc.). Now logic formulas can be used for the description of properties of agent or environment states. The first order formulas with quantifiers and some temporal modalities can be used as formulas defining the properties of environment states.

The sets of *local descriptions* of environments are used in IMS for the definition of insertion functions. The general form of local descriptions used in IMS is the following:

$$\forall x(\alpha(x, r) \rightarrow P < x, r > \beta(x, r)),$$

where x is a list of typed parameters, r is a list of attributes, $\alpha(x, r)$ and $\beta(x, r)$ are logic formulas, $P < x, r >$ is a process - finite behavior of an environment. Local descriptions can be considered as formulas of dynamic logic, or Hoare triples, or productions - the most popular units of procedural memory in AI and BICA. In any case they describe local dynamic properties of environment behavior: for all possible values of parameters, if precondition is true then a process of a local description can start and after successful termination of this process a postcondition must be true.

Equations in behavior algebra are used as local descriptions of low level agents (agents which are not environments).

The goal of BICA is the construction of cognitive agents, which model human-like mind and behavior. They must have some internal structure, and this structure must agree with our knowledge about human brain. We believe that insertion models of interaction of agents and environments are adequate for the development of BICA. The notion of agent can be used as an abstract model of the group of neurons joined for the performing of some functionalities. High level neurons which control and coordinate the behavior of low level groups can be considered as environment for these groups.

3 Mutilevel Environments and the Leyers of Neocortex

The *type of environment* is defined by two action sets: the set of environment actions and the set of agent actions. The last defines the type of agents which can be inserted into this environment: if the set of agent actions is included in the set of agent actions of environment then this agent can be inserted into this environment. This relation is called *compatibility* relation between environments and agents (environment is compatible with agent if this agent can be inserted into this environment).

Multilevel environment is a family of environments with distinguished the most external environment. The compatibility relation on the set of environments defines a directed graph and we demand for multilevel environment that the most external environment would be reachable from any environment of the family in this graph. Some agents have the actions that describe the moving of agents from one environment to another compatible with these agents (mobility actions).

Multilevel environments are represented in IMS by means of *environment descriptions* and local descriptions for insertion functions. Environment description contains the signature of environment (types of attributes and types of inserted agents). Local descriptions are organized as a knowledge base with special data structures providing efficient access to the needed local descriptions (procedure memory). Anther knowledge base contains the description of different types of agents and environments together with the histories of their use (episodic memory).

Our **first hypothesis** is that the multilevel environments can be used for modeling of the six leyers of neocortex. Moving from low levels to higher ones we increase the levels of abstraction using more and more abstract symbolic models. This means that the agents inserted into the higher levels correspond to the classes of low level agents.

4 Insertion Machines and the Models of Mind

To implement the models of multilevel environment different kinds of insertion machines are used in IMS. The general architecture of insertion machines and the detailes of its functioning can be found in [9]. The main component of insertion machine is model driver, the component which controls the machine movement along the behavior tree of a model. The state of a model is represented as a text in the input language of insertion machine and is considered as an algebraic expression. The computation of the next state of a model is performed on a base of symbolic manipulation and logic inference.

Two kinds of insertion machines are distinguished: *real time* or *interactive* and *analytical* insertion machines. The first ones exist in the real or virtual environment, interacting with it in the real or virtual time. Analytical machines intended for model analysis, investigation of its properties, solving problems etc. The drivers for two kinds of machines correspondingly are also divided into interactive and analytical drivers. Interactive driver must select exactly one alternative and perform the action specified as a prefix of this alternative. Insertion machine with interactive driver operates as an agent inserted into external environment with insertion function defining the laws of functioning of this environment. Cognitive interactive driver has criteria of successful functioning in external environment, it accumulates the information about its past in episodic memory, develops the models of external environment, uses some learning algorithms to improve the strategy of selecting actions and increase the level of successful functioning.

Analytical insertion machine as opposed to interactive one can consider different variants of making decision about performed actions, returning to choice points (as in logic programming) and consider different paths in the behavior tree of a model. The model of a system can include the model of external environment of this system, and the driver performance depends on the goals of insertion machine. In the general case analytical machine solves the problems by search of states using different strategies to reduce the search spsce. Analytical machine in IMS are enriched by logic and deductive tools used for generating traces of symbolic models of systems. The state of symbolic model is represented by means of properties of the values of attributes rather than their concrete values

Our **second hypothesis** is that in human mind there must exist some mechanisms like drivers of insertion machines to activate and use different kinds of models of external and internal environments produced during the life experience, learning and inherited genetically.

We are working now on the development of insertion cognitive architecture on a base of IMS. Our **third hypothesis** is that cognitive architecture can be constructed as real time insertion machine which realizes itself as a high level internal environment. It has a center to evaluate the success of its behavior. This center has the form of a special agent that can observe the interaction of a system with external environment. The main goal of a system is achieving maximum success repeated. As an agent, this machine is inserted in its external environment and has the means to interact with it. The internal environment of cognitive agent creates and develops its own model and the model of external environment. If the external environment contains other agents, they can be modeled by internal environment which creates corresponding machines and interprets those machines using corresponding drivers, comparing the behaviors of models and external agents.

References

1. Baranov, S., Jervis, C., Kotlyarov, V., Letichevsky, A., Weigert, T.: Leveraging UML to deliver correct telecom applications in UML for Real: Design of Embedded Real-Time Systems. In: Lavagno, L., Martin, G., Selic, B. (eds.), pp. 323–342. Kluwer Academic Publishers (2003)
2. Bergstra, J.A., Klop, J.W.: Process algebra for synchronous communications. Information and Control 60(1/3), 109–137 (1984)
3. Bergstra, J.A., Ponce, A., Smolka, S.A. (eds.): Handbook of Process Algebra. North-Holland (2001)
4. Gilbert, D., Letichevsky, A.A.: A universal interpreter for nondeterministic concurrent programming languages. In: Gabbrielli, M. (ed.) Fifth Compulog Network Area Meeting on Language Design and Semantic Analysis Methods (September 1996)
5. Letichevsky, A.A., Gilbert, D.: A general theory of action languages. Cybernetics and System Analyses 1 (1998)
6. Letichevsky, A.A., Gilbert, D.: A Model for Interaction of Agents and Environments. In: Bert, D., Choppy, C., Moses, P. (eds.) WADT 1999. LNCS, vol. 1827, pp. 311–328. Springer, Heidelberg (2000)
7. Glushkov, V.M.: Automata theory and questions of design structure of digital machines. Cybernetics 1, 3–11 (1965)
8. Glushkov, V.M., Letichevsky, A.A.: Theory of algorithms and descrete processors. In: Tou, J.T. (ed.) Advances in Information Systems Science, vol. 1, pp. 1–58. Plenum Press (1969)
9. Letichevsky, A.A., Letychevskyi, O.A., Peschanenko, V.S.: Insertion Modeling System. In: Clarke, E., Virbitskaite, I., Voronkov, A. (eds.) PSI 2011. LNCS, vol. 7162, pp. 262–273. Springer, Heidelberg (2012)
10. Hoare, C.A.R.: Communicating Sequential Processes. Prentice Hall (1985)
11. Kapitonova, J., Letichevsky, A.: Mathematical theory of computational systems design. Moscow, Science, 295 (1988) (in Russian)
12. Kapitonova, J., Letichevsky, A., Volkov, V., Weigert, T.: Validation of Embedded Systems. In: Zurawski, R. (ed.) The Embedded Systems Handbook. CRC Press, Miami (2005)
13. Letichevsky, A.: Algebra of behavior transformations and its applications. In: Kudryavtsev, V.B., Rosenberg, I.G. (eds.) Structural theory of Automata, Semigroups, and Universal Algebra. NATO Science Series II. Mathematics, Physics and Chemistry, vol. 207, pp. 241–272. Springer (2005)
14. Letichevsky, A., Kapitonova, J., Letichevsky Jr., A., Volkov, V., Baranov, S., Kotlyarov, V., Weigert, T.: Basic Protocols, Message Sequence Charts, and the Verification of Requirements Specifications. In: ISSRE 2004, WITUL (Workshop on Integrated Reliability with Telecommunications and UML Languages), Rennes, November 4 (2005)

15. Letichevsky, A., Kapitonova, J., Letichevsky Jr., A., Volkov, V., Baranov, S., Kotlyarov, V., Weigert, T.: Basic Protocols, Message Sequence Charts, and the Verification of Requirements Specifications. Computer Networks 47, 662–675 (2005)
16. Letichevsky, A., Kapitonova, J., Volkov, V., Letichevsky Jr., A., Baranov, S., Kotlyarov, V., Weigert, T.: System Specification with Basic Protocols. System Specification with Basic Protocols. Cybernetics and System Analyses 4 (2005)
17. Milner, R.: A Calculus of Communication Systems. LNCS, vol. 92. Springer, Heidelberg (1980)
18. Milner, R.: Communication and Concurrency. Prentice Hall (1989)
19. Milner, R.: The polyadic π-calculus: a tutorial. Tech. Rep. ECS–LFCS–91–180, Laboratory for Foundations of Computer Science, Department of Computer Science, University of Edinburgh, UK (1991)
20. Minsky, M.: The Society of Mind, 339 p. Touchstone Book (1988)
21. Park, D.: Concurrency and Automata on Infinite Sequences. In: Deussen, P. (ed.) GI-TCS 1981. LNCS, vol. 104, pp. 167–183. Springer, Heidelberg (1981)

A Parsimonious Cognitive Architecture for Human-Computer Interactive Musical Free Improvisation

Adam Linson, Chris Dobbyn, and Robin Laney

Faculty of Mathematics, Computing and Technology,
Department of Computing,
The Open University, Milton Keynes, UK

Abstract. This paper presents some of the historical and theoretical foundations for a new cognitive architecture for human-computer interactive musical free improvisation. The architecture is parsimonious in that it has no access to musical knowledge and no domain-general subsystems, such as memory or representational abilities. The paper first describes some of the features and limitations of the architecture. It then illustrates how this architecture draws on insights from cybernetics, artificial life, artificial intelligence and ecological theory by situating it within a historical context. The context presented consists of a few key developments in the history of biologically-inspired robotics, followed by an indication of how they connect to James Gibson's ecological theory. Finally, it describes how a recent approach to musicology informed by ecological theory bears on an implementation of this architecture.

1 Introduction

This paper presents some of the historical and theoretical foundations for a new cognitive architecture for human-computer interactive musical free improvisation. The paper first describes some of the features and limitations of this architecture. It then situates the architecture within a historical context by presenting a few key developments in the history of biologically-inspired robotics. This is followed by an indication of how these developments connect to James Gibson's ecological theory. Finally, it describes how a recent approach to musicology informed by ecological theory bears on an implementation of this architecture.

2 Features and Limitations of the Architecture

Although the architecture seeks neither to model human-internal cognitive structure, nor to closely emulate human musical behaviour, it does aim to provide musical behaviour that supports engaging improvisational interaction with an expert human performer. This is achieved with domain-specific low-level subsystems for dealing separately with pitch, loudness and timing, respectively, in both input and output. Subsystem output is combined into discrete monophonic note

A. Chella et al. (Eds.): Biologically Inspired Cognitive Architectures 2012, AISC 196, pp. 219–224.
springerlink.com

streams, which are in turn combined into continuous polyphonic note streams. The resulting musical behaviour is open-ended and highly responsive to a collaborative human co-performer. The architecture allows for reciprocal influence between human and computer, and it ultimately facilitates engaging musical collaboration, according to preliminary findings from on-going empirical research. The architecture is parsimonious in that it has no access to musical knowledge and no domain-general subsystems, such as long-term memory or representational abilities (for more on the relation between cognitive architecture and knowledge, see Laird [9]; for more on the interaction between domain-general and domain-specific cognitive abilities, see Gerrans and Stone [4]).

Two key limitations of the parsimonious architecture, as implemented in a current prototype agent, Adam Linson's *Odessa*, are counterbalanced as described below. The first key limitation, that the agent lacks any knowledge, is addressed by drawing on principles of the 'subsumption architecture' (Brooks [2]). In an agent with a subsumption architecture, intelligent behaviour can arise from the interaction dynamics between agent and environment using neither representation nor reasoning, but rather, via hardwired "competing" behaviours, organised into layers. In the case of *Odessa*, the agent's layers are organised in order to give it the appearance of an intentional agent to a human co-performer (for more on the role of intentional agency in this context, see Linson, et al. [11]). Although the agent does not merely mirror the human input, it does, at times, reorganise some of the input elements, providing a perceptible degree of resemblance. While stored knowledge could be expected to enhance the agent's musicality, the current implementation serves as an existence proof that a musical agent can function effectively without musical knowledge.

The second key limitation is that, without long-term memory, the agent lacks the internal capacity for developing the music in the course of a performance. However, this does not mean that no musical development takes place in a given human–computer collaboration. Although here, too, an internal mechanism for musical development could be expected to enhance the agent's performance, in the absence of one, the agent's continuous short-term responsiveness serves to offload long-term musical development onto the human co-performer. This offloading takes place while nonetheless preserving a constant (non-hierarchical) reciprocal influence between human and computer, made possible by the fact that, within the dynamic musical collaboration, the human and computer performers function as tightly coupled subsystems. Thus, a performance may exhibit musical development due to the reflective practice of the expert human improvisor, despite the computer agent's inability to handle long-term dependencies.

This parsimonious architecture, implemented in *Odessa*, draws on insights from cybernetics, artificial life, artificial intelligence and ecological theory. The remainder of this paper presents a few key developments in the history of biologically-inspired robotics. It then connects these developments to James Gibson's ecological theory. Finally, it describes how a recent approach to musicology informed by ecological theory bears on the design of *Odessa*.

3 Cybernetics, Nouvelle AI and Ecological Psychology

According to Alan Turing's view, as presented in his well-known account of the so-called "Turing test", intelligence can be assessed behaviourally in an unstructured, open-ended interactive context (Turing [15]). This view has close parallels to the work of cybernetics pioneer W. Grey Walter. Walter is perhaps best known for his autonomous robotic tortoises built in 1948-49 (Holland [8]), and first published about in the same year as Turing's 1950 paper (Walter [16]). Walter's ideas were partly based on his own empirical research on the human brain, and partly on the work of his colleague and fellow cybernetics pioneer, W. Ross Ashby. Ashby proposed a dynamic systems model of a brain, based on the premise that a small collection of simple subsystems, tightly coupled but capable of temporary independence, could produce a wide range of complex behaviours (Ashby [1]). He maintained that despite the simplicity of his model, nothing prevented it from being theoretically scaled up to human-level behaviour. Although Walter's robotic tortoises were far from human-level intelligence, they did exhibit basic lifelike behaviour through their interactions with each other, with humans and with their physical environment.

Several decades later, Rodney Brooks developed the 'subsumption architecture', a concept he applied to building insect-like robots and other biologically analogous models (Brooks [2]). For Brooks, like Walter before him, the basic premise of his robot design is that a small number of simple interconnected components can lead to complex emergent behaviour. The mechanically embodied robots built with the subsumption architecture were rapidly prototyped with "real world" empirical testing. This meant that at an early stage of development, they were deployed, for example, in crowded offices, and could be iteratively adjusted depending on how well they negotiated such everyday environments. This approach stood in contrast to the elaborate pre-deployment development of earlier GOFAI robots, such as the Stanford Research Institute's *Shakey*, and their evaluation in carefully constructed specialised environments.

Another hallmark of the approach to robotics evident in both Walter and Brooks is that, rather than focusing solely on an agent's internal construction, they emphasise its adaptive behaviour in terms of feedback loops between perception and action, or, more broadly, between the agent and its environment. This theoretical orientation bears a strong resemblance to that of ecological psychology, a field which largely emerged from the work of James Gibson. Ecological psychology fundamentally concerns itself with the tight coupling of perception and action, and of agent and environment, in the context of adaptive behaviour; Gibson is also the source of the well-established notion of *affordances* (Gibson [5], [6]). Simply put, an affordance is a feature of an environment that presents a possible action to an agent, relative to the agent's capabilities.

4 Traditional, Critical and Ecological Musicology

With respect to the study of music, traditional musicological theory generally regards classical western notation as sufficient to grasp the salient aspects of music. More recently, however, critical approaches to musicology have served to identify, for example, aspects of musical improvisation that previously escaped traditional musicological theory. Significantly, it has been shown that there are psychosocial dynamics at work in improvisation that are not reducible to musical structure defined solely in terms of traditional western musical notation (Sansom [14]). This definition of structure is not only the primary object of traditional musicological analysis, but also the predominant focus of computer music research. In the critical mode of analysis, musical structure is not denied, but rather, it is viewed as a vehicle for socioculturally situated and context-dependent meaning production and exchange.

Drawing on the work of Gibson, Eric Clarke has advanced a new strain of musicology and music psychology research which he describes as "an ecological approach to the perception of musical meaning" (Clarke [3]). In this context, he addresses improvised music, pointing out how performers of collective improvisation forcefully demonstrate the ecological link between perception, action and meaning (see Clarke [3], p. 152-154). Additional writings on improvisation within critical musicology, though not explicitly ecological, can be viewed in terms of this perception-action-meaning linkage. For example, in the context of jazz improvisation, Ingrid Monson examines performative musical irony, looking at what can be understood as high-level affordances for creative meaning production (Monson [13]). In another account of jazz improvisation, Benjamin Givan examines ways in which the low-level physicality of performers' bodies and instruments can become central affordance-rich vehicles that facilitate creative exploration (Givan [7]). And, in a direct consideration of free improvisation, Matthew Sansom focuses on general interconnections between musical structure and social dynamics, highlighting the ways in which nominally extramusical structure can afford new directions in underlying musical structure, as it unfolds in real time (Sansom [14]). These qualitative approaches to analysis identify salient features of music that remain invisible (or, at best, marginal) to traditional musicology, computational musicology and most quantitative approaches to musical perception (see Linson, et al. [12]).

5 Deployment Context

By design, *Odessa* does not impose limitations on the behavioural and sonic complexity of the human performance. While its output may not conform to the expectations of traditional musicological analysis, its open-ended complex interactive musical behaviour provides affordance-rich material for an expert human improvisor to engage with in real time. In these respects, the design draws, in part, on techniques pioneered by Walter and Brooks, as well as on insights from Clarke's ecological musicology.

On one hand, considerations of agent–environment interaction apply here, with the idea that an interactive agent for musical improvisation should be able to cope with the real-time dynamics of its unstructured musical environment, like the physical environments of Walter's and Brooks' robots. On the other hand, for human–computer musical free improvisation, the interaction model relates more to Turing's test, because the musical environment – like a conversation – consists in an open-ended agent–agent interaction (see Lewis [10]). In designing an agent for freely improvised music, both modes of interaction are relevant.

In such music, even in the limited case of duets, there are simultaneous sonic-musical streams between the players in both directions. Everything in these streams is heard by each co-performer either as a potential response to what was played, or as a potential affordance for a new response – or as both of these at once. Musical streams from each agent – whether human or computer – may sometimes support one another, sometimes provide contrast or resistance, and sometimes steer the interaction in a new direction.

6 Conclusion

Contrary to a frequently encountered assumption, it is not necessary for a computer co-performer to be mistaken for a human in order for the human–computer interaction to be engaging. An expert human improvisor can collaborate with an agent that has a unique way of making music, thus producing a shared musical outcome. The computer's musical output and, indeed, even the collaborative musical result, are an entirely separate matter from the unfolding dynamics of the collaborative mutual engagement. The on-going empirical studies of *Odessa*'s collaborative abilities seek to demonstrate the effectiveness of a parsimonious cognitive architecture in this context.

Acknowledgments. Adam Linson thanks Allan Jones of the Open University's Department of Communication and Systems for the opportunity to present some of this material to the Society and Information Research Group. Additional thanks to those who attended for the interesting discussion following the presentation.

References

1. Ashby, W.R.: Design for a brain: The origin of adaptive behaviour. Chapman and Hall (1960)
2. Brooks, R.A.: Cambrian intelligence: The early history of the new AI. MIT Press (1999)
3. Clarke, E.F.: Ways Of listening: An ecological approach to the perception of musical meaning. Oxford University Press (2005)
4. Gerrans, P., Stone, V.: Generous or parsimonious cognitive architecture? Cognitive neuroscience and theory of mind. The British Journal for the Philosophy of Science 59(2), 121–141 (2008)
5. Gibson, J.J.: The theory of affordances. In: Shaw, R., Bransford, J. (eds.) Perceiving, Acting, and Knowing: Toward an Ecological Psychology, pp. 67–82. Lawrence Erlbaum (1977)
6. Gibson, J.J.: The ecological approach to visual perception. Mifflin and Company, Houghton (1979)

7. Givan, B.: Thelonious Monk's pianism. Journal of Musicology 26(3), 404–442 (2009)
8. Holland, O.: Exploration and high adventure: The legacy of Grey Walter. Philosophical Transactions of the Royal Society of London 361(1811), 2085–2121 (2003)
9. Laird, J.: The Soar cognitive architecture. MIT Press (2012)
10. Lewis, G.: Improvising tomorrow's bodies: The politics of transduction. E-misférica 4.2 (2007)
11. Linson, A., Dobbyn, C., Laney, R.: Interactive intelligence: Behaviour-based AI, musical HCI and the Turing test. In: Müller, V., Ayesh, A. (eds.) Revisiting Turing and His Test: Comprehensiveness, Qualia, and the Real World (AISB/IACAP Symposium, Alan Turing Year 2012), pp. 16–19 (2012a)
12. Linson, A., Dobbyn, C., Laney, R.: Critical issues in evaluating freely improvising interactive music systems. In: Maher, M., Hammond, K., Pease, A., Pérez, R., Ventura, D., Wiggins, G. (eds.) Proceedings of the Third International Conference on Computational Creativity, pp. 145–149 (2012b)
13. Monson, I.: Doubleness and jazz improvisation: Irony, parody, and ethnomusicology. Critical Inquiry 20(2), 283–313 (1994)
14. Sansom, M.J.: Musical meaning: A qualitative investigation of free improvisation. PhD thesis, University of Sheffield (1997)
15. Turing, A.M.: Computing machinery and intelligence. Mind 59(236), 433–460 (1950)
16. Walter, W.G.: An imitation of life. Scientific American 182(5), 42–45 (1950)

Cognitive Integration through Goal-Generation in a Robotic Setup

Riccardo Manzotti[1], Flavio Mutti[2], Giuseppina Gini[3], and Soo-Young Lee[4]

[1] Institute of Communication and Behaviour, IULM University, Milano, Italy
riccardo.manzotti@iulm.it
[2,3] Dipartimento di Elettronica e Informazione, Politecnico di Milano, Italy
{mutti,gini}@elet.polimi.it
[4] Brain Science Research Center, KAIST, Daejeon, Republic of Korea
sy-lee@kaist.ac.kr

Abstract. What brings together multiple sensory, cognitive, and motor skills? Experimental evidences show that the interaction among thalamus, cortex and amygdala is involved in the generation of elementary cognitive behaviours that are used to achieve complex goals and to integrate the agent skills. Furthremore, the interplay among these structure is likely to be responsible of a middle level of cognition that could fill the gap between high-level cognitive reasoning and low-level sensory processing. In this paper, we address this issue and we outline a goal-generating architecture implemented on a NAO robot.

Keywords: Thalamo-Cortex, ICA, Hebbian Learning, Development, Goals.

1 Introduction

What brings together different skills and modules in a biological agents? In simpler animals like insects the integration is likely either the result of a genetic blueprint or embodiment [1]. Yet, in more complex cognitive agents, there seems to be some yet-to-be-understood architectural principle that encourages a seamless integration of different sensory capabilities and cognitive skills. It is long well-known, from both psychological and neuroscientific data, that severe anatomical damage does not prevent the brain from achieving a working unity [2]. Somewhat suprisingly, to a large extent, the unity of the mind is still one of the most puzzling and unresolved aspects of cognition and – although many models have been proposed[3-5] [6] – there are no universally accepted model for sensory-cognitive-motor integration.

This paper addresses the issue of integration through a goal-generating architecture targeted to that middle level of cognition investigated by Jackendoffion [7]. He suggests that there is a cognitive bridge that fills the gap between a high-level reasoning (e.g. task planning, etc.) and low-level sensorimotor control. It is tempting to wonder whether this intermediate level could be the key to cognitive integration in mammals-like cognitive systems. In fact, it is fair to maintain that most human actions are neither the result of sensorimotor contingencies nor the output of logical reasoning.

A. Chella et al. (Eds.): Biologically Inspired Cognitive Architectures 2012, AISC 196, pp. 225–231.
springerlink.com

The cortex shows an almost universal capability to adapt to different kind of stimuli, taking into consideration the particular input statistics [6,8]; so it makes sense to look at a general approach. Moreover, experimental evidences show each brain area can potentially develop any cognitive or computational skill [9,10]. On the other hand, it is well know that the mammal brain is able both to learn how to achieve goals and what goals have to pursued [11]. Putting together goal generation and generality leads to the idea the goal-generation arise by means of interplay between thalamus and cortex [12]. In fact, the thalamus is closely coupled with the cortex and each partition of the thalamus seems to provide control information to the corresponding cortical areas [13]. A further evidence of this fact is the massive neural connections from the cortex to the thalamus whereas the backward connections are less data intensive [14]. As working hypothesis it may be assumed that the cortex provides memory storage and input stimuli classification whereas the thalamus evaluates to what extent the current stimuli are related with previous data. The hypothesis is consistent with the experimental evidence showing that the brain bootstraps the generation of new goals taking advantage of hardwire criteria likely located in the amygdala [15,16].

In the robotic fields, many of the robotic setups focus either on the sensorimotor level or on the high-level reasoning. A lot of them are designed for very specific goal, since designer aim to solve specific sensorimotor, relational or logic issues. Of course, there are also cognitive agents able to develop new skills and to adapt to novel environments, such as those exploiting embodiment [17-19].

In this paper we present an intermediate cognitive system able to generate new goals and process the incoming sensory information. The cognitive framework can be implemented using either bioinspired algorithms or not. The main contributions of this paper are: to extend a computational framework of the thalamus-cortex interaction and validate some preliminary results presented in [20].

2 Goal-Generating Architecture

In this section we sketch the Goal-generating Architecure (IA) which ought to help fleshing out a cognitive middleware. Specifically, the IA is composed of a set of Intentional Modules (IM) and a single Global Phylogenetic Module (GPM). To a certain extent, the IM ought to model the interaction between the thalamus and the cortex whereas the GPM models the amygdala, providing bootstrap innate criteria whose output is projected to each IM in the network. Such approximation is justified by neuroanatomical evidence showing that the amygdala and other parts of core affective circuitry modulate the attentional matrix in the brain. It is also worth to remember that the amygdala also projects to other areas notably the forebrain and the thalamus [15]. The input layer of the IA receives the sensory data whereas the output layer sends command to the actuators. Here, we focus on the internal details of the IM. The IM is the key module of the network and represents the basic computational unit that generates goal during the sensory acquisition. The IM receives two inputs: sensory information and a control signal. The control signal is the maximum between an external control signal $e_s(t-1)$ coming from another IM (if any) and a signal from the GPM. The GPM control signal is broadcasted to all IMs in the archtiecture. The IM has two symmetrical outputs: a category signal $\mathbf{y}$ and a new control signal r_s computed internally by the model. r_s is important since expresses the relevance of the sensory input at time t with respect to the previous history of the IM. Furthremore, r_s plays an important role during both the learning phase and the runtime activity.

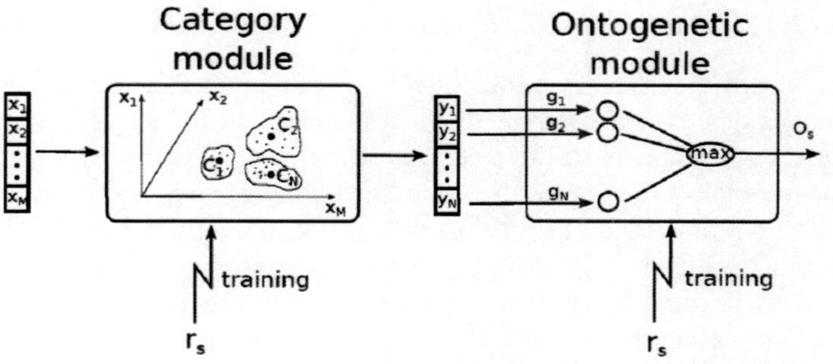

Fig. 1. (Left pane) The category module (Right pane) The ontogenetic module

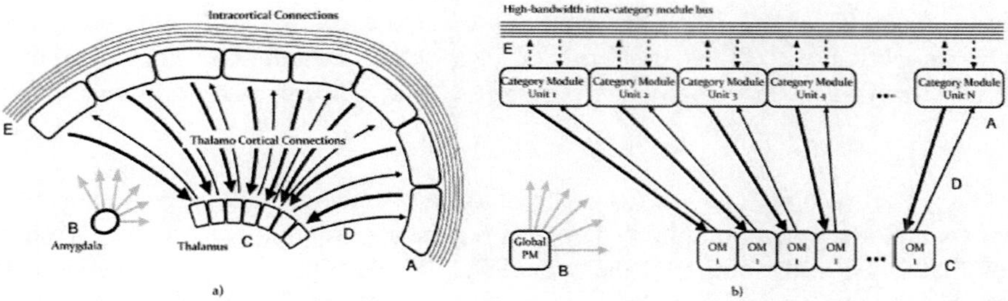

Fig. 2. A sketchy comparison between the thalamocorticalsystem and the presented architecture: A. Category Modules vs cortical areas, B. GPM vs. amygdala, C. Ontogenetic modules vs. thalamus, D. Control signals, E. high bandwidth data bus vs. intracortical connections

The IM is composed by two different submodules:the Category Module (CM) and the Ontogenetic Module (OM). The CM deals with the categorization of the sensory input, a sketchy aproximation of a cortical subarea. It is worth noting that the input vector could be generated by a single sensory source or could be a mix of different sensor modalities. The CM must be able to categorize the input, i.e. to *represent* the sensory input as correlation measure of a set of basis. On the other hand, the OM models a thalamus subareas and it must be able to associate the sensory information with the relevance of the sensory input itself. The relevance may be the result either of a natively phylogenetic signal (in the model this is GPM outcome) or of an internal association between the representation of the sensor values (e.g. the output of the CM) and a past experience with a *similar* input. The signal propagation through the IM can be summarized by the following steps:

1. Acquisition of the sensory input $\boldsymbol{x}$, and the global relevant signal g_s.
2. Given the input $\boldsymbol{x}$, the CM computes the category signal $\boldsymbol{y}$ (see below for details).
3. The OM models the associative memory by means of Hebbian Learning.
4. IM computes r_s representing the relevance of the incoming input $\boldsymbol{x}$ at time t.
5. Training phase: the IM adds new categories update the the OM.

The CM deals with the representations of the sensory input that can be generated by different sources or by a mixature of them. It implies that the internal algorithm can deal with different sensory modalities regardless the unity of measure. Using independent component analysis (ICA) each CM extracts the relevant basis given its input. The input is thus projected on the corresponding set of components and thus normalized and dimensionally simplified. It is interesting to observe that each IM has to shut down, from time to time, in order to improve or extend its set of ICA components. This activity is time-consuming and it presents, very speculatively, a strong resemblance with the memory consolidation role of sleep which, particularly in mammals, has a function not completely defined [21,22]. The outcome is then fed to a clustering function with an ad-hoc distance measure (Fig. 2, left pane). The sensory input can be defined as a vector $\boldsymbol{x} \in R^K$. This input is projected on the set of ICA M components. The projection $\mathbf{w} \in R^M$ is used to define a set of centers $C_1, C_2, ..., C_N$ in the M-dimensional space. Eventually, the vector $\boldsymbol{y} \in R^N$ may be computed as follows:

$$y_i = \rho(w, C_i) \tag{1}$$

where $\mathbf{y}=[y_1, y_2, ..., y_N]$ and $\rho(x, C_i)$ is the positive correlation index. Each element of $\mathbf{y}$ shows how the corresponding center well represents the input vector. It is worth noting that y is also the output of the overall IM. We consider the correlation measure $\mathbf{y}$ good enough to represent the sensory information. During the execution, the sensory inputs are grouped in few clusters, showing a certain statistics of the sensory information. However, one of the key features of the model is to dynamically generate the centers with respect to the relevance of the incoming sensory values.

The OM ideally represents the IM associative memory. It receives as input the category vector $\mathbf{y}$, the relevant signal r_s during the training phase, and returns as ouput a signal $o_s \in R$ representing the relevance of the sensory input with respect to IM past history. To accomplish this task, the OM contains a set of internal state variables, called gates, ideally representing the weights of neurons with one input and one output (Fig. 2, right pane). The gates are grouped in a vector $\mathbf{g}= [g_1, g_2, ..., g_N]$ where N is $\mathbf{y}$ dimension. The signal o_s is defined as:

$$o_s = max(y_i \cdot g_i) \tag{2}$$

where y_i is the element value of the CM output vector and g_i is the value of the single gate. o_s shows which y_i is associated with a past relevant information. During the training phase, to take into account the memory information within **g**, a stable Hebbian rule (Oja's version) is used:

$$g_i(t) = g_i(t-1) + \Delta\, g_i(t), \tag{3}$$

$$\Delta g_i(t) = \eta[r_s(t) \cdot y_i(t) - g_i(t-1) y_i(t)^2]$$

where η is the learning rate, r_s is the relevant signal that acts as desidered output of the neurons i and y_i is the input of the gate. The training phase implements an associative learning between the relevant signal and the incoming inputs. The gates implicitly encode new generated goals.

3 Robotic Setup

The goal-generating architecture can potentially be applied to control any robot regardless of the kind of sensors and actuators. Moreover, the incoming sensory information can be either raw or filtered data since the CM generates categories using the sensory values without any a priori knowledge as the nature of incoming data. For the same reason, the cognitive agent may control any kind of actuator, after a development phase in which the categories and the gates are generated.

We tested our architecture on a NAO robot. It has 21 degrees of freedom (DOF). Given the rich endowment of sensors, only a few of them have been connected to the developing goal-generating architecture. The aim of the experiment is to verify wheter the robot is able to generate a new goal starting from a hardwired innate criterion. At the beginning, the robot generates a positive phylogenetic signal when it see a colored object regardless of its shape. The objective of the experiment, after the interaction with a structured environment, is the acquisition of a new criterion for a specific object shape regardless of its color.

The overall architecture is composed by a single IM. The IA receives colored logpolar image frames from the NAO camera and produces in output the pan and tilt commands where the angular amplitude of a movement is chosen randomly by a Gaussian distribution in such a way that the lower the relevant signal the greater the movement. It is worth noting that, in principle, the output signal of the IM could potentially drive directly the motors but in this first experiment we focused on the capability of the ontogenetic module to infer a new goal with respect to the category algorithm. The experiment is divided in three different steps:

1. a set of black geometrical objects are shown in front of the camera.
2. a single colored object with star shape is shown in front of the NAO.
3. the same set of black objects is shown (including a gray star-shaped object.

In the first step the robot does not focus on black objects, whereas in the second step the phylogenetic signal becomes immediatly high. After a while, also the ontogenetic signal becomes high due to the gates saturation. In the third step the agent shows interest, through the relevant signal, for star-shaped black objects.

Although this experiment may look very close to previous work [20], there are important differences. First, the system is now able to generalize to a whole network of IM (previously the system was limited to a single IM). Second, the system implements memory consolidation and invariant extraction by means of ICA in each IM. Third, a theoretically more satisfying version of Hebbian Learning has been used.

In the near future, some improvements are planned: first, the action-control model (based on action for perception paradigm) will be integrated in the intentional architecture; finally the dynamical generation of Intentional Modules will be implemented.

Acknowledgements. This work has been possible thanks to the partial support provided by the 2010-2012 Italian-Korean bilateral project (ICT-CNR and KAIST). The authors gratefully acknowledge the contribution the NVIDIA Academic Partnership.

References

1. Pfeifer, R., Lungarella, M., Fumiya, L.: Self-Organization, Embodiment, and Biologically Inspired Robotics. Science 5853, 1088–1093 (2007)
2. Sperry, R.: Cerebral organization. Science 133(3466), 1749–1757 (1961)
3. Shanahan, M.: Embodiment and the Inner Life. Cognition and Consciousness in the Space of Possible Minds. Oxford University Press, Oxford (2010)
4. Baars, B.J.: A Cognitive Theory of Consciousness. Cambridge Univ. Press (1988)
5. Tononi, G.: An information integration theory of consciousness. BMC Neuroscience 5, 1–22 (2004)
6. Dileep, G.: How the brain might works. Stanford University (2008)
7. Jackendoff, R.S.: Consciousness and the computational mind. MIT Press (1987)
8. Hawkins, J., Blakeslee, S.: On Intelligence. Times Books, New York (2004)
9. Sharma, J., Angelucci, A., Sur, M.: Induction of visual orientation modules in auditory cortex. Nature 404, 841–849 (2000)
10. Sur, M., Rubenstein, J.L.R.: Patterning and Plasticiy. Science 310, 805–810 (2006)
11. Manzotti, R.: A Process-oriented Framework for Goals and Motivations. In: Poli, R. (ed.) Causality and Motivation, pp. 105–134. Ontos-Verlag, Frankfurt (2010)
12. Ward, L.M.: The thalamic dynamic core theory of conscious experience. Consciousness and Cognition 20, 464–486 (2011)
13. Modha, D.S., Singh, R.: Network architecture of the long-distance pathways in the macaque brain. Proc. of the Nat. Acad. of Sciences 107(30), 13485–13490 (2010)
14. Nieuwenhuys, R., Voogd, J., van Huijzen, C.: The Human Central Nervous System: A Synopsis and Atlas. Steinkopff, Amsterdam (2007)
15. Duncan, S., Feldman Barrett, L.: The role of the amygdala in visual awareness. Trends in Cognitive Sciences 11, 190–192 (2007)
16. de C. Hamilton, A.F., Grafton, S.T.: Goal Representation in Human Anterior Intraparietal Sulcus. The Journal of Neuroscience 26(4), 1133–1137 (2006)
17. Lungarella, M., Metta, G., Pfeifer, R., Sandini, G.: Developmental robotics: a survey. Connection Science 15, 151–190 (2003)

18. Vernon, D., Metta, G., Sandini, G.: A Survey of Artificial Cognitive Systems: Implications for the Autonomous Development of Mental Capabilities in Computational Agents. IEEE Trans. on Evolutionary Computation 11(2), 151–180 (2007)
19. Asada, M., Hosoda, K., Kuniyoshy, Y., Ishiguro, H., Inui, T., Yoshikawa, Y., Ogino, M., Yoshida, C.: Cognitive developmental robotics: A survey. IEEE Transactions on Autonomous Mental Development 1(1), 12–34 (2009)
20. Manzotti, R., Tagliasco, V.: From behaviour-based robots to motivation-based robots. Robotics and Autonomous Systems 51, 175–190 (2005)
21. Diekelmann, S., Born, J.: The memory function of sleep. Nature Neuroscience 11, 114–126 (2010)
22. Stickgold, R.: Sleep-dependent memory consolidation. Nature 437(7063), 1272–1278 (2005), doi:10.1038/nature04286

A Review of Cognitive Architectures for Visual Memory

Michal Mukawa and Joo-Hwee Lim

IPAL (UMI CNSR 2955), Institute for Infocomm Research, 1 Fusionopolis Way
#21-01 Connexis A*STAR, Singapore 138632
{stumam,joohwee}@i2r.a-star.edu.sg

Abstract. Numerous cognitive architectures have been proposed for human cognition, ranging from perception, decision making, to action and control. These architectures play a vital role as foundation for building intelligent systems, whose capabilities may one day be similar to that of the human brain. However, most of these architectures do not address the challenges and opportunities specific to visual perception and memory, which form important parts of our daily tasks and experiences. In this paper, we briefly review some of the cognitive architectures related to perception and memory. As studies of visual perception and memory are active research areas in cognitive science, we summarized what has been done so far, how neurobiology and psychology have identified different memory systems, and how different stages of memory processing are performed. We described different types of visual memory, namely short-term memory, long-term memory, episodic and semantic memories. Finally, we tried to predict what should be done towards a visual memory architecture for enabling autonomous visual information processing systems.

Keywords: Visual memory, Cognitive vision, Cognitive architectures, Short-term memory, Long-term memory.

1 Introduction

Much has been described about the visual pathways in animals' eyes and brain. However, little has been said about high-level cognition involving visual memories, especially reconciling the visual information processing flow in the human brain. In unfettered environments human vision is considered better than computer vision in many ways. Therefore, possibilities arise for cognitive and computer researchers to design and implement computer vision systems that will be able to extend our visual abilities [7]. Cognitive architectures provide necessary information for building intelligent systems whose capabilities should be similar to those of humans. Artificial general intelligence (AGI) aims to create a system which will go beyond human capabilities in many areas [3]. To achieve that, there needs to be progress in several other domains. One of these domains is memory modeling. Here, we briefly review different models for visual memory, as well as relevant cognitive architectures which use visual inputs.

A. Chella et al. (Eds.): Biologically Inspired Cognitive Architectures 2012, AISC 196, pp. 233–238.
springerlink.com

2 Cognitive Architectures with Visual Inputs

The main task for cognitive architectures is to model human performance in different task situations [3]. Newell [10] provided 12 criteria for evaluation of cognitive systems: adaptive behaviour, dynamic behaviour, flexible behaviour, development, evolution, learning, knowledge integration, vast knowledge base, natural language, real-time performance, and brain realization. For each cognitive architecture, progress in memory development can be considered as one of the most important factors, which allows us to create more advanced architectures [3]. We can categorize cognitive architectures into three main groups based on the different memory model used: symbolic, emergent, and hybrid architecture. Symbolic architectures use rule-based or/and graph-based representation of memory. In emergent systems globalist memory and localist memory are used. For hybrid architectures, localist-distributed or symbolic-connectionist memory model is used. In Table 1 simplified taxonomy of different memory types used in cognitive architectures is presented.

Table 1. Memory types used in different cognitive architectures

Symbolic architecture	**Emergent architecture**	**Hybrid architecture**
Rule-based memory	*Globalist memory*	*Localist-distributed memory*
Graph-based memory	*Localist memory*	*Symbolic-connectionist memory*

SOAR is a classic example of symbolic cognitive architecture. Laird [6] presented an extended version of SOAR system, this improved cognitive architecture gained, among other things: semantic memory, episodic memory, a set of processes and memories to support visual imagery, and clustering. In SOAR, semantic memory is built up from structures that occur in working memory. Information from semantic memory is retrieved by receiving a cue, which was generated in working memory, after finding best match semantic memory sends data to working memory. New modules added to SOAR support visual imagery. Short-term memory was added, where images are constructed and manipulated; long-term memory that contains images that can be retrieved into the short-term memory; processes that manipulate images in short-term memory, and processes that create symbolic structures from the visual images. With the addition of visual imagery, Laird [6] demonstrates that it is possible to solve spatial reasoning problems orders of magnitude faster than without it, and by using significantly less procedural knowledge.

DeSTIN is an emergentist cognitive architecture. This system addressed the problems of general intelligence by using hierarchical spatiotemporal networks. So far practical use of DeSTIN has focused on visual and auditory processing. This architecture allows for powerful unsupervised classification of different visual inputs [4]. It can create structures adequate for different categories of objects located in the scenes (categories like: beds, lamps, pets), this system is also capable of creating categories for actions (e.g. reaching, falling) which it sees [4].

NOMAD (Neurally Organized Mobile Adaptive Device) is another emergent architecture. It is also known as Darwin automata. The main task of this architecture is pattern recognition in real time. NOMAD is equipped with vision sensors, as one of the inputs to the system. As for now this cognitive architecture is controlled by simulated nervous system ($\sim 10^5$ neurons with $\sim 10^7$ synapses) [3], which run on powerful computers. One of task of Darwin automata systems VI and VII is the ability for invariant visual object recognition.

DUAL is probably the most impressive hybrid architecture. This system uses both symbolic and neural network mechanisms. This combination can be considered to be DUEL's most significant feature. Nestor and Kokinov [9] present model of DUAL system where they connect raw visual input with systems semantic memory, this model takes advantage of this hybridity by combining massively parallel activation-based computations with a serial attention-based symbolic processing mechanism instantiating the principle of active vision. Due to this change, DUAL becomes capable of processing its visual input.

3 Visual Memory

As memory and vision concern both the processes of the memory and nature of the stored representations, research in these domains seems to be especially interesting [8]. Tulving and Craik [14] provide a general definition of memory as the "neurocognitive capacity to encode, store, and retrieve information", they also suggested that there may be many separate memory systems that fit this definition. Nowadays, memory systems are characterized by differences in timing, storage capacity, conscious access, active maintenance, and mechanisms of operation.

Two main types of memory can be distinguished as follows: short-term memory (in this paper also referred as working memory), and long term memory. When we are describing visual processes in the human brain, we often denote working memory as visual working memory, and long-term memory as visual long-term memory. Visual working memory has a very limited capacity (only a few items), on the other hand visual long- term memory can retain thousands of items [2].

We, as humans, are only conscious of what is currently stored in our working memory. We are unconscious of things stored in the long-term memory[13]. Through manipulation of data stored in our working memory we can perform cognitive tasks [1]. All information currently stored in the working memory can be moved to long-term memory [13].

How is data stored in working memory? Brady, Konkle, and Alvarez [2] present one of the possibilities that initial encoding process is object-based (or location-based), but that the "unit" of visual working memory is a hierarchically structured feature bundle: at the top level of individual "unit" is an integrated object representation; at the bottom level of an individual "unit" are low-level feature representations. In this same publication, authors provide evidence that individual items are not represented independent of other items on the same display, visual system efficiently extracts and encodes structure from the spatial and featural information across the visual scene.

Wood [15] presents a system that uses three representations: spatio-temporal representation for object tracking, object property/kind representation for object recognition, and view-dependent snapshot representation for place recognition. Each of these systems may use its own buffer to keep information. Wood refers to this idea as “core knowledge architecture“ of visual working memory.

As humans are not directly conscious of their long term memory, the same situation occurs for visual long-term memory [13]. Visual long-term memory can be divided into two parts: semantic visual memory, and episodic visual memory. According to Nuxoll and Laird [11] semantic memory is what you “know“, and consists of isolated facts that are decontextualized. They are not organized in a specific experience and are useful in reasoning about general properties of the world. This memory is also called “general knowledge“. This storehouse of knowledge is built over our lifetime of visual experience. Many models assumed that this knowledge is stored hierarchically. Lower levels in this hierarchy consist of basic features, which are shared across many categories. Higher levels consist of visual features, this features are more category specific [2].

Episodic memory is what you “remember“, and includes contextualized information about specific events, such as a memory of a vacation or the last time you had dinner [11]. This memory is capable of storing thousands of events. Episodic visual memory consists of multiple levels, low levels contain individual items, and higher levels store conceptual representation [2]. In Figure 1, a simplified model of information flow in human visual memory is presented.

To store any knowledge in long-term memory, we need processes such as adaptation and learning, but this topic is not discussed in this paper. A "forgetting" process in memory models is also needed to preserve frequently-used information, while removing less important or rarely used ones. This topic is also omitted.

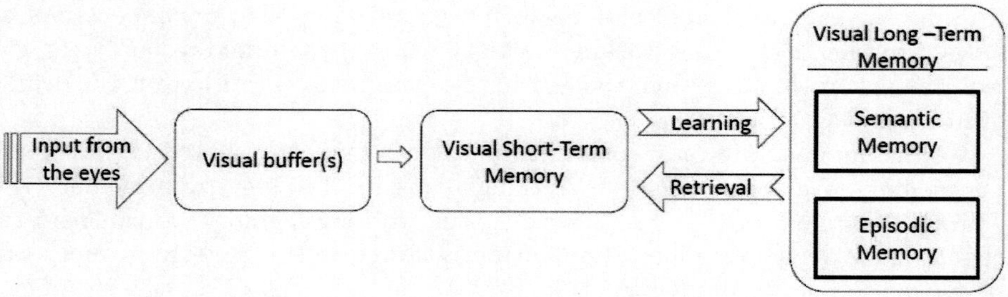

Fig. 1. Data flow in human visual memory

4 Towards Cognitive Architecture with Visual Memory

So far, cognitive architectures with visual memory are used in very few real-world applications. These architectures require constant evaluation. Humans are only conscious of the final processing stage of their visual system. This is one of the main reasons why more research about human visual memory is needed. Solving problems for which vision skills of cognitive architecture are needed may require addition of specialized computer vision modules, and specialized memory structures, like visual working memory, semantic memory, and episodic memory. Some interactions in cognitive architectures require deeper research, like different methods of moving knowledge between episodic and semantic memories.

Starzyk [12] introduces a new computational model of cognitive architecture, where attention switching mechanism is based on internal motivations and mental saccades. Mental saccades provide mechanism for selection of the current attention spotlight. This proposed mechanism can be use in future cognitive architectures for creating visual memory models.

In real-world scenes, humans encode the gist and also some detailed information about specific objects [5]. This may indicate that we need more than one visual memory buffer when trying to model the visual memory. For example one buffer could be responsible to store general information about the scene, and the second for processing highly detailed information about objects. Moreover as Wood presented in his work [15], we may even need three buffers for visual working memory. One for spatiotemporal information, second for object identity information, and third for snapshot information.

Brady et al. [2] suggest that models of memory must go beyond characterizing how individual items are stored and move toward capturing the more complex, structured nature of memory representation.

Future research may also focus on how memory systems interact in the same moment in time, operating over similar structured representations.

Learning processes for visual memory should be continuously developed. Lim and Liu [7] believe that there are two main mechanisms by which a visual scene is learned. The first one is attention-driven. This mechanism is used when our attention is directed on some specific object in the scene. In this process, eye saccades movements play huge role when we search important "items" in the scene. The second mechanism is so called frequency-based. It presumes that information about the object will be stored finally in the visual long term memory when we see this object many number of times.

5 Conclusion

Numerous cognitive architectures have been proposed for human cognition. These architectures play a vital role as foundation for building intelligent systems. However, none of these architectures addresses the challenges and opportunities specific to visual perception and memory. In this paper, we briefly reviewed some cognitive architectures with visual input. We summarized what was done so far in visual memory research. We also briefly described different types of human visual memory. Finally, we proposed some ideas about what can be done next to create cognitive architecture with visual memory.

References

1. Baddeley, A.: The episodic buffer: a new component of working memory? Trends in Cognitive Sciences 4(11), 417–423 (2000)
2. Brady, T.F., Konkle, T., Alvarez, G.A.: A review of visual memory capacity: Beyond individual items and toward structured representations. Journal of Vision 11(5) (2011)
3. Duch, W., Oentaryo, R.J., Pasquier, M.: Cognitive architectures: Where do we go from here? In: Proceeding of the 2008 Conference on Artificial General Intelligence 2008: Proceedings of the First AGI Conference, pp. 122–136. IOS Press (2008)
4. Goertzel, B., Lian, R., Arel, I., de Garis, H., Chen, S.: A world survey of artificial brain projects, part ii: Biologically inspired cognitive architectures. Neurocomputing 74(1), 30–49 (2010)
5. Hollingworth, A.: Constructing visual representations of natural scenes: the roles of short-and long-term visual memory. Journal of Experimental Psychology: Human Perception and Performance 30(3), 519 (2004)
6. Laird, J.E.: Extending the soar cognitive architecture. In: Proceeding of the 2008 Conference on Artificial General Intelligence 2008: Proceedings of the First AGI Conference, pp. 224–235. IOS Press (2008)
7. Lim, J.H., Liu, S.: Extended visual memory for computer-aided vision. In: Proceedings of the 33rd Annual Meeting of the Cognitive Science Society. Cognitive Science Society (2011)
8. Luck, S.J., Hollingworth, A.R.: Visual memory. Oxford University Press, USA (2008)
9. Nestor, A., Kokinov, B.: Towards active vision in the dual cognitive architecture (2004)
10. Newell, A.: Unified theories of cognition, vol. 187. Harvard Univ. Pr. (1994)
11. Nuxoll, A.M., Laird, J.E.: Extending cognitive architecture with episodic memory. In: Proceedings of the National Conference on Artificial Intelligence, vol. 22, p. 1560. AAAI Press, MIT Press, Menlo Park, Cambridge (1999/2007)
12. Starzyk, J.A.: Mental saccades in control of cognitive process. In: The 2011 International Joint Conference on Neural Networks (IJCNN), pp. 495–502. IEEE (2011)
13. Sweller, J., Van Merrienboer, J.J.G., Paas, F.G.W.C.: Cognitive architecture and instructional design. Educational Psychology Review 10(3), 251–296 (1998)
14. Tulving, E., Craik, F.I.M.: The Oxford handbook of memory. Oxford University Press, USA (2005)
15. Wood, J.N.: A core knowledge architecture of visual working memory. Journal of Experimental Psychology: Human Perception and Performance 37(2), 357 (2011)

A Model of the Visual Dorsal Pathway for Computing Coordinate Transformations: An Unsupervised Approach

Flavio Mutti[1], Hugo Gravato Marques[2], and Giuseppina Gini[3]

[1] Dipartimento di Elettronica e Informazione, Politecnico di Milano, Italy
mutti@elet.polimi.it
[2] University of Zurich, Institute for Informatics, AI Lab, Zurich 8050, Switzerland
hgmarques@gmail.com
[3] Dipartimento di Elettronica e Informazione, Politecnico di Milano, Italy
gini@elet.polimi.it

Abstract. In humans, the problem of coordinate transformations is far from being completely understood. The problem is often addressed using a mix of supervised and unsupervised learning techniques. In this paper, we propose a novel learning framework which requires only unsupervised learning. We design a neural architecture that models the visual dorsal pathway and learns coordinate transformations in a computer simulation comprising an eye, a head and an arm (each entailing one degree of freedom). The learning is carried out in two stages. First, we train a posterior parietal cortex (PPC) model to learn different frames of reference transformations. And second, we train a head-centered neural layer to compute the position of an arm with respect to the head. Our results show the self-organization of the receptive fields (gain fields) in the PPC model and the self-tuning of the response of the head-centered population of neurons.

1 Introduction

A coordinate transformation (CT) is the capability to compute the position of a point in space with respect to a specific frame of reference (FoR), given the position of the same point in another FoR. The way the mammal brain solves the problem of CTs has been largely studied. Nowadays it is fairly well known from lesion studies [10] that the main area involved in this type of computation is the Posterior Parietal Cortex [1] [6].

The computation of CT seems to exploit two widespread properties of the brain, namely, population coding [7], and gain modulation [2] [9]. Population coding is a general mechanism used by the brain to represent information both to encode sensory stimuli and to drive the body actuators. The responses of an ensemble of neurons encode both sensory or motor variables in such a way that can be further processed by the next cortical areas, e.g. motor cortex. There are at least two main advantages of using a population of neurons to encode information: robustness to noise [7] and the capability to approximate nonlinear

A. Chella et al. (Eds.): Biologically Inspired Cognitive Architectures 2012, AISC 196, pp. 239–246.
springerlink.com

transformations [8]. Gain modulation is an encoding strategy for the amplitude of the response of a single neuron that can be scaled without changing the response selectivity of the neuron. This modulation, also know as *gain field*, can arise from either multiplicative or nonlinear additive responses [2] [3].

Several computational models of the PPC address the problem of CTs using three-layer feed-forward neural networks (FNNs) [11], recurrent neural networks (RNNs) [9], or basis functions (BFs) [8]. The FNNs and the BFs models are trained with supervised learning technique whereas the RRNs model uses a mix of supervised and unsupervised approaches to train the neural connections, encoding multiple FoRs transformation in the output responses.

It is worth noting that gain modulation plays an important role in the computation of the coordinate transformations but it is still unclear if this property emerges in the cortex from statistical properties of the afferent (visual) information. Recently, [5] shows evidence to support that gain fields can arise through the self-organization of an underling cortical model called Predictive Coding/Biased Competition (PC/BC) . It demonstrates that the gain modulation mechanism arises through the competition of the neurons inside the PC/BC model, and comments on the feasibility of such system to compute CTs.

These computational models of the PPC could be particularly suitable for the robotics community to solve the well-known problem of CT. In the recent past, an architecture was proposed that explicitly includes a PPC model composed by a set of radial basis functions trained with supervised learning techniques [4]. However, most of the approaches in robotics address the problem of FoR transformation inside the more general sensorimotor mapping approach, without explicitly exploit the features of PPC models [6].

Following these ideas, we present a biologically inspired model for CTs. First we describe the training of a PPC model with an unsupervised learning approach; and second we introduce the computation of the arm position with respect to the head position. We hypothesise that gain modulation mechanisms can emerge in the PPC neurons, and that basis functions, encoding parallel CTs, can emerge after the training phase. The main contributions of this paper are: first to show an unsupervised approach to the learning of sensorimotor mapping; second to exploit the synergy between a biologically inspired neural network and the population coding paradigm; and third to introduce quantitative evaluation of the sensorimotor mapping performance.

This paper is organized as follows. In Section 2 we design the neural network model that performs the implicit sensorimotor mapping, in Section 3 we present the performed experiments and in Section 4 we derive our conclusions.

2 Model

In this section we present the neural model used for computing CTs between an arm and the head FoR. We define a simple mechanical structure composed by an eye, a head and an arm with the same origin. We assume the same origin because the fixed translations among these FoRs can be neglected due to their

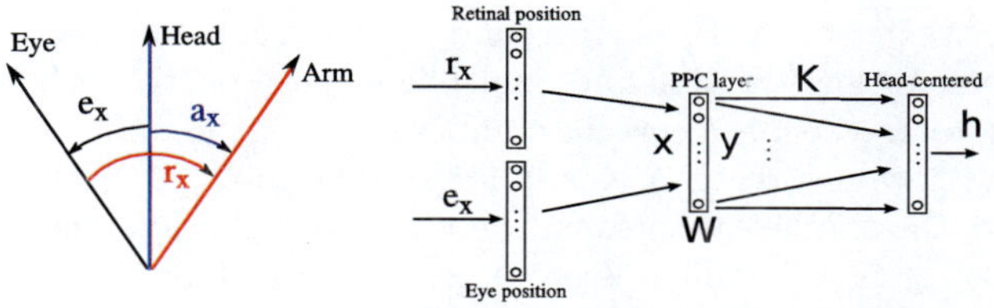

Fig. 1. (Left pane) Body definition composed by an eye, a head and an arm with the same origin. (Right pane) Neural Network model. The first layer encodes the sensory information into a neural code, the second layer models the posterior parietal cortex and it performs the multi sensory fusion and the third layer encodes the arm position with respect to the head frame of reference.

known contribution in the computation of the CTs (Figure 1, left pane). The eye position is defined by the angle e_x with respect to the head FoR, the retinal stimuli position of the arm is defined by the angle r_x with respect to the eye FoR; the head-centered position of the arm is defined by $a_x = r_x + e_x$ angle (see Figure 1, left pane). The neural architecture is divided in three layers: the first is composed by two populations of neurons which represent the information of the retinal position of the arm, r_x, and the eye position with respect to the head, e_x. The second is composed by PPC population of neurons that encode the position of the arm in different FoRs. The third is a population of neurons that encodes the arm position with respect to the head FoR.

The first layer of the network model receives as input the eye position with respect to the head FoR (e_x) as well as the arm position with respect to the retinal FoR (r_x). We define the eye angle e_x in degrees and the retinal position of the target r_x both in degrees and in pixels (see Section 3). These *numeric* values are encoded in a *population coding* paradigm, where a given sensor value is represented as a population of neural responses [8]. The response of a population neuron is defined as a Gaussian as follows:

$$n_i = A \exp\left(-\frac{(v - \mu_i)^2}{2\sigma^2}\right) \tag{1}$$

where n_i is the response of neuron i, μ_i is the neuron preferred sensor value, v is the *numeric* input angle (in degrees), and σ is the standard deviation of the Gaussian.

The PPC layer is based on the Predictive Coding/Biased Competition model (PC/BC) proposed in [5]. The model is trained with a unsupervised approach that is based on Hebbian learning. The system equations are:

$$s = x \oslash (\epsilon_2 + \hat{W}^T y) \qquad y = (\epsilon_1 + y) \otimes W s \tag{2}$$

where s is the internal state of the PC/BC model, $x = [n_0, \ldots, n_{L-M}]$ is the neural population input vector defined by M retinal neurons and L neurons encoding the eye position, W is the weight matrix, $\hat{W}$ is the normalized W,

y is the output vector of the PPC layer, and ϵ_1, ϵ_2 are constant parameters; $\oslash$ and $\otimes$ indicate element-wise division and multiplication respectively. These equations are evaluated iteratively for a certain number of time steps; after a certain period of time, y and e values reach a steady state. The internal state s is self-tuned and represents the similarity between the input vector x and the reconstruction of the input $\hat{W}^T y$ ($s \approx 1$ indicates an almost perfect reconstruction).

The unsupervised training rule is given by:

$$W = W \otimes \{1 + \beta\, y(s^T - 1)\} \tag{3}$$

where β is the learning rate. This training rule minimizes the difference between the population responses x and the input reconstruction $W^T y$; the weights increase for $s > 1$ and decrease for $s < 1$.

Let's consider the output vector $y = [y_0, \ldots, y_T]$ as the population responses of the PPC model. Each neuron response y_i should be compatible with the gain modulation paradigm, according to the experimental results of [5], in such a way that the response exhibit a multiplicative behaviour, as a function of both eye and retinal positions. The weight matrix, which encodes the response properties, is *internal* to the PPC model and the training phase is independent with respect to the unsupervised training phase that will involve the head-centered network layer.

The population of neurons associated to the head-centered frame of reference deals with the estimation of the arm position a_x given the eye angle e_x and the projection of the arm in the retina r_x. The synapses between the PPC layer and head-centered frame are trained with an Hebbian learning, taking into account the arm position, a_x. Estimating a_x means identifying the maximum response inside a population of neuron that encodes a_x with the population coding paradigm. The head-centered population responses are given by $h = K\, y$, where y is the output vector of the PPC model, K is the weight matrix representing the fully-connected synapses between the PPC model and the head-centered layer and h is a vector that contains the population responses, encoding the estimated a_x. The dimension of h depends on the granularity of the a_x encoding. The training phase is performed using Hebbian learning:

$$K = K + \delta\, h\, p_a^T \qquad \delta = \frac{1}{N} \tag{4}$$

where p_a is the vector that contains the proprioceptive population responses encoding a_x, and δ is the learning rate depending by N, the number of samples.

3 Experimental Results

In this section we present the results obtained in two experiments; in the first experiment we train and analyse the network where either the eye angle and the retinal position are encoded in degrees and in the second experiment we introduce a simple camera model to encode the retinal information in pixels.

The training phase is carried out in two steps: (1) train the PPC layer and (2) train the head-centered layer. The PPC layer is trained following the method described in Section 2 (Equation 3) and the synapses between the PPC and head-centered layer are trained using Hebbian learning as described in Section 2 (Equation 4).

In the first experiment, we encode both r_x and e_x in degrees and for the PPC layer, we use the same parameter values as in [5]. The y consists of a 64-element vector with a range for the sensors values defined as follows: $r_x \in [-30°, 30°]$, $e_x \in [-30°, 30°]$, $a_x \in [-60°, 60°]$. We encode the sensory input with a population of 61 neurons with a gaussian response and with a standard deviation $\sigma = 6°$. The σ value is chosen taking into account the experiment described in [5] whereas the neuron preferred values are equally distributed inside the range value.

After the training of the PPC layer, we train the head-centered layer with a population of 121 neurons, defining h as a 121-elements vector. With 121 neurons representing a_x the coding resolution (1°) can be analytically derived. The standard deviation of the neuron responses associated to the arm position a_x is equal to 6°. The population of neurons, encoding the proprioceptive position of the arm, has the same number of neurons of the head-centered layer (121) and each neuron has the same standard deviation (6°). The proprioceptive responses vector p_a drives the Hebbian learning for the head-centered neural layer (Equation 4).

Figure 2 shows the analysis of the trained network: top left pane shows the responses of the trained network that represents the arm position a_x with respect to the head frame of reference. The red solid line represent the response for $a_x = 20°$, the green dashed-dot line represent the response for $a_x = 0°$ and the blue dash line represent the response for $a_x = -20°$. Top right pane shows the error distribution (in degrees) of the estimated a_x with respect to the arm position, the eye position and the retinal position respectively. The solid lines represent the mean error and the dashed lines represent the standard deviation limits. The error distributions are quite similar and, in general, the error is quite low with a global mean error equal to 1.93° with a global standard deviation equal to 1.89°. Bottom left pane shows the receptive field after the training phase of the PPC layer: it is shown the global shape of the gain modulation. As expected, the curves shapes are compatible with the gain modulation paradigm, supporting the evidence that an unsupervised method can effectively learn a multiplicative behaviour. Bottom right shows the contours at half the maximum response strength for the 64 PPC neurons: it is worth noting the different color of the contours that represent different level of activations. A qualitative analysis points out that the population responses are stronger where the correspondent neuron receptive fields are slightly overlapped. Moreover, the PPC neurons receptive fields almost cover the whole subspace in the e_x-r_x plane, indicating that there is at least a neuron firing for each combination of e_x and r_x.

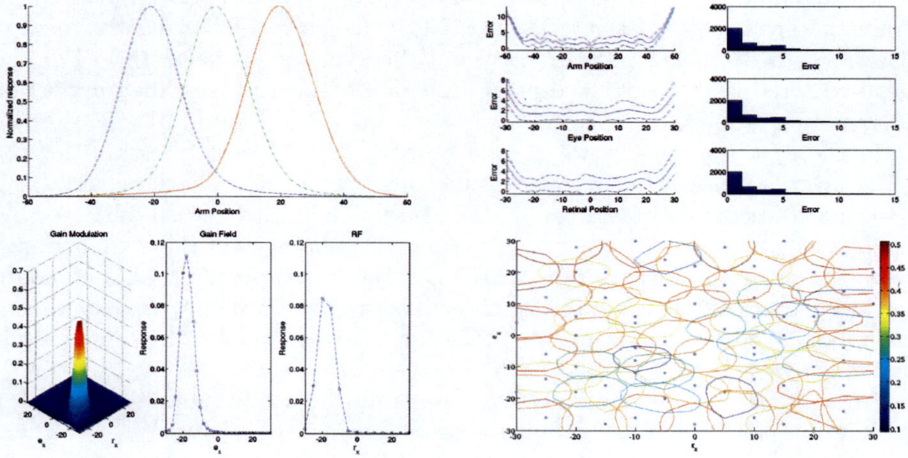

Fig. 2. Experimental results with r_x in degrees.(Top left) It shows the responses of the trained network that represents the arm position a_x with respect to the head frame of reference for $-20°$,$0°$ and $20°$, respectively. (Top right) The error distribution (in degrees) of the estimated a_x with respect to the arm position, the eye position and the retinal position respectively. The solid lines represent the mean error and the dashed lines represent the standard deviation limits. (Bottom left) It represents a receptive field after the training phase of the PPC layer. (Bottom right) Contours at half the maximum response strength for the 64 PPC neurons.

In the second experiment we investigate a more realistic scenario where the retinal position is a pixel position in the image plane. We just consider only the horizontal component of the image position of the arm. To compute the real a_x value we exploit some geometrical constraints, given by the camera model. In the specific:

$$a_x = e_x + \tan^{-1}\left(\frac{r_x}{f}\right) \qquad [°] \tag{5}$$

where r_x is the retinal position in pixels of the arm and f is the focal length of the camera. For our purposes, we choose a focal length equal to 120 pixels that represents a camera with a open lens of about 140°.

The PPC layer contains 64 neurons but the input range are $r_x \in [-320, 320]$, $e_x \in [-25°, 25°]$, $a_x \in [-94°, 94°]$ where r_x is defined in pixels; it follows that we suppose to have a image plane with an horizontal component that has a size equal to 641 pixels. The range of a_x follows the maximum value that the a_x can reach. We use 101 and 51 neurons to represent r_x and e_x, respectively. We use the standard deviation σ of gaussian representing r_x equal to 60 pixels. Also in this case the standard deviation of the proprioceptive neurons encoding a_x is equal to 6°.

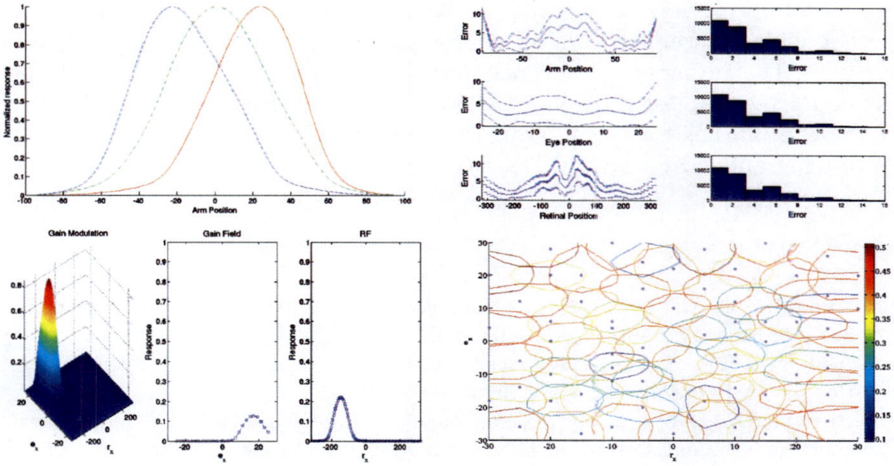

Fig. 3. Experimental results with r_x in pixels.(Top left) It shows the responses of the trained network that represents the arm position a_x with respect to the head frame of reference for $-20°$,$0°$ and $20°$, respectively. (Top right) The error distribution (in degrees) of the estimated a_x with respect to the arm position, the eye position and the retinal position respectively. The solid lines represent the mean error and the dashed lines represent the standard deviation limits. (Bottom left) It represents a receptive field after the training phase of the PPC layer. (Bottom right) Contours at half the maximum response strength for the 64 PPC neurons.

Figure 3 shows the results from the analysis of the trained network. The overall performance is lower than that obtained in the previous experiments: the top right pane shows the error distribution with respect to the arm, eye and retinal position, respectively. In this set of experiments, during the PPC learning, the system is able to learn PPC receptive fields that are compatible, in a qualitative way, with the gain modulation principle (see Figure 3, bottom left pane). The bottom right pane shows the receptive fields distribution in the space r_x-e_x where we have the same qualitative features of the previous experiment. The estimation of a_x has a global mean error equal to 3.36° with a global standard deviation equal to 2.90°.

4 Conclusions

This work described an unsupervised approach to learn coordinate transformations. The results show how the system is able to correctly compute the position of a target with respect to the stable head frame of reference knowing only the projection of the target onto the image plane and the eye position with respect to the head. Further experiments are foreseen to validate the model for more realistic scenarios, trying the method on a real robotic system and extending the model for complex physical architectures.

Acknowledgments. This work has been possible thanks to the partial support provided by the 2010-2012 Italian-Korean bilateral project (ICT-CNR and KAIST). The authors gratefully acknowledge the contribution the NVIDIA Academic Partnership for providing GPU computing devices. The research leading to these results has received funding from the European Community's 7th Framework Programme FP7 no. 207212 - eSMCs.

References

1. Andersen, R.A., Cui, H.: Intention, action planning, and decision making in parietal-frontal circuits. Neuron 63, 568–583 (2009)
2. Andersen, R.A., Essick, G.K., Siegel, R.M.: Encoding of spatial location by posterior parietal neurons. Science 230, 456–458 (1985)
3. Brozovic, M., Abbott, L.F., Andersen, R.A.: Mechanism of gain modulation at single neuron and network levels. Journal of Computational Neuroscience 25, 158–168 (2008)
4. Chinellato, E., Antonelli, M., Grzyb, B.J., del Pobil, A.P.: Implicit sensorimotor mapping of the peripersonal space by gazing and reaching. IEEE Transactions on Autonomous Mental Development 3(1), 43–53 (2011)
5. De Meyer, K., Spratling, M.W.: Multiplicative gain modulation arises through unsupervised learning in a predictive coding model of cortical function. Neural Computation 23, 1536–1567 (2011)
6. Hoffmann, M., Marques, H., Arieta, A., Sumioka, H., Lungarella, M., Pfeifer, R.: Body schema in robotics: A review. IEEE Transactions on Autonomous Mental Development 2(4), 304–324 (2010)
7. Knill, D.C., Pouget, A.: The bayesian brain: the role of uncertainty in neural coding and computation. Trends in Neurosciences 27(12), 712–719 (2004)
8. Pouget, A., Sejnowski, T.J.: Spatial transformations in parietal cortex using basis functions. Journal of Cognitive Neuroscience 9(2), 222–237 (1997)
9. Salinas, E., Abbott, L.F.: Coordinate transformations in the visual system: How to generate gain fields and what to compute with them. Progress Brain Research 130, 175–190 (2001)
10. Shadmehr, R., Krakauer, J.W.: A computational neuroanatomy for motor control. Experimental Brain Research 185, 359–381 (2008)
11. Xing, J., Andersen, R.A.: Models of the posterior parietal cortex which perform multimodal integration and represent space in several coordinate frames. Journal of Cognitive Neuroscience 12(4), 601–614 (2000)

Multiagent Recursive Cognitive Architecture*

Zalimkhan V. Nagoev

Institute of Computer Science and Problems of Regional Management
of Kabardino-Balkar Scientific Center of Russian Academy of Sciences, Russia
zaliman@mail.ru

Abstract. The hypothesis of invariant of organizational structure of intelligent decision making process based on cognitive functions, the *multiagent recursive cognitive architecture* (MuRCA) is proposed. Its application for the creation of self-organizing emergent systems, capable of goal-setting and adaptive behavior based on the semantic models of reality is substantiated.

Keywords: cognitive architectures, multiagent systems, artificial intelligence.

1 Introduction

Our approach to the decision of the problem of formalizing the semantics of rational thinking bases on theoretical foundation of cognitive psychology and cognitive neuroscience [1], [2], [5]. Minsky structured an intelligent cognitive architecture (CA) on a basis of a community of internal agents [4]. Basing on cognitive modeling, we try to find the architectural and functional solutions that would enable such agents to collaborate, thus, emulating the mental functions. We propose the idea of building a CA based on multi-agent system, each agent of which, in turn, implements a CA.

2 An Intelligent System Based on the Invariant of MuRCA

The aim of this work is the development of the CA of an intelligent agent that implements the principle of the emergent formation of intelligent behavior on the basis of self-organization in a complex symbiotic system of neuron cells. The latter are considered autonomous intelligent systems, interacting with each other basing on the principles of collective optimization of the parameters that are critical to preserve the integrity of the entire system. Specialization of neurons, the structure and function of the internal cognitive tract are genetically determined. However, each functional center within the cognitive tract has an opportunity

* The paper is written with the support of Russian Foundation for Basic Research grants ## 12-01-00367-a, 12-07-00624-a, 12-07-00744-a.

A. Chella et al. (Eds.): Biologically Inspired Cognitive Architectures 2012, AISC 196, pp. 247–248.
springerlink.com

to learn during the lifetime of the system. Learning is provided on the basis of ontoneuromorphogenesis and energy optimality principles [3].

According to these principles, a neuron, in general, approximates a nonlinear function of the choice of counterparties (neurons that may contribute to the joint solution of the problem) from a variety of other neurons. The knowledge base of a neuron is represented as a production system. We suggest that unlike traditional artificial neurons, the model of such a proactive neuron cell should include a number of specialized data processing centers, which we call the cognitive blocks (CBs) (their special functions we call the cognitive functions) (Fig. 1). The neuron through the synapses located on the axon interacts with other neurons, which, in their turn, each specializing in specific cognitive functions, constitute the cognitive architecture of the upper level by themselves (Fig. 2).

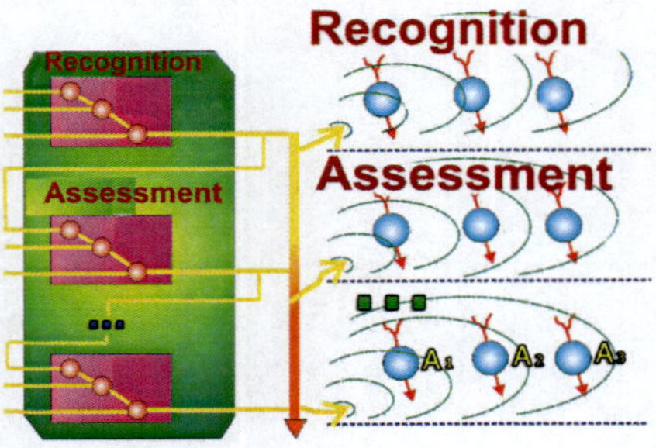

Fig. 1. Structural scheme of MuRCA

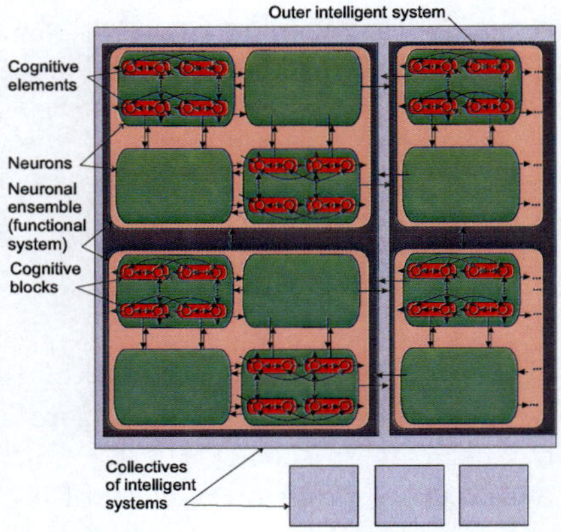

Fig. 2. The invariant of CA

The invariant of the MuRCA consists of 6 CBs: *recognition*, *assessment*, *goal formation*, *synthesis of an action plan*, *proactive modeling*, *control of actions*.

References

1. Chomsky, N.A.: A Review of Skinner's Verbal Behavior. In: Jakobovits, L.A., Miron, M.S. (eds.) Readings in the Psychology of Language. Prentice-Hall (1967)
2. Gazzaniga, M.S. (ed.): Conversations in the Cognitive Neurosciences. MIT Press, Cambridge (1999)
3. Nagoev, Z.V.: An intelligent system on a basis of fractal multiagent neuronal plastic cognitive architecture. In: Proceedings of the International Congress on Intelligent Systems and Information Technologies IS&IT 2011, vol. III, pp. 5–10. TTI Press, Taganrog (2011)
4. Minsky, M.: The Society of Mind. Simon and Schuster, New York (1988)
5. Newell, A.: Unified Theories of Cognition. Harvard University Press, Cambridge (1990)

A Biologically-Inspired Perspective on Commonsense Knowledge

Pietro Perconti

University of Messina,
Dept. of Cognitive Science
Messina, Italy

Since the seminal papers by John McCarthy [1,2], the problem to design intelligent systems able to handle common sense knowledge has become a real puzzle [3,6,7]. According to the McCarthy and Hayes suggestion, "The first task [to construct a general intelligent computer program] is to define even a naive, common-sense view of the world precisely enough to program a computer to act accordingly. This is a very difficult task in itself" [5]: 6. Perhaps the frame problem, i.e., how can a representational system deal with the enormous amount of knowledge that is necessary to everyday behaviour, needs nowadays a new account. The BICA challenge, that is, the challenge to make a general purpose and computational equivalent of the human intelligence by means of an approach based on biologically inspired cognitive architectures, can be considered as an example of this kind of new perspective [1,8].

In my talk I would like to show the possibilities of a "quasi-pragmatist" and a "biologically inspired" attitude on commonsense knowledge. It would be a step toward the BICA roadmaps commitment to embodiment and biological plausibility of computational architecture. My attempt is inspired by the idea that common sense has not an unitary structure and that its cognitive architecture depends on a set of biological constraints. In particular, I will argue that commonsense knowledge is articulated at two different levels: a deep and a superficial level of common sense.

The deep level consists of know-how procedures, of metaphorical frames based on the imaginative bodily representation, and of a set of adaptive behaviour, like disgust and the feelings of pain and pleasure. There are body schematic processes which generate an ongoing pre-reflective experience of the body as it performs and moves in the environment. This embodied experience is imaginative in kind and it works as a set of biological constraints for the human capacity to shape concepts [2]. The "quasi-pragmatistic" account is based on the idea that the deepest level of common sense is oriented toward the efficacy, and not the truth. Furthermore, deep common sense is unavailable for any fast change because it depends on human biology more than on culture conventions. Cognitive science can only appreciate this state of affair; it is useless to attempt to change the deep level of common sense, as in the case one would like to change the way human eyes perceive the world.

On the contrary, superficial level includes beliefs and judgments. They are emendable by means of further propositional knowledge and can be true or false. In a word, they are culture-dependent creatures. Differently from the deep

A. Chella et al. (Eds.): Biologically Inspired Cognitive Architectures 2012, AISC 196, pp. 249–250.
springerlink.com

level, superficial common sense is really challenged by the findings of cognitive science and we should are interested in these changes that are advantageous to deal with new technological and bioethical issues.

References

1. Chella, A., Lebiere, C., Noelle, D.C., Samsonovich, A.V.: On a Roadmap to Biologically Inspired Cognitive Agents. In: Samsonovich, A.V., Johannsdottir, K.R. (eds.) Biologically Inspired Cognitive Architectures 2011 - Proceedings of the Second Annual Meeting of the BICA Society. Frontiers in Artificial Intelligence and Applications, vol. 233, pp. 453–460 (2011)
2. Gallese, V., Lakoff, G.: The Brain's Concepts: The Role of the Sensory-Motor System in Conceptual Knowledge. Cognitive Neuropsychology 22(3/4), 455–479 (2005)
3. Haselager, W.F.G.: Cognitive Science and folk psychology: The right frame of mind. Sage, London (1997)
4. McCarthy, J.: Programs With Common Sense. In: Proceedings of Teddington Conference on the Mechanization of Thought Processes (1958)
5. McCarthy, J., Hayes, P.J.: Some philosophical problems from the standpoint of artificial intelligence. In: Meltzer, B., Michie, D. (eds.) Machine Intelligence. Edinburgh University Press, Edinburgh (1969)
6. Minsky, M.: Commonsense-based interfaces. Communications of the ACM 43(8), 67–73 (2000)
7. Pylyshyn, Z.W. (ed.): The robots dilemma. Ablex, Norwood (1987)
8. Samsonovich, A.V.: On a roadmap for the BICA Challenge. Biologically Inspired Cognitive Architectures 1, 100–107 (2012)

Coherence Fields for 3D Saliency Prediction

Fiora Pirri, Matia Pizzoli, and Arnab Sinha

ALCOR Lab, Department of Computing, Control and Management Engineering Antonio Ruberti, Sapienza, University of Rome
{pirri,pizzoli,sinha}@dis.uniroma1.it

Abstract. In the coherence theory of attention [26] a coherence field is defined by a hierarchy of structures, supporting the activities across the different stages of visual attention. At the interface between low level and mid level attention processing stages are the *proto-objects*, generated in parallel and collecting features of the scene at specific location and time. These structures fade away if the region is not further attended by attention. We introduce a method to computationally model these structures on the basis of experiments made in dynamic 3D environments, where the only control is due to the Gaze Machine, a gaze measurement framework that can record pupil motion at the required speed and project the point of regard in the 3D space [25],[24]. We show also how, from these volatile structures, it is possible to predict saliency in 3D dynamic environments.

1 Introduction

The main goal of computational attention is to predict saccade directions in tasks such as visual search of targets in a complex scene ([14], [34], [12], [20], [4], [6]). The ability by which humans can find targets in crowded scenes without overloading memory has stimulated a large body of research on pre-attentive processing to model the rapid gaze shifts characterizing visual search. Approaches have exploited either the simulation of saccades by active camera(s) [5], [17] or biologically founded prior models of saliency [23], [1], [11], [7], [30], [16], to cite some of the works from the wide literature on saliency prediction. The main motivation on this rich literature on saliency prediction to model visual search is twofold: on the one hand the complexity of searching the visual field is too high to be managed by processing the whole visual input [34] at the same resolution of the fovea; on the other hand feature detectors (such as Harris [10]) and orientation filters can handle pre-attentive processing, by partially discarding the visual input, but cannot handle the further integration processing required to lift up the low-level structures to focused attention.

The main problems in artificial models for saliency prediction, with respect to the psychophysical, neurophysiological and psychological studies (PNP) on pre-attentional and attentional processing, reside in the need to model also mechanical, electronic, and software limitations and noise of an (ecological) artifact which, despite these flaws, has to interpret, learn and model saliency in real world, where it performs its tasks. Therefore the tacit assumption is that artificial computational models transform those obtained in the PNP literature to cope with these limitations.

Since Treisman's [33] foundational work on feature integration, one of the main challenges in the computational studies of visual search is modeling how attention is guided

A. Chella et al. (Eds.): Biologically Inspired Cognitive Architectures 2012, AISC 196, pp. 251–263.
springerlink.com

towards distinctive items. In the pre-attentive, early vision phase, primitive visual features can be rapidly accessed in searching tasks. For example colors, motion, orientation can be processed in parallel and effortlessly, and the underlying operations occur in within hundreds of milliseconds. So the pre-attentive level of vision is based on a small set of primitive visual features organized in maps, that are extracted in parallel while the attentive phase serves to group these features into coherent descriptions of the surrounding scene. When attention is thus focused on a particular region of the field of view, processing passes from parallel to serial. A great amount of work, since Treisman's feature integration theory, has been done to understand which are the basic features and how visual strategies affect their integration over a limited portion of the visual field.

Several models have been further provided in the literature for feature integration; among those that led to a concept of representation we consider [8] who have observed that there is a large differentiation *in search difficulty, observed across different stimulus material*. On this basis Duncan introduced the theory of visual selection in three stages: the parallel one, that produces an internal structured representation; a selective one, matching the internal representation and a last one, providing the input of selected information to the visual short term memory. This theory relies on the evidence of less efficiency of parallel processing of basic features in the presence of heterogeneous distractors, which led to the hypothesis of internal structure representation given to the visual input, that Duncan calls *structural units* (close to the 3-D model by Marr and Nishihara [19]). Further Wolfe, closely working on visual search [36], supported the concept of structural units, by noting that visual search might need grouping and categorization. In [38] Wolfe suggests that categorization is a strategy that is invoked when it is useful and that it could affect different features of the visual input. In [37] he makes clear that attentional deployment is guided by the output of earlier parallel processes, but its control can be exogenous, *based on the properties of the visual stimulus*, or endogenous, based on the subject task. He introduces the notion of feature maps (see also Treisman [32]) as independent parallel representations for a set of basic, limited visual features. Finally activation maps, both bottom-up and top-down, serve in his model [37] to guide attention towards distinctive items in the field of view. In summary, Wolfe suggests that information extracted in parallel, with loss of details, serves to create a representation for the purpose of guiding attention.

The huge amount of literature that has studied how from parallel processing, across large areas of the visual field, focused attention emerges (see also [21] and [13]) has led to the quest for a virtual representation that could explain the way input is discarded and focused attention is guided, leading to a coherent view of the scene.

In this paper we propose a methodology, suitable for computational artificial-attention, to study saliency for visual search in dynamic complex scenes, motivated by the concept of virtual representation originated in the coherence theory of attention of [29], [26], [27]. Rensink introduces the concept of *proto-object* as a volatile support for focused attention, which is actually needed to see changes [28]. He assumes that proto-objects are formed in parallel across the visual field [26] and form a continuously renovating flux that is accessed by focused attention. Proto-objects are collected by focused attention to form a stable object temporally and spatially coherent, which provides a structure for perceiving changes.

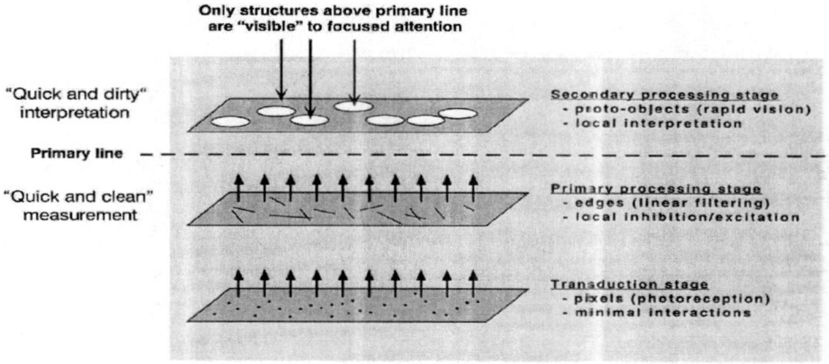

Fig. 1. The image above, taken from [26], illustrates Rensink low-level vision architecture whose output are proto-objects, *that become the operands for attentional objects* [26]

In Figure 1 the two stages of early visual operations, as suggested by Rensink, are illustrated. In this architecture the lower level corresponds to the retinotopic mapping and, going up, proto-objects are structures for more complex feature configurations formed in parallel across the visual field and lying at the interface between low-level vision and higher attentional operations. These structures are said to be volatile, and fading away as new stimuli occur, within "few hundreds of milliseconds" [29]. Focused attention, in Rensink's triadic architecture [26], accesses some of the generated proto-objects to stabilize them and form individual objects "with both temporal and spatial coherence" [26]. Proto-objects are linked within a coherence field to the *nexus*, a structure coarsely summarizing the properties of the stabilized ones. Proto-objects have been explored in computational attention for modeling how object recognition can use their representation and generation, thus at the high-level interface, in [35],[22]. Here, instead, we are interested in the other side of the interface, namely we model their generation and study their spatial and time persistence across the visual fields in visual search tasks, carried out in real dynamic environments. Furthermore we show that these structures can be used to learn the parameters of the underlying process and predict saliency distribution across the scene.

The paper is organized as follows. In the next section we illustrate the setting and the motivations for modeling proto-objects. In Section 3 we present the motion field and the generation of proto-objects via the two-dimensional wave equation. In Section 4 we discuss the proposed experiments and finally we conclude the paper with a discussion on how proto-objects can be further used within the coherence field for saliency prediction.

2 Experiments Setting vs. Coherence Theory

Saliency prediction is about obtaining an elementary model of the features that are meant to provide a schema of the visual field on which a search strategy is deployed. These features are induced by the mental representation of both what is to be searched and what is in general a source of stimuli, even if the subject has no exact representation of what she should search for. In our experiments we have used toy digits and animals.

We aim at modeling the features that are selected during a search task, whether these specify general properties that are preserved across tasks or local properties closely related to the target. These properties may characterize the motion field of the selected regions, or an alternation between spatial and temporal relations, triggered by stimuli. As highlighted in [31] the V4 area displays neural activity with features similar to the target, and these is the area involved in the formation of a coherence field, according to the coherence theory of attention. Furthermore it is noted that the interaction between stimuli-driven and voluntary factors [31] is further and further relevant in the later stages of attentional processing, where more complex coherent fields of features configurations are formed. From the stand point of computational attention a *proto-object* can be described as a *configuration of features having relative time and spatial coherence, directly affected by attention, and having a motion field orienting the gaze towards the target.*

Proto-objects in this sense are dynamic and relatively volatile feature structures related both to fast eye movements, namely saccades, and to saliency. These feature structures are precursors of attention and further used by attention to drive recognition – this is the double face of proto-objects between pre-attentive and selective attention [8], [26] – and can be localized in time and space: proto-objects may last few milliseconds up to hundreds of milliseconds. The relation of proto-objects to saccades, within computational attention, is due to the fact that we can identify these structures via the PORs of the subject in the real world scene and their relation to saliency is due to the intrinsic value, in terms of features, of the stimuli inducing the gaze shift.

We recall that the POR, namely the Point of Regard, is *the point on the retina at which the rays coming from an object regarded directly are focused.* In particular we assume that PORs are the point on the fovea, sub-tending a visual angle of about 1.7 degrees.

Saccades are fast eye movements that can reach peak velocities of $1000°/s$. While a subject is moving, like in our framework, saccades do not exceed 30°, but the velocity follows an exponential function. According to Bahill [3] the range in the duration of 30° saccades can be up to 100 *ms*. Saccade models (see [2], [3], [39] and for a review [15] and the references therein) rarely explain the role of saliency, being mainly motivated by the need to model the motion control. These models do not contribute to the interpretation of proto-objects, although saccades direction and speed are substantial to explain the motion-field of a proto-object and how it fades away.

Similarly, saliency models not grounded in the 3D visual scene lack to explain the reduced coherence of a specific proto-object, hence its fading away. The 3D dynamic structure of the scene and its motion field is necessary to explain the proto-object dynamics. In this section we overview the experiments with the Gaze Machine (GM), see [25] and [24], and the acquisition and data model. The Gaze Machine enables good controlled experiments, as the device can be well fitted on the head, the pupil rate acquisition can reach 180 Hz – conditions depend on the support – ensuring to get good saccades approximation, while the visual field can be acquired up to 20-30 Hz. The model of the POR, described in [25] and [24], namely of the projection of the gaze on the visual field, is very precise as it relies on a robust localization of the subject and structure from motion (see Figure 2). The above statements are endorsed by several experiments

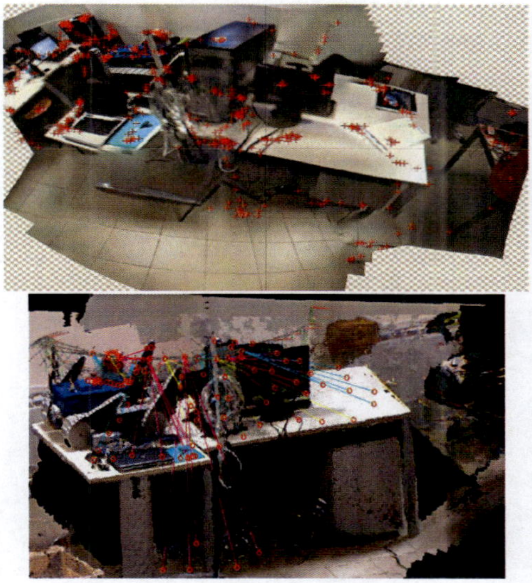

Fig. 2. On the left panel, a panoramic stitching and the PORs collected in $20s$; the stitching has been realized with 30 images over a collection of 600 left images of the scene. The acquisition of the scene is at 30 Hz while the acquisition of the eye is at 120 Hz. The PORs are measured on the scene via dense structure from motion and further reprojected on the image. On the right the head poses of the subject during the experiment *searching for the T*, computed with a robust localization algorithm, and the rays joining the head pose with the PORs (the red circles) projected on the dense depth map of the scene. The lines represent, ideally, the intersection of the visual axes.

evaluating the error between land-marked targets, measured by calibrated cameras, and POR localization, also while the subject is moving, namely walking and moving his/her head. Experiments are at the basis of our experimental model of saliency, whose main stages are shown in Figure 3.

To begin with we give some definitions on the acquisition process that are meant to introduce the computational nature of proto-objects. We do not specify all the geometrical background used to handle the POR projection, including the subject localization, the 3D reconstruction and obviously the pupil tracking. We regard an experiment as a certain collection of raw and elaborated data, that are necessary to compute the head pose, the visual axes, hence the PORs vectors in space, and their re-projections on the images.

An experiment, begins with a calibration phase, in which the subject moves her/his eyes, head and body while fixating a specified target. This phase is needed to calibrate the wearable device with the subject eye motion manifold [24] and scene cameras. Thereafter, according to the search task, the search experiment lasts a certain amount of time T, $120s \leq T \leq 180s$ and it collects the frame sequence F, of the left and right images, at a frequency of $f_T \in [15,30]Hz$; frames are gathered in bundles specifying

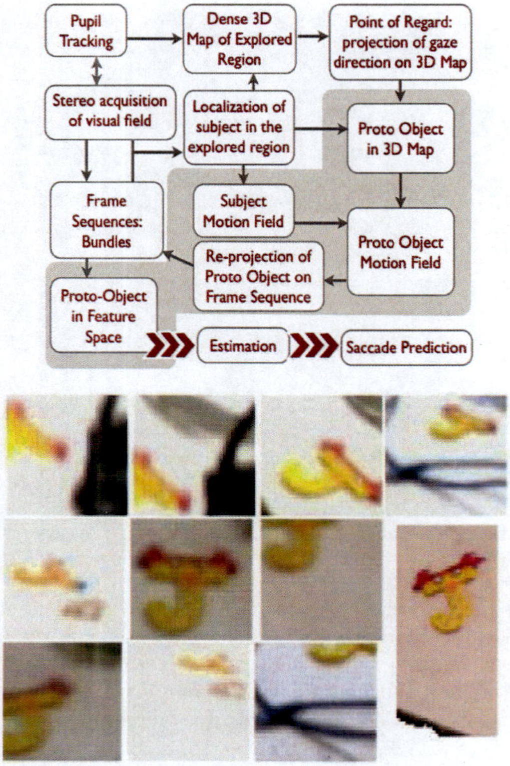

Fig. 3. The left panel shows the stages of saliency prediction according to our *experimental saliency model*. We use the term *experimental* as it is based on 3D measurements of the gaze in natural scenes and of its motion field. The model copes with the coherence theory of attention with respect to the interpretation of Proto-Objects in early attention stages. On the right the re-projection of proto-objects during the task *looking for* **J**, the last image in the right panel is a proto-object in the 3D dense map.

the local coherence of the gaze motion. Further it collects the pupil sequence P at a frequency $f_t \in [120,\ 180]Hz$ and the head motion H via a tiny inertial device part of the acquisition device. Data are processed off-line and the following set of data is returned together with a synchronization of images, visual axes and head poses: the head pose in global coordinates $\mathscr{H}$, via the localization [25], the dense depth map $\mathscr{M}$ in global coordinates, the visual axes of the eye manifolds, namely the PORs directions projected as points in the global coordinates of the scene $\mathscr{P}$, the re-projections of the PORs to the images $\mathscr{R}_{POR}$, synchronized so that in each image a certain amount of PORs, between 7 up to 15 is reprojected. Finally, $\mathscr{B}$ are the relative positions of the observer with respect to the scene. An experiment, therefore, comes with the following formal structure:

$$E = \langle \mathscr{H}, \mathscr{M}, (\mathscr{B}, \Delta T), (\mathscr{P}, \Delta t), \mathscr{R}_{POR} \rangle \tag{1}$$

Here ΔT is the time lapse between two measurements of the scene, $\Delta T \approx 60ms$; Δt is the time lapse between two measurements of the PORs direction in the scene, $\Delta t \approx 8ms$ exploiting the scene constancy – namely, the speed of the eyes is faster than any meaningful motion in the scene and of the head and body motion. To these data we add the model of the proto-objects and their feature space that we describe in this work. The principal outcomes of an experiment E are the PORs and their localization in the 3D space together with the localization of the head pose in the dense map reconstruction of the scene. These are illustrated in Figure 2, right panel, showing the dense map, the path of the head poses, together with PORs as located in the natural scenes, and in Figure 2, left panel showing a meaningful part of an experiment, via a stitched panorama, with the PORs re-projection to the images.

3 Generating Proto-objects

Before the target is reached there is a competition among PORs to survive. For instance, some are the outcome of micro-saccades and endogenous saccades, driven by the search task, and therefore they are parts of aggregates of PORs while some others are the outcomes of reflexive saccades, driven by stimuli that can be distractors. To determine the volatility of the proto-objects in terms of the amount of surface processed by the gaze at each time step t of pupil measurements, we consider the velocity field of the gaze. In other words, if we consider that the high resolution region is $\rho = 2z_B \tan(1.7\pi/180)$, where z_B is the distance of the attended spot from the observer, the effective surface imaged depends on how these spots are tessellated by the motion of the gaze. Likewise processing of the visual field is inversely dependent on the velocity of the gaze. Namely the generation of a proto-object and its volatility is specifically connected to the response of the gaze to the stimulus. The incoming stimulus from the visual field is either caught and discarded, or is processed to decide whether it is the target or not, or it is processed with the purpose to determine whether it is in a place where the target can be found, and so on. In other words the strategy emerges specifically from the way proto-objects are generated and fade away, and thus the feature associated to more coherent proto-objects can lead to an understanding of the visual search.

According to these hypotheses we assume that the gaze velocity field reflects the interest towards a certain region of the visual field. Therefore we model the motion in the velocity field to account for these different cues of the gaze in visual search tasks. Here we give simply a hint of the model.

As said in the previous section, we represent the visual field as a surface where at each point $(x,y,z)^\top$ a label is attached, which is different from 0 only if a POR has been identified at that position at time t. The set of these labels constitutes the PORs map $(\mathcal{P},\Delta t)$. As PORs are on the surface of the reconstructed space, the generation of the proto-object centered at the position $(x,y,z)^\top$ is according to the propagation of a wave, since we can observe only the velocity of the gaze and the single point $(x,y,z)^\top$. We want to infer the size and interest of the region from which features can be sampled, along with the coherence and persistence of the induced proto-object.

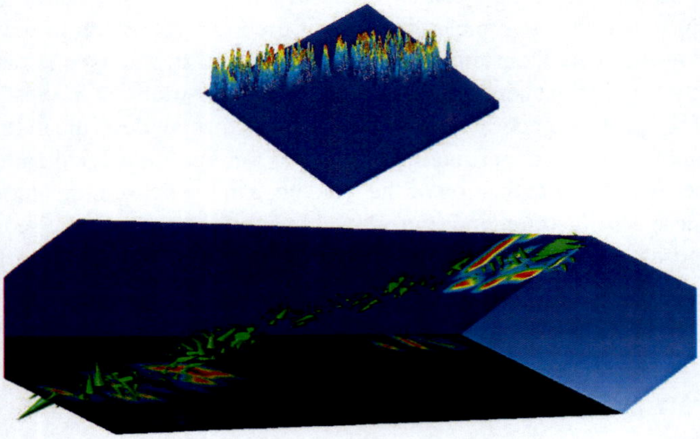

Fig. 4. The figure shows on the left the proto-objects generated in $1s$ using the velocity field of the PORs. On the right the velocity field of the PORs within $\Delta t = 1s$, this amounts of about 20 measurements of the scene and about 120 measurement of the PORs directions. The wave propagation is projected on the slice.

To this end we first expand the single cell, labeled by a 1, with an isotropic Gaussian kernel $g = N((x,y,z)^\top, \rho)$ of size 5, centered on the cell and scaled according to the associated resolution region determined by ρ. The convolution is applied to the map of labels $M = \mathscr{P} \star g$. This operation determines a denser tessellation of the PORs, and on this basis a new contour of the gazed region is drawn, so as to determine the displacement $\mathbf{u} = (u(x,y,z,t), v(x,y,z,t), w(x,y,z,t))^\top$ of the gaze in the direction of the neighbors. Here t is related to the neighbor cells reached by the convolution. Note that this displacement is different from the one determined by the velocity of the gaze.

Let λ and μ be two constants related to the point-wise stimulus. The equation of motion for the region centered at $(x,y,z)^\top$ is:

$$(\lambda + \mu)\nabla(\nabla \cdot \mathbf{u}) + \mu\nabla^2\mathbf{u} + \mathbf{f}. \tag{2}$$

Here $\mathbf{f} = (f_1(x,y,z,t), f_2(x,y,z,t), f_3(x,y,z,t))^\top$ is the coherence in time and space of the stimulus, hence of the induced proto-object. The dilatation of the wave is, on the other hand, determined by the gaze displacement:

$$\Delta^{-1} = (x_{t+\Delta t} - x_t, y_{t+\Delta t} - y_t, z_{t+\Delta t} - z_t)^\top \tag{3}$$

Since high speed must be related to saccades, hence to shifts towards new stimuli, the wave dies away quickly.

The equation for dilatation becomes:

$$(\lambda + \mu)\nabla \cdot \nabla\Delta + \mu\nabla^2\Delta = (\lambda + 2\mu)\nabla^2\Delta \tag{4}$$

Writing $(\lambda + 2\mu)$ as c^2 we obtain the propagation of the wave at speed $(\lambda + 2\mu)$. Letting $\omega = \nabla \times \mathbf{u}$ shows that also the rotation propagates as a wave.

The goal is to obtain **f** from the above equations, in order to determine the coherence and permanence of the proto-object.

Observe that whenever there is no POR the value is zero, and for each point **x** in the neighbor of a POR the value is according to eq. (4). The expansion in $\Delta t = 1s$ of the wave is illustrated in Figure 4, left panel. Proto-objects have thus a coherence in time and space according to their dimension and intensity, the surface of the 3D objects considered will coincide with the high-values of the proto-objects.

Figure 4 shows the velocity field of the PORs, within $\Delta t = 1s$, generating the proto-objects. We have been using the coneplot facilities of Matlab and the slice facility to show the waves expansion, namely the proto-objects, in terms of space time values. More specifically, the wave function is used to localize the regions in the 3D world which have a *slightly* coherent structure.

4 Experimental Validation

Gaze localization and mapping. Investigating the accuracy of the proposed acquisition model involves different aspects. Localization and mapping of the PORs in the 3D scene both rely on the estimation of each POR relative position and the localization of the subject in a reference frame for the experiment.

A first evaluation focuses on investigating the accuracy of the proposed method in localizing and mapping the PORs. The ground truth has been produced as follows: 5 visual landmarks have been placed in the experimental scenario and their position has been measured with respect to a fixed reference frame; 6 subjects have been instructed to fixate the visual landmarks while freely moving in the scenario, annotating (by speaking) the starting and ending of the landmark observations. In each sequence, an average of 60 PORs were produced for each landmark. The validation sequences comprise about 6000 frames each. After registration of the subject initial pose with the fixed reference system, the PORs in the annotated frames were computed and compared with the ground truth, producing a Root Mean Square (RMS) value of 0.094 meters.

For a quantitative analysis of the strategy to select coherent subsequences we relied on a manual coding to produce ground truth data: after the acquisition, subjects were shown the scene sequence overlapped with the POR re-projections to the image plane and used their innate human pattern recognition skill to select coherent subsequences, annotating for each one the starting and ending frames. The performance measure is the *agreement*, defined as the ratio between the number of coherent subsequences recognized by the system over the number of subsequences identified by the subject. Experiments on sequences characterized by a number of frames in the range 4000-6000, yielding a number of keyframes in the range 120-200 produced an average agreement of 85%.

Velocity field and feature set. POR generation is related to the value a point on the surface has according to equation (4), which depends on the velocity and kind of motion of the gaze. To validate the method introduced, we quantify the extent of the POR

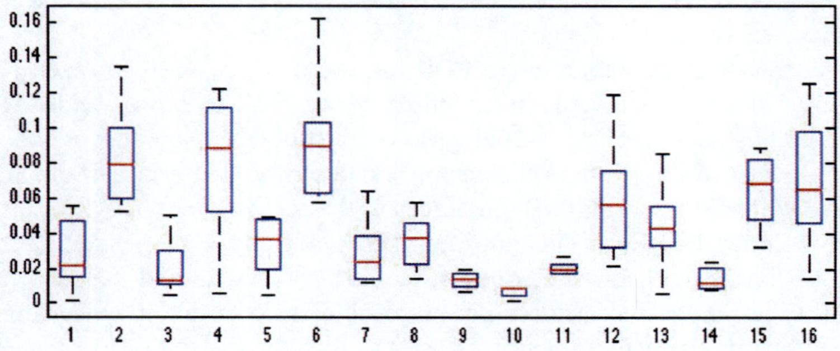

Fig. 5. Box plot for the extent of the re-projections of 16 POR fields to the related bundle images (percentages)

re-projections in each of the related bundle images. The result for an experiment producing 16 image bundles and POR velocity fields, with distances ranging from 1.8 and 8 meters from the observer, is shown in Figure 5. For each bundle, the extent of the re-projections of the POR fields to the bundle images is evaluated as the ratio between the image coordinates that are re-projections of the POR fields and the total number of image pixels. The Figure shows the median values, the boxes representing the 25th and 75th percentiles, the minimum and maximum values. The validation confirms that the extent of the re-projections is mostly confined between 1% and 10% of the image area, and is thus suitable for the proposed feature model.

Validation of feature set and saliency model. From the proto-objects generation, illustrated in the previous section, it is possible to obtain the associated image intensity features, by projecting the surface region, generated according to equation (4), to the images where the proto-objects are imaged. Thus a sample is made of the set of points on the selected surface (the proto-object) and of the imaged pixels, resulting from its re-projection. The complete feature set, of the specific sample, is made of the 3D position of the proto-object in the global map, the composed velocities of the associated PORs, the RGB values, their gradient, and the relative position of the subject w.r.t. the proto-object. The feature set has been chosen according to the feature selection method proposed by [18]: an embedded-method [9] performing variable selection in the process of training. Given the set of m observations in the input feature space $\mathbb{R}^n$, represented by the matrix $\mathbf{V} \in \mathbb{R}^{m \times n}$, each point V_i belongs to the projection of the bundle in the image, namely it lies in the neighborhood of a fixation, if the diagonal matrix D_{ii} is 1 and it does not if it is -1. Discrimination is achieved by a non-linear kernel evaluated on the input feature patterns. Mangasarian and Wild idea [18] is to use a mixed-integer non linear program to determine the non-linear separating surface $k(\mathbf{VE}, \mathbf{EV}')\mathbf{w} - b = 0$, with $\mathbf{E}$ a diagonal matrix in $\mathbf{R}^{n \times n}$ having on the diagonal 1 if the feature is selected and 0 if it is not. To make the mixed-inter programming feasible the matrix $\mathbf{E}$ is held fixed at each cycle of testing at which the objective function is computed, we refer the reader

Table 1. Results from the *k-fold cross validation of the maximum margin classification using the complete image+bundle feature set*

iteration	**number of positives**	wp_+	wp_-	**Accuracy**
1	44707	0.0127	0.0318	95.334%
2	46881	0.01883	0.0206	93.591%
3	420034	0.0093	0.0157	93.019%

to [18] for the method and its converging properties. We have implemented a slightly modified version in order to focus on sets of features and to obtain the *balanced error rate*, namely

$$ber = \frac{1}{2}\left(\frac{wp_+}{|D|_+} + \frac{wp_-}{|D|_-}\right). \tag{5}$$

Here $|D|_+$ are the positive instances of **VE** and $|D|_-$ are the negative ones, while wp_+ and wp_- are, respectively, the false negatives and false positives. In fact, to keep trace of the decrease of the objective function on feature groups, we generate $k!/(k-m)!m!$ m-tuples of even features, up to $k=5$, so as to assign a *ber* value to each feature group. Given an observation task, a model trained on the complete set of features is able to predict if a new sample point is likely to be attended, namely if it belongs to a POR field. To validate this assumption, we ran maximum margin classification experiments. A *K-fold cross-validation* strategy has been followed: we divided the available data comprising more than 6 million points in 3 subsets; in turn, 2 of the three subsets have been used to train the classifier and the remaining one for validation. The process is iterated until every subset is used for validation. As expected, classification accuracy is very high, as reported in Table 1.

5 Conclusions

The computational theory of visual attention aims at mimicking the human capability to select, among stimuli acquired in parallel, those that are relevant for the task at hand. Similar to the biological counterpart, artificial systems can accomplish this by orienting the vision sensors towards regions of space that are more promising. 3D saliency prediction resides in defining a quantitative measure of how attention should be deployed in the three-dimensional scene. Current state-of the art does not model the integration of features in space and time, which is required when dealing with a three-dimensional, dynamic scene. In the coherence theory of attention the concept of proto-object emerged to explain how focused attention collects features to form a stable object that is temporally and spatially coherent. In this work we addressed modeling the process of formation of proto-objects and their relative spatial and temporal coherence according to characteristic that we measure in the velocity field of the gaze. More coherent proto-objects encapsulate the information about the search task and we show how it is possible, from three-dimensional gaze tracking experiments, to extract features that are relevant to predict saliency.

References

1. Ackerman, C., Itti, L.: Robot steering with spectral image information. IEEE Transactions on Robotics 21(2), 247–251 (2005)
2. Bahill, T.A., Bahill, K.A., Clark, M.R., Stark, L.: Closely spaced saccades. Investigative Ophtalmology 14(4), 317–321 (1975)
3. Bahill, T.A., Stark, L.: The trajectories of saccadic eye movements. Scientific American 240(1), 1–12 (1979)
4. Belardinelli, A., Pirri, F., Carbone, A.: Bottom-up gaze shifts and fixations learning by imitation. IEEE Trans. Syst., Man and Cyb.B 37, 256–271 (2007)
5. Butko, N.J., Zhang, L., Cottrell, G.W., Movellan, J.R.: Visual saliency model for robot cameras. In: ICRA, pp. 2398–2403 (2008)
6. Carmi, R., Itti, L.: Visual causes versus correlates of attentional selection in dynamic scenes. Vision Research 46(26), 4333–4345 (2006)
7. Cerf, M., Harel, J., Einhäuser, W., Koch, C.: Predicting human gaze using low-level saliency combined with face detection. In: NIPS (2007)
8. Duncan, J., Humphreys, G.W.: Visual search and stimulus similarity. Psychological Review 96(3), 433–458 (1989)
9. Guyon, I., Elisseeff, A.: An introduction to variable and feature selection. JMLR 3(7-8), 1157–1182 (2003)
10. Harris, C., Stephens, M.: A combined corner and edge detector. In: Proc. of Fourth Alvey Vision Conference, pp. 147–151 (1988)
11. Hügli, H., Jost, T., Ouerhani, N.: Model Performance for Visual Attention in Real 3D Color Scenes. In: Mira, J., Álvarez, J.R. (eds.) IWINAC 2005. LNCS, vol. 3562, pp. 469–478. Springer, Heidelberg (2005)
12. Itti, L., Koch, C., Niebur, E.: A model of saliency-based visual attention for rapid scene analysis. IEEE PAMI 20(11), 1254–1259 (1998)
13. Julesz, B.: Texton gradients: The texton theory revisited. Biological Cybernetics 54, 245–251 (1986)
14. Koch, C., Ullman, S.: Shifts in selective visual-attention: towards the underlying neural circuitry. Hum. Neurobiol. 4(4), 219–227 (1985)
15. Kowler, E.: Eye movements: The past 25years. Vision Research, 1–27 (January 2011)
16. Mahadevan, V., Vasconcelos, N.: Spatiotemporal saliency in dynamic scenes. IEEE PAMI 32, 171–177 (2010)
17. Mancas, M., Pirri, F., Pizzoli, M.: From saliency to eye gaze: Embodied visual selection for a pan-tilt-based robotic head. In: ISVC (1), pp. 135–146 (2011)
18. Mangasarian, O.L., Wild, E.W.: Feature selection for nonlinear kernel support vector machines. In: IEEE-ICDM Workshops, pp. 231–236 (2007)
19. Marr, D., Nishihara, H.K.: Representation and Recognition of the Spatial Organization of Three-Dimensional Shapes. Proceedings of the Royal Society of London. Series B. Biological Sciences 200, 269–294 (1978)
20. Minato, T., Asada, M.: Image feature generation by visio-motor map learning towards selective attention. In: Proc. of IROS 2001, pp. 1422–1427 (2001)
21. Neisser, U., Becklen, R.: Selective looking: Attending to visually specified events. Cognitive Psychology 7(4), 480–494 (1975)
22. Orabona, F., Metta, G., Sandini, G.: A proto-object based visual attention model. In: Paletta, L., Rome, E. (eds.) Attention in Cognitive Systems. Theories and Systems from an Interdisciplinary Viewpoint, pp. 198–215. Springer, Heidelberg (2008)
23. Pichon, E., Itti, L.: Real-time high-performance attention focusing for outdoors mobile beobots. In: Proceedings of AAAI Spring Symposium (AAAI-TR-SS-02-04), p. 63 (2002)

24. Pirri, F., Pizzoli, M., Rudi, A.: A general method for the point of regard estimation in 3d space. In: CVPR 2011, pp. 921–928 (2011)
25. Pizzoli, M., Rigato, D., Shabani, R., Pirri, F.: 3d saliency maps. In: 2011 IEEE Computer Society Conference on Computer Vision and Pattern Recognition Workshops (CVPRW), pp. 9–14 (2011)
26. Rensink, R.A.: The dynamic representation of scenes. Visual Cognition 7, 17–42 (2000)
27. Rensink, R.A.: Change detection. Annual Review of Psychology 53, 245–277 (2002)
28. Rensink, R.A., O'Regan, J.K., Clark, J.J.: To see or not to see: The need for attention to perceive changes in scenes. Psychological Science 8, 368–373 (1997)
29. Rensink, R.A., O'Regan, J.K., Clark, J.J.: On the failure to detect changes in scenes across brief interruptions. Visual Cognition 7, 127–145 (2000)
30. Sala, P.L., Sim, R., Shokoufandeh, A., Dickinson, S.J.: Landmark selection for vision-based navigation. IEEE Transactions on Robotics 22(2), 334–349 (2006)
31. Serences, J.T., Yantis, S.: Selective visual attention and perceptual coherence. Trends in Cognitive Sciences 10(1), 38–45 (2006)
32. Treisman, A.: Preattentive processing in vision. Comput. Vision Graph. Image Process. 31(2), 156–177 (1985)
33. Treisman, A., Gelade, G.: A feature-integration theory of attention. Cognitive Psychology 12, 97–136 (1980)
34. Tsotsos, J.K., Culhane, S., Wai, W., Lai, Y., Davis, N., Nuflo, F.: Modeling visual attention via selective tuning. Artifical Intelligence 78, 507–547 (1995)
35. Walther, D., Koch, C.: Modeling attention to salient proto-objects. Neural Networks 19(9), 1395–1407 (2006)
36. Wolfe, J.M.: The parallel guidance of visual attention. Current Directions in Psychological Science 4, 124–128 (1992)
37. Wolfe, J.M.: Guided search 2.0. a revised model of visual search. CPsychonomic Bulletin and Review 2, 202–238 (1994)
38. Wolfe, J.M., Friedman-Hill, S.R., Stewart, M.L., O'Connell, K.M.: The role of categorization in visual search for orientation. Journal of Experimental Psychology: Human Perception and Performance 18, 34–49 (1992)
39. Zhou, W., Chen, X., Enderle, J.: An updated time-optimal 3rd-order linear saccadic eye plant model. International Journal of Neural Systems 19(5) (2009)

Principles of Functioning of Autonomous Agent-Physicist

Vladimir G. Red'ko

Scientific Research Institute for System Analysis, Russian Academy of Sciences, Moscow, Russia
vgredko@gmail.com

Abstract. An interesting approach towards human-level intelligence has been proposed at BICA 2011 [1]. Namely, it is proposed that an intelligent artificial BICA agent would be able to win a political election against human candidates.

The current work proposes another approach. Our approach is based on the fact that *the most serious cognitive processes are processes of scientific cognition.* The background of this approach is the report by Modest Vaintsvaig at the Russian conference "Neuroinformatics-2011" [2]; that report considers the models of an autonomous agent that tries to cognize elementary laws of mechanics. The agent observes movements and collisions of rigid bodies, forms its own knowledge about interactions of bodies. Basing on these observations, the agent can generalize its knowledge and cognize regularities of mechanical interactions. So, modeling of such autonomous agents, we can try to analyze, how agents could discover (by themselves, without any human help) elementary laws of mechanics. Ultimately, such agents could discover three Newton's laws of mechanics. Thus, we can investigate autonomous agents that could come to the discovery of the laws of nature. It is natural to think that these agents have human-level intelligence.

Using our knowledge about scientific activity of Isaac Newton, we can represent intelligence of such investigating agent in some details. The agent should have an aspiration for the acquisition of the new knowledge and for the transforming of its knowledge into compact form. The agent should have the curiosity that directs the agent to ask the questions about the external world and to resolve these questions by executing real physical experiments. The agent should take into account the interrelations between different kinds of the scientific knowledge. It is natural to assume that a certain society of cognizing agents exists; the agent of the society informs other agents about its scientific results. For example, considering Isaac Newton as a prototype of the main agent, we can consider also agents that are analogous to Galileo Galilei, Rene Descartes, Johannes Kepler, Gottfried Wilhelm Leibniz, Robert Hooke. The agent should have the self-consciousness, the emotional estimation of the results of its cognition activity and the desire to reach the highest results within the society. Agents should have the tendency to get the clear, strong and compact knowledge, such as Newton's laws or Euclidean axioms.

Keywords: Towards human-level intelligence, critical mass, autonomous agent-physicists.

A. Chella et al. (Eds.): Biologically Inspired Cognitive Architectures 2012, AISC 196, pp. 265–266.
springerlink.com

References

1. Chella, A., Lebiere, C., Noelle, D.C., Samsonovich, A.V.: On a roadmap to biologically inspired cognitive agents. In: Samsonovich, A.V., Johannsdottir, K. (eds.) Biologically Inspired Cognitive Architectures 2011. Proceedings of Second Annual Meeting of the BICA Society, pp. 453–460. IOS Press, Amsterdam (2011)
2. Vaintsvaig, M.N.: Learning to control of behavior in the world of objects of space-time. In: Tyumentsev, Y.V. (ed.) XIII All-Russian Scientific Engineering Conference "Neuroinformatics 2011". Lectures on Neuroinformatics, pp. 111–129. NRNU MEPhI, Moscow (2010)

Affect-Inspired Resource Management in Dynamic, Real-Time Environments

W. Scott Neal Reilly, Gerald Fry, and Michael Reposa

Charles River Analytics, Cambridge, MA,
Richard West,
Computer Science Department, Boston University, Boston, MA

Abstract. We describe a novel affect-inspired mechanism to improve the performance of computational systems operating in dynamic environments. In particular, we designed a mechanism that is based on ideas from fear in humans to dynamically reallocate operating system-level resources to processes as they are needed to deal with time-critical events. We evaluated this system in MINIX and Linux in a simulated unmanned aerial vehicle (UAV) testbed. We found the affect-based system was not only able to react more rapidly to time-critical events as intended, but since the dynamic processes for handling these events did not need to use significant CPU when they were not in time-critical situations, the simulated UAV was able to perform even non-emergency tasks at a higher level of efficiency and reactivity than was possible in the standard implementation.

1 Introduction

Many modern computational systems, such as operating systems for real-time applications, need to operate effectively in complex, dynamic environments. Current systems are limited in their ability to dynamically modify their behavior to suit the changing environment and user needs. Humans and other biological creatures, while far from perfect, are considerably better at adapting to dynamic environments than are computational systems. Therefore, the study of human adaptation can provide insights into mechanisms to improve the performance of computational systems.

While it might not be obvious at first why adaptation mechanisms that work for humans should be suitable for computational systems, we believe there are enough architectural similarities that such a transfer of effective adaptation mechanisms is plausible. For instance, they both have multiple, concurrent threads/goals competing for time/attention resources; both are limited in terms of resources, time to respond, and ability to perceive the environment, both have limited short and long term memory; and both act in a social/networked environment. We, therefore, believe that computational systems, such as modern operating systems being used for real-time applications, can be a useful application area for biologically inspired cognitive architectures.

Self-Aware Computing provides a conceptual, metacognitive approach for generating dynamic capabilities for computational systems inspired by the adaptive, introspective capabilities of biological systems [1;2]. These metacognitive approaches are

A. Chella et al. (Eds.): Biologically Inspired Cognitive Architectures 2012, AISC 196, pp. 267–272.
springerlink.com

designed to observe and adapt their own processing, as well as other processing within the self-aware system, in an effort to achieve their specified goals even in dynamic environments. Agarwal and Harrod [1] identify a number of desirable properties for self-aware systems, each of which is associated with the key goal of making the processing system adaptive.

Current metacognitive approaches (e.g., [3-5] and IBM's autonomic computing program) have shown promising results in this area, but none of these approaches makes any effort to utilize the rich conceptual resources available to humans through *affective* processes. While many might think of emotions and moods as being irrational, psychologists and neuroscientists have largely come to believe that affect is an evolutionarily adaptive aspect of human behavior that is useful for living in dynamic, resource-bounded, dangerous, social environments (e.g., [6]). Computational scientists and philosophers have come to believe that computational systems with multiple, competing motivations, operating with limited resources (including time, memory, and computational power), and operating in complex, dynamic, and social environments will *require* affect-like mechanisms to be effective [7-10].

2 Approach

Human affect moderates attentional, cognitive, and physical resources based on the state of current goals and the relevant aspects of the environment. For instance, fear results in attentional/perceptual focusing that filters out elements of the environment that would normally be attended to by resources that are needed by the threatened goal.

We believe computational systems can benefit from a similar mechanism. For instance, operating systems, which are responsible for allocating resources (e.g., CPU, memory, network access) to processes, can more effectively allocate these resources if they are aware when the processes are being threatened due to a lack of those resources. For instance, an unmanned aerial vehicle's (UAV) navigation process is threatened when the UAV is in danger of colliding with another object but does not normally require large number of resources for normal flight. Obviously, the success of such a mechanism relies on a level of self-awareness on the part of the processes to recognize threats, but one of the results of this effort was to demonstrate that it is possible to create such self-aware processes.

We implemented this concept as the *Affective Process Management Module* (APMM) illustrated in Figure 1. The basic APMM architecture is adapted from Neal Reilly [11], where it was used to model the generation and influence of affect in software agents for games. The architecture is, in turn, based on the cognitive-appraisal theory of emotion put forth by Ortony et al. [12].

In this approach, System Processes (e.g., operating system threads that correspond to software applications) are extended to support *Affect Attributes* (e.g., likelihood of success) in addition to standard annotation (e.g., name, status). The *Affective Process Management Module* looks for *Affect Patterns*, which are inspired by the affective literature and which match affective situations such as a lowered likelihood of success for a high-importance process. The patterns correspond to process-specific "emotions." These emotions have a type (e.g., fear), and intensity (which in the case of

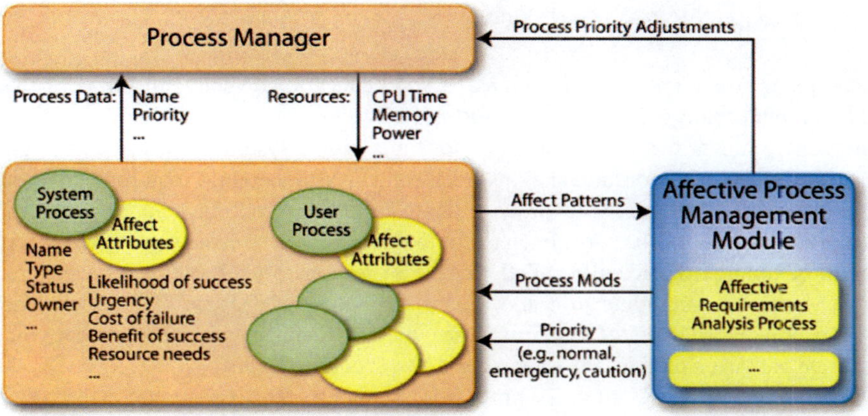

Fig. 1. APMM Conceptual Design

fear, is based on the importance of the process not failing and the likelihood that it will if not tended to), and the cause of the emotion (e.g., fear of failing due to lack of CPU resources or network bandwidth). The latter is used to provide more specific and appropriate emotional response, just as in humans fear of failing a test and fear of a mugger produce different responses that are suited to the situation [13]. The emotions are used by the Affective Process Management Module to reallocate resources (e.g., CPU) via the standard *Process Manager* for the OS.

Self-aware applications are then designed for use with the APMM, which enables them to update their affective attributes as needed to improve performance at key junctures. For instance, the self-aware UAV will know that the success of its avoid-obstacles objective is threatened when an obstacle in seen to be in its flight path, and the urgency of this threat is based on the distance to this obstacle. Such application-specific knowledge is most easily encoded in the applications themselves. Those applications that do not use any affective attribute management will execute under their standard priority with no modifications made by the APMM. When an affect-enhanced application updates its affective attributes, however, through a library the APMM provides, the APMM analyzes those updates and determines changes in the application's priority and resource requirements. Based on these changes, the APMM will direct adjustments to the process's priority through the OS Process Manager.

Consider the practical example of an "affective" UAV gathering information in a particular region. While "content," it distributes resources based on standard process priorities, just as a normal UAV. However, when this affective UAV notices a potential collision, its affective processes react to ensure high priority goals are achieved. In this case, the key goals are to avoid the collision and to ensure that all data collected is downloaded to a network resource. Thus, the affective navigation might report a dramatic increase in urgency which is combined with the high cost of failure associated with this objective not being met. This combination of factors would cause the APMM to significantly increase the priority of the navigation process, which, in turn, would cause the UAV to shift additional resources to that process.

3 Results

We implemented the APMM in both MINIX (an educational version of Linux that let us build a rapid prototype) and then in a full Linux system. We created a basic UAV simulation environment that enables a variety of time-critical system events (e.g., navigation obstacle avoidance, communications requirements, processing requirements, sensor tasks provided by simulated ground agents) to be defined and played back in the specified order to a set of processes. Each event requires a system response within a specified period of time before it fails. Failed navigation-responses lead to crashes; other failed events, such as communication and sensor responses, are logged but not considered system failures. In cases of system failures, the simulation is continued at that point to continue gathering data for the remainder of the scenario events.

The simulated UAV runs its processes as either affective or non-affective, dispatches the time-critical events to the processes, and collects profiling information. When started as affective processes, each process registers an initial affective process profile and then, during execution, uses information from the time-critical events it handles to update the values of its affective profile properties. For instance, the navigation process updates its likelihood of failure to be higher the closer it is to an obstacle (and, therefore, the less time there is to respond before failing).

The scenario was designed to mostly fully occupy the simulated UAV, with transient periods of under-utilization and over-utilization. Thus, our simulation experiment evaluates the affective and non-affective scheduling policies under a variety of system environments.

The primary metric used for this experiment was whether the system responded within the time-limits allocated for each simulated event. Our preliminary results showed that both the MINIX and Linux APMMs provide dramatic improvements in efficiency over the non-affective versions. This is shown in Table 1. The improvements come both in terms of the ability to respond to non-catastrophic, time-critical events (Failures in the table are failures to respond in time to events such as sensor or communications tasks; there were 60 such events in the simulation) and catastrophic, time-critical events (Crashes in the table; there were two such events in the simulation).

Table 1. Performance Statistics for APMM

	Affective Linux	Non-affective Linux	% Improvement	Affective MINIX	Non-affective MINIX	% Improvement
Total Processing Time (ms)	28471	184135	85%	63550	156551	59%
Failures	0	49	100%	23	50	54%
Crashes	0	1	100%	0	1	100%

One concern with affective processing was that it might help in the extreme, emergency cases, but it might not be effective otherwise due to overhead associated with computing affective priorities. This would imply that some non-emergency tasks that would be handled otehrwise would be missed by an affective system. We found, however, that this was not the case. In fact, we found just the opposite: because the emergency-response tasks could be reduced in priority when there were no emergencies, the entire system performed more efficiently and that non-emergency, time-critical event responses were handled successfully more often. In fact, none of the 23 Failures of the affective MINIX system were handled successfully by the non-affective system.

4 Discussion

To date we have only used this architecture to implement and evaluate a fear-inspired method of reallocating CPU resources to threads. The point of this initial fear-focused effort has been to demonstrate that using ideas from human emotional mechanisms as inspirations for computational system design is a fruitful path for generating effective resource-allocation mechanisms at all, but we believe that there are many other such mechanisms that might be developed using this same motivation. For instance, we already have the infrastructure in place to support "hope," which we believe would tend to provide additional resources to threads that are nearing completion, thus potentially reducing context switching. Similarly, using the existing mechanism, we could use fear or hope-based mechanisms to allocate non-CPU resources, such as network bandwidth. Other mechanism corresponding to, say, anger or pity are also possible, and could be used to adjust to threads that regularly over-demand or under-receive CPU resources. We also believe reallocating physical resources, much as emotion-driven adrenaline is used to allocate physical resources in humans, could be used to make more power-efficient systems. This could be used to improve the performance of smart phones or other power-limited devices.

Acknowledgments. This work was performed under DARPA contract number Contract W31P4Q-09-C-0469.

References

1. Agarwal, A., Harrod, W.: Organic Computing. Self Aware Computing Concepts Paper, Cambridge, MA, MIT CSAIL/DARPA IPTO (2006)
2. Ganek, A.G., Corbi, T.A.: The dawning of the autonomic computing era. IBM Systems Journal 42(1), 5–18 (2003)
3. Rhea, S., Eaton, P., Geels, D., Weatherspoon, H., Zhao, B., Kubiatowicz, J.: Pond: the OceanStore prototype. In: 2nd USENIX Conference on File and Storage Technologies (FAST 2003), San Francisco, CA (2003)
4. Patterson, D.A., Brown, A., Broadwell, P., Candea, G., Chen, M., Cutler, J., Enriquez, P., Fox, A., Kiciman, E., Merzbacher, M., Oppenheimer, D., Sastry, N., Tetzlaff, W., Traupman, J., Treahaft, N.: Recovery Oriented Computing (ROC): Motivation, definition, techniques, and case studies. High Performance Transition Systems Workshop (HTPS). Technical Report UCB // CSD-02-1175, Berkeley, CA, U.C. (March 15, 2002)

5. Anderson, M.L., Perlis, D.R.: Logic, self-awareness, and self-improvement: The metacognitive loop and the problem of brittleness. Journal of Logic and Computation 15(1), 21–40 (2005)
6. Damasio, A.R.: Descrates' Error: Emotion, Reason and the Human Brain. G.P. Putnam's Sons, New York (1994)
7. Sloman, A., Croucher, M.: Why Robots Will Have Emotions. In: 7th International Joint Conference on Artificial Intelligence, Vancouver (1981)
8. Toda, M.: The Design of the Fungus Eater: A Model of Human Behavior in an Unsophisticated Environment. Behavioral Science 7 (1962)
9. Pfeifer, R.: The New Age of the Fungus Eater. In: Second European Conference on Artificial Life (ECAL), Brussels (1993)
10. Minsky, M.: The Emotion Machine. Simon & Schuster, New York (2006)
11. Neal Reilly, W.S.: Believable Social and Emotional Agents (PhD Thesis), CMU-CS-96-138. Carnegie Mellon University, Pittsburgh, PA (1996)
12. Ortony, A., Clore, G.L., Collins, A.: The Cognitive Structure of Emotions. Cambridge University Press, NY (1988)
13. Neal Reilly, W.S.: Modeling What Happens Between Emotional Antecedents and Emotional Consequents. In: Trappl, R. (ed.) Cybernetics and Systems 2006, Vienna, Austria (2006)

An Approach toward Self-organization of Artificial Visual Sensorimotor Structures

Jonas Ruesch, Ricardo Ferreira, and Alexandre Bernardino

Instituto Superior Técnico, Institute for Systems and Robotics, Computer and Robot Vision Laboratory, Av. Rovisco Pais 1, 1049-001 Lisbon
{jruesch,ricardo,alex}@isr.ist.utl.pt

Abstract. Living organisms exhibit a strong mutual coupling between physical structure and behavior. For visual sensorimotor systems, this interrelationship is strongly reflected by the topological organization of a visual sensor and how the sensor is moved with respect to the organism's environment. Here we present an approach which addresses simultaneously and in a unified manner i) the organization of visual sensor topologies according to given sensor-environment interaction patterns, and ii) the formation of motor movement fields adapted to specific sensor topologies. We propose that for the development of well-adapted visual sensorimotor structures, the perceptual system should optimize available resources to accurately perceive an observed phenomena, and at the same time, should co-develop sensory and motor layers such that the relationship between past and future stimuli is simplified on average. In a mathematical formulation, we implement this request as an optimization problem where the variables are the sensor topology, the layout of the motor space, and a prediction mechanism establishing a temporal relationship. We demonstrate that the same formulation is applicable for spatial self-organization of both, visual receptive fields and motor movement fields. The results demonstrate how the proposed principles can be used to develop sensory and motor systems with favorable mutual interdependencies.

Keywords: sensorimotor coupling, morphological adaptation, self-organization.

1 Introduction

Visual perception is often considered a one-way process which passes a recorded stimulus along a sensory pathway at the end of which a conclusion is reached regarding the observed scene. However, by simply observing an animal relying on visual perception in its natural environment, it becomes immediately clear that the animal's motor apparatus is permanently engaged in supporting perception by orienting and relocating the visual sensory organs. Motor and sensory systems are working in a very close relationship where not only visual input affects future actions, but motor actions also actively contribute to the process of perception by "shaping" the sequence of recorded stimuli. Thus, in living organisms, the process of visual perception does not merely consist of visual stimuli being analyzed along the sensory pathway, but must be considered as a closed sensorimotor loop in which the animal's body plays an important role [13]. In developmental psychology, this point of view has most prominently been advocated by

A. Chella et al. (Eds.): Biologically Inspired Cognitive Architectures 2012, AISC 196, pp. 273–282.
springerlink.com

Gibson [5]. But also more recently, O'Regan and Noë argue that visual percepts are acquired through training and execution of so-called sensorimotor skills [11]. In their view, a visual percept is "created" through exploration of sensorimotor contingencies during interaction with the observed environment. From this perspective, it is clear that sensor and motor systems must closely function together to support perception and consequently, from a developmental point of view, must also evolve together. In this work we follow this line of thinking and propose an approach where sensor and motor structures develop conjointly into a well concerted sensorimotor system.

1.1 Related Work

The exploration of the advantages of temporally extended sensorimotor loops for perception and their implementation in artificial agents is still on-going work. On the one hand, sensorimotor learning traditionally focuses on learning a generally nonlinear coordinate transformation from a sensor related reference frame to motor space such that sensory input can be translated into a motor action appropriate for a task at hand [10,14]. Moreover, with the advent of motor theories of perception [2,3], a mapping of sensor and motor systems in the opposite direction – i.e. forward models predicting stimuli from motor commands – has gained attention in the robotics community too [18]. On the other hand, work which also considers structural adaptations of sensor and motor systems to obtain adequate sensorimotor maps, as addressed in this paper, is less widespread. This fact seems opposed to the importance of the induced coupling between a given motor apparatus and the physical structure of a sensory system which has been described by roboticists early on; e.g. for visual perception in [4], but also in general for embodied agents with different sensor modalities in [13]. A notable exception is a structurally adaptive sensorimotor system described in [8]. There a robot evolves a 1-dimensional visual sensor such that projected stimuli undergo a uniform translation during straight locomotion.

In a broader context, [9] analyses the causal structure present in the information flow induced by sensorimotor activity using information theoretic measures. The results in essence confirm that the characteristics of recorded stimuli have strong ties to spatiotemporal relationships defined by the physical embodiment and the movement strategies executed by the considered artificial agent.

The authors of this article investigated in previous work the structure of linear stimulus prediction models for visual sensors [15]. It was found that the pairing of a particular sensor topology and sensor actuation strategy has a profound impact on the complexity of a visual stimulus prediction model. The adaptation of motor primitives with respect to a given visual sensor topology and, vice versa, the adaptation of a sensor topology given a particular interaction pattern were previously addressed in [16] and [17] considering these two problems independently.

1.2 Contribution

In this work, we develop a computational approach to conjointly synthesize visual sensor topologies and visual motor layers according to a given agent-environment interaction. The presented method takes as input an agent's interaction pattern with its environment and evolves a spatial layout for both, light receptive fields and motor movement fields. The resulting sensorimotor structure is tuned to the characteristics of the agent's interaction with its environment. We show that visual receptive fields and motor movement fields can evolve simultaneously when minimizing a simple error measure which contemplates the reconstruction error for recorded stimuli with respect to given input signals, and the prediction error for stimuli resulting from self-initiated actions. Driven by the predominantly low spatial frequency of natural images, spatially coherent and smoothly overlapping receptive fields organize on the sensor side without any further constraint on spatial shape. At the same time on the motor side, individual movement fields evolve such as to displace the sensor ensuring high temporal coherence of visual stimuli. Compared to [16], we additionally relax here the constraint that movement fields must implement a Gaussian model and instead allow them to evolve freely. At the beginning of the adaptation process, both, visual receptive fields and motor movement fields can be initialized with randomly chosen activation functions and eventually develop into compact fields.

2 Self-organization of Visual Sensorimotor Structures

A common line of thinking in biology proposes that evolutionary adaptation implicitly optimizes some underlying criterion which is related to the fitness of an organism [12]. From an abstract perspective, it can be argued that similarly any autonomous artificial system should optimize a certain overall cost function in order to temporally maximize its resource-efficiency, task completion rate, or in general its functional subsistence. In the remainder of this work, we consider an artificial agent inhabiting a given world or ecological niche (N) developing so as to optimize an underlying cost function $c_{\texttt{agent}}$. Clearly, the function $c_{\texttt{agent}}$ strongly depends on the agent's body and behavior, where the body of the agent can be further decomposed into its perceptual abilities (S) and its motor apparatus (M). The behavior (Q) is defined as a lifelong sequence of motor actions which depends on the agent's particular survival strategy. We propose that a developmental process for the considered artificial agent should implicitly strive to optimize a loosely defined optimization problem

$$\min_{(S,M,Q)} c_{\texttt{agent}}(S,M,Q;N), \tag{1}$$

which can always be separated into

$$\min_{Q}\left[\min_{(S,M)} c_{\texttt{agent}}(S,M;Q,N)\right]. \tag{2}$$

Note that in this form, the full problem can be locally solved by iteratively optimizing first variables S and M while keeping Q constant and then optimizing Q while keeping S and M constant. Here, we are interested in optimizing sensorimotor structures (S,M), and hence we address only the inner problem in (2) and consider Q fixed. In this case, the agent's interaction with its environment can be recorded as a set of efferent and afferent signals experienced during lifetime according to Q.[1] In line with observations made in living organisms, the first hypothesis in this work is that the characteristics of such lifelong sensorimotor activity is the principal driving force for the co-development of sensorimotor structures (S, M). With this hypothesis the inner optimization problem given in (2) can be rewritten as

$$\begin{aligned} \min_{(S,M)}\ & c_{\mathrm{sm}}\,(S,M;I_0,I_1,a) \quad , \\ \text{s.t.}\ & (I_0,I_1,a) \sim B(Q,N) \end{aligned} \tag{3}$$

where the agent's behavior Q and environment N enter the problem as overall experienced before-and-after signals (I_0, I_1) when executing actions a. The function B defines how triplets (I_0, I_1, a) are sampled from Q and N.

In problem (3), S and M describe the agent's sensorimotor structure. Again, from an abstract perspective, both, sensory and motor systems can be considered a physical implementation which reduces in a specific way the dimensionality of perceivable stimuli and possible actions. In this sense, S can be thought of as a descriptor of the sensor's structure which defines how the agent records a stimulus from available signals I. For visual sensors, such a structure is typically implemented as a 2-dimensional spatially non-uniform distribution of light sensitive receptors which linearly integrate luminance through receptive fields. In motor systems, a reduction in dimensionality can be considered to be present when lower level actions are organized into directly addressable higher level movements with some added value for the acting agent. For an example on how such a reduction in dimensionality is implemented in a biological system, see e.g. [6]. In this work, we address such a reduction in dimensionality for a very early motor layer. Similarly as for the sensory system, the structure of the motor system M is composed of discrete motor movement fields covering the available motor space non-uniformly in a way which provides an advantage for the considered agent. An example of such first layers of motor structures in biology are the motor layers in the optic tectum or superior colliculus as found in mammalian species [7].

With S and M encoding the structure of the agent's sensorimotor system, we now incorporate the second hypothesis of this paper which addresses the co-development of S and M. We propose that sensory and motor systems organize so as to minimize the expected error between available signals I and stimuli which the agent actually records via $S(I)$. Reducing such an error directly relates to the request for the sensorimotor system to optimize available resources in favor of accurate perception. To measure a distance between $S(I)$ and I, a reconstructed signal $S^{+}(S(I))$ is compared to the original signal I, where S^{+} projects the low-dimensional signal back into the original sensor space. Furthermore, for the perceptual process to work as a continuous sensorimotor loop, we not only want an accurate spatial relationship of the agent to its environment, but also maintain this relationship in a coherent manner over time. We thus include a coupling of sensory and motor systems via a prediction mechanism (P) capable of

[1] In neuroscientific terms, a motor signal sent from the central nervous system to the periphery of an organism is called *efference*. Conversely, a sensory signal traveling from the periphery of an organism to the central nervous system is called *afference*.

predicting future sensory stimuli from executed motor actions a.[2] Hence, the second hypothesis proposes that c_{sm} is of the form

$$\min_{(S,M,P)} \mathbb{E}\left[\left\|S^{+}\left(P\left(M,a,S(I_0)\right)\right)-I_1\right\|^2\right], \quad \text{s.t. } (I_0,I_1,a)\sim B(Q,N) \tag{4}$$

where the norm is used to measure the reconstruction error of a predicted stimulus and the actually experienced signal. Other distance measures could be considered instead. The interested reader can find an excellent review on the ubiquity of stimulus prediction in living organisms e.g. in [3].

2.1 Realization

To solve problem (4), sensor and motor spaces are discretized as regular grids which yield sensor signals and motor activity as vectors $\mathbf{I}_0$, $\mathbf{I}_1$, and $\mathbf{a}$, sampled according to $B(Q,N)$. Note that similarly, as $\mathbf{I}$ represents recorded activation on the given sensor surface, a motor action $\mathbf{a}$ is a vector describing an activation profile on the motor space. Here it is assumed that the considered agent possesses a given motor system which transforms activation profiles $\mathbf{a}$ into specific motor actions. Such a transformation can e.g. be thought to be a weighted vector sum of activated locations in the motor space, compare also [7] for an example in biological systems.

Sensor and motor structures S and M are represented as positive matrices $(\mathbf{S},\mathbf{M})$ which when applied to $\mathbf{I}$ and $\mathbf{a}$ yield visual stimuli $\mathbf{SI}$ and motor movement field activations $\mathbf{M}^\top\mathbf{a}$. We choose $\mathbf{S}$ and $\mathbf{M}$ to be positive since on the sensor side $\mathbf{S}$ represents light integrating receptive fields, and $\mathbf{M}$ on the motor side encodes movement fields integrating activation from the underlying motor space (Fig. 1). To predict sensory stimuli, we consider P to be a stateless predictor represented as a mixture of linear predictors of the form $P(\mathbf{M},\mathbf{a},\mathbf{SI}_0) = \left[\sum_i \lambda_i(\mathbf{M},\mathbf{a})\mathbf{P}_i\right]\mathbf{SI}_0$, as introduced in [16]. Additionally, we relax the constraint that λ_i must be composed solely of Gaussian receptive fields as in the previous work and instead allow for arbitrary field shapes $\lambda_i(\mathbf{M},\mathbf{a}) = \mathbf{m}_i{}^\top\mathbf{a}$, where $\mathbf{m}_i$ is the i-th column of $\mathbf{M}$. After prediction, the signal is reconstructed using the adjoint operator $\mathbf{S}^\top$. In this sense, we rewrite (4) as

$$(\mathbf{S}^*,\mathbf{M}^*,\mathbf{P}^*) = \operatorname{argmin}_{(\mathbf{S},\mathbf{M},\mathbf{P})} \sum_a \left\|\mathbf{S}^\top \sum_i \left[\left(\mathbf{m}_i{}^\top\mathbf{a}\right)\mathbf{P}_i\right]\mathbf{SI}_0 - \mathbf{I}_1\right\|^2. \quad \text{s.t. } \mathbf{S}\geq\mathbf{0},\quad \mathbf{M}\geq\mathbf{0},\quad \mathbf{P}\geq\mathbf{0} \tag{5}$$

The savvy reader will notice that the apparent ambiguity which arises by the interaction between $\mathbf{P}$ and $\mathbf{M}$ nearly disappears with the positivity constraints.

2.2 Method

We consider the organization of $N_S = 16$ visual receptive fields taking place on a sensor surface in the shape of a disk discretized at $n_S = 481$ locations in a grid-like layout.

[2] Note that for a complex agent, the consequences of an action a might depend on the current state of the agent in which case the predictor P must be state aware. In this paper, as described in Sect. 2.1, we consider only cases where prediction is state independent.

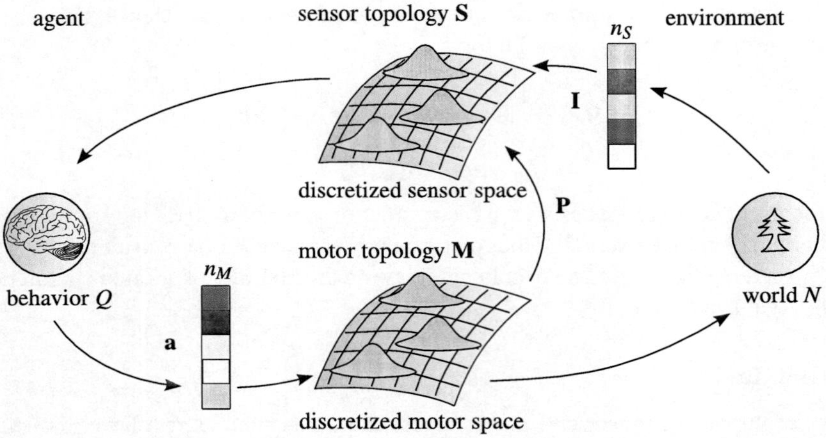

Fig. 1. The sensorimotor loop considered when organizing lower-dimensional sensory and motor topologies **S**, **M** and learning the stimulus predictor **P**. On the motor side, the agent generates motor commands **a** of size n_M according to a given behavior Q. On the sensor side, the agent experiences input signals **I** of size n_S which represent a projection of the world onto the sensor. When executing the full sensorimotor loop, each action **a** changes the input signal **I** and generates a triplet $(\mathbf{I}_0, \mathbf{I}_1, \mathbf{a})$. During learning, the lower-dimensional sensor and motor topologies **S** and **M** evolve according to given triplets $(\mathbf{I}_0, \mathbf{I}_1, \mathbf{a})$. At the same time, the prediction operator **P** is learned such as to predict future sensory stimuli $\mathbf{SI}_1$ from previous stimuli $\mathbf{SI}_0$ for any action **a**.

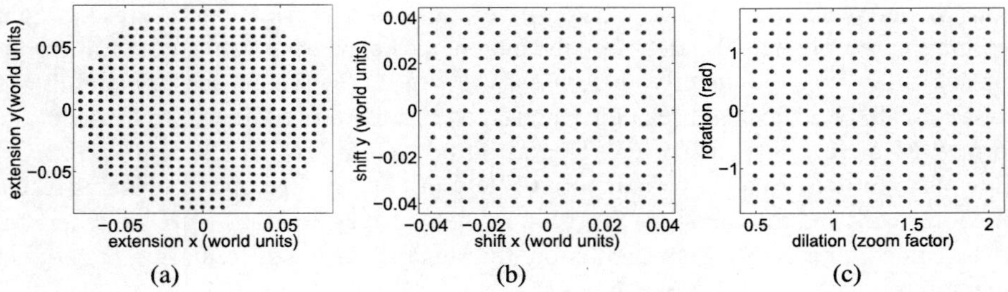

Fig. 2. (a) Discretization of the given sensor space; (b) discretization of the given motor space for a behavior with horizontal and vertical translation actions; (c) discretization of the given motor space for a behavior with dilation and rotation actions. Sensor area and translation distances are specified in world coordinates ranging from -1 to 1 in x- and y-direction.

Similarly, experiments presented in Sect. 3 consider $N_M = 16$ motor movement fields evolving on 2-dimensional motor spaces discretized at $n_M = 15 \times 15$ locations in a grid-like layout, see also Fig. 2.

The environment is given as a plane textured by a very high resolution image (2448×2448 pixels) depicting a real world scene. In this article we assume the sensor surface to be parallel to the plane recording grayscale images **I**. The sensor can interact with the environment through four types of actions, translations in x- and y-directions, rotations and changes in distance to the plane (zoom). A set of 22 500 triplets $(\mathbf{I}_0, \mathbf{I}_1, \mathbf{a})$

is obtained via $B(Q,N)$, where for the presented experiments the underlying Q selects actions $\mathbf{a}$ with sharp activation profiles (all entries in $\mathbf{a}$ are zero except one) according to a uniform distribution over the discretized action space. Each triplet is obtained by positioning the agent in a random position on the environment and taking the chosen action a. To find $(\mathbf{S}^*, \mathbf{M}^*, \mathbf{P}^*)$, we iteratively improve the optimization problem given in Eq. (5) using a projected gradient descent method [1]. While it is no problem to find a solution with an online method, convergence is much slower, we therefore choose here the batch approach for practical reasons. However, we note that under different circumstances an online implementation might be preferable, e.g. for a purely biologically inspired implementation in a robot with stronger memory constraints and a longer exploration phase. The experiments presented in Sect. 3 were initialized as follows: the motor layout $\mathbf{M}$ randomly according to a uniform distribution between zero and one; $\mathbf{S}$ randomly such that each discrete sensor location belongs to exactly one receptive field (row of $\mathbf{S}$), scaled so as to obey $\mathbf{SS}^\top = \mathbb{I}$. The prediction matrices $\mathbf{P}_i$ were initialized with given random $\mathbf{S}$ and $\mathbf{M}$ to the least squares solution to predict $\mathbf{SI}_1$ with $[\sum_i \lambda_i(\mathbf{M}, \mathbf{a})\mathbf{P}_i]\,\mathbf{SI}_0$ and subsequently projected according to $\mathbf{P} \geq \mathbf{0}$. It is important to note that with a randomized initialization, nothing prevents the adaptation process from converging to a locally optimal solution. However, from a biological point of view, we accept these solutions as possible branches of evolutionary development.

2.3 Implication

The presented approach for the co-development of visual sensor and motor structures is based on two main hypotheses. The first states that sensorimotor structures can be decoupled within problem (2) in the sense of (3), and the second proposes that S and M evolve such as to optimize i) the reconstruction of higher dimensional signals, and ii) stimulus predictability. Per se, it is not clear if these hypotheses are justifiable. However, if the proposed framework is capable of reproducing some characteristics of in nature observed sensorimotor structures, then an indication is provided that the implementation captures some inherent principles present in phylogenetic and or ontogenetic development of biological systems. In this case, even though we are not aware of the true evolutionary cost function, we might claim that the made assumptions could hold, and that the proposed framework with its simple underlying principles has explanatory power.

3 Results

Two different behaviors Q_1 and Q_2 were considered to co-develop sensor and motor topologies $\mathbf{S}_1^*$, $\mathbf{S}_2^*$ and $\mathbf{M}_1^*$, $\mathbf{M}_2^*$. In a first setup, $B(Q_1,N)$ samples sensor translation actions from a 2-dimensional motor space of a given range as shown in Fig. 2(b). Triplets $(\mathbf{I}_0, \mathbf{I}_1, \mathbf{a})$ are sampled choosing actions $\mathbf{a}$ with uniform probability from the available discrete actions. This scenario relates to translational unbiased oculomotor control causing random stimulus displacements. The second behavior is composed of mixed zoom and rotation actions where $B(Q_2,N)$ samples combined sensor rotations and stimulus dilations from a 2-dimensional motor space as shown in Fig. 2(c). As for Q_1, triplets

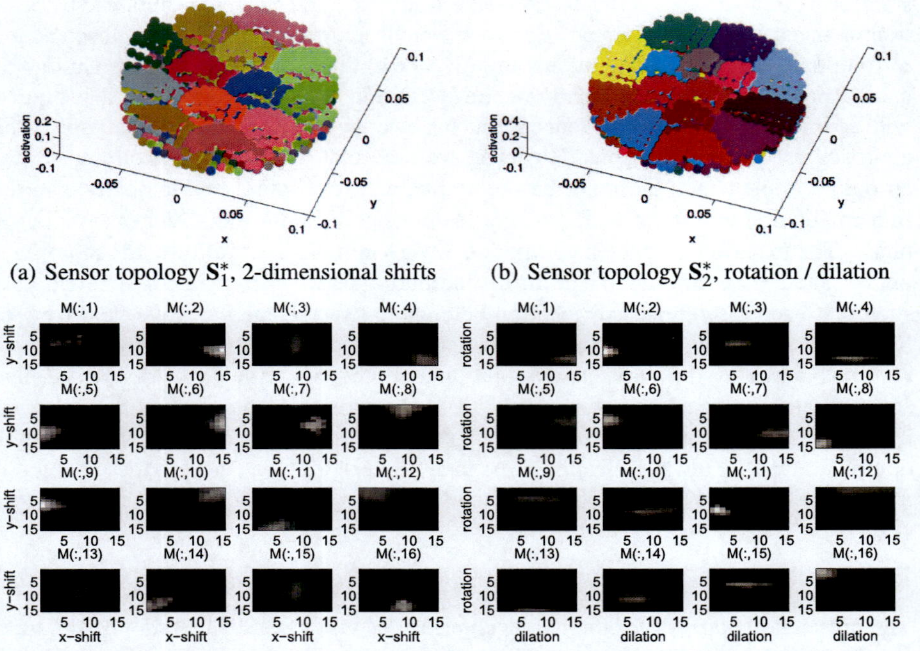

(a) Sensor topology $\mathbf{S}_1^*$, 2-dimensional shifts (b) Sensor topology $\mathbf{S}_2^*$, rotation / dilation

(c) Motor topology $\mathbf{M}_1^*$, 2-dimensional shifts (d) Motor topology $\mathbf{M}_2^*$, rotation / dilation

Fig. 3. Sensor and motor topologies obtained under behaviors with actions uniformly sampled from motor spaces as shown in Fig. 2(b) and Fig. 2(c). Resulting sensor layouts are shown in (a) and (b) where each color denotes a different visual receptive field, and each dot shows the activation of that field at the respective location on the sensor surface. In the translation only case we can identify a tendency for hexagonal tiling structures, whereas in the rotation and dilation case the receptors organize more radially. In (c) and (d) the evolved motor movement fields are shown. For the translation only case, motor fields organize as compact Gaussian-like areas, whereas in the rotation and dilation case, elongated elliptic fields develop reflecting the higher axial resolution of sensor $\mathbf{S}_2^*$ compared to its radial resolution. Note that some motor fields happen to overlap and therefore appear less pronounced as their contribution is combined according to Eq. (5).

$(\mathbf{I}_0, \mathbf{I}_1, \mathbf{a})$ were sampled with uniform probability from the available discrete actions. Behavior Q_2 mimics an object manipulating agent where the oculomotor system stabilizes the sensor on target, mechanically compensating for image translations but not image rotations or scaling. The resulting sensor and motor topologies $\mathbf{S}_1$, $\mathbf{S}_2$ and $\mathbf{M}_1$,$\mathbf{M}_2$ are shown in Fig. 3. The results demonstrate that different behaviors Q induce sensorimotor structures of different macroscopic nature. Note that even though the proposed algorithm is unaware of the topological order present in recorded stimuli $\mathbf{I}$, visual receptors cluster as smoothly overlapping receptive fields and motor primitives appear as spatially coherent Gaussian-like areas.

4 Conclusions and Outlook

This paper investigated how the behavior of an artificial agent can shape the sensorimotor structure of its visual system. We proposed that well adapted sensor and motor layouts organize such as to accurately represent given input signals not only spatially but also temporally for a set of motor actions characteristic for the behavior of the considered agent. We showed that this criterion is captured by comparing the reconstruction of a predicted future stimulus and the actually experienced signal and can be used to conjointly develop visual receptive fields and motor movement fields. In living organisms, comparable structures mapping visual sensory input to motor output can be found in the optic tectum or superior colliculus in mammals.

In future work, we intend to address larger scale problems with an optimized version of the current implementation, which eventually could also serve as a design tool for synthesizing behavior-specific sensorimotor structures for artificial agents. Furthermore, we plan to apply the introduced principles to other sensory modalities, e.g. in a frequency domain for auditory perception.

Acknowledgements. This work was supported by the European Commission proj. FP7-ICT-248366 RoboSoM, by the Portuguese Government – Fundação para a Ciência e Tecnologia (FCT) proj. PEst-OE/EEI/LA0009/2011, proj. DCCAL PTDC/EEA-CRO/105413/2008, and FCT grant SFRH/BD/44649/2008.

References

1. Absil, P.-A., Mahoney, R., Sepulchre, R.: Optimization Algorithms on Matrix Manifolds. Princeton University Press (2008)
2. Craighero, L., Fadiga, L., Rizzolatti, G., Umiltà, C.: Action for perception: A motor-visual attentional effect. Exp. Psych.: Human Perception and Performance 25, 1673–1692 (1999)
3. Crapse, T.B., Sommer, M.A.: Corollary discharge across the animal kingdom. Nat. Rev. Neuroscience 9, 587–600 (2008)
4. Franceschini, N., Pichon, J.M., Blanes, C.: From insect vision to robot vision. Phil. Trans. Biological Sciences 337, 283–294 (1992)
5. Gibson, J.J.: The theory of affordances. In: Shaw, et al. (eds.) Perceiving, Acting, and Knowing: Toward an Ecological Psychology, pp. 67–82 (1977)
6. Kazuya, S., Ménard, A., Grillner, S.: Tectal control of locomotion, steering, and eye movements in lamprey. J. Neurophysiol. 97, 3093–3108 (2007)
7. Lee, C., Rohrer, W.H., Sparks, D.L.: Population coding of saccadic eye movements by neurons in the superior colliculus. Nature 332, 357–360 (1988)
8. Lichtensteiger, L., Eggenberger, P.: Evolving the morphology of a compound eye on a robot. In: Proc. 3rd Europ. Worksh. on Adv. Mobile Robots, pp. 127–134 (1999)
9. Lungarella, M., Sporns, O.: Mapping information flow in sensorimotor networks. PLoS Computational Biology 2, 1301–1312 (2006)
10. Massone, L.L.E.: Sensorimotor learning. In: Arbib, M.A. (ed.) The Handbook of Brain Theory and Neural Networks, pp. 860–864. The MIT Press (1995)
11. O'Regan, J.K., Noë, A.: A sensorimotor account of vision and visual consciousness. Behavioral and Brain Sciences 24(5), 939–1031 (2001)
12. Parker, G.A., Maynard Smith, J.: Optimality theory in evolutionary biology. Nature 348, 27–33 (1990)

13. Pfeifer, R., Bongard, J.: How the Body Shapes the Way We Think. MIT Press (2006)
14. Pouget, A., Snyder, L.: Computational approaches to sensorimotor transformations. Nature Neuroscience 3, 1192–1198 (2000)
15. Ruesch, J., Ferreira, R., Bernardino, A.: A measure of good motor actions for active visual perception. In: Proc. Int. Conf. on Development and Learning (ICDL), Frankfurt, Germany. IEEE (2011)
16. Ruesch, J., Ferreira, R., Bernardino, A.: Predicting visual stimuli from self-induced actions: an adaptive model of a corollary discharge circuit. IEEE Transactions on Autonomous Mental Development (accepted 2012)
17. Ruesch, J., Ferreira, R., Bernardino, A.: Self-organization of Visual Sensor Topologies Based on Spatiotemporal Cross-Correlation. In: Ziemke, T., Balkenius, C., Hallam, J. (eds.) SAB 2012. LNCS, vol. 7426, pp. 259–268. Springer, Heidelberg (2012)
18. Wolpert, D.M., Diedrichsen, J., Flanagan, J.R.: Principles of sensorimotor learning. Nature Reviews Neuroscience 12, 739–751 (2011)

Biologically Inspired Methods for Automatic Speech Understanding

Giampiero Salvi

KTH (Royal Institute of Technology), School of Computer Science and Communication, Dept. for Speech, Music and Hearing
giampi@kth.se

Abstract. Automatic Speech Recognition (ASR) and Understanding (ASU) systems heavily rely on machine learning techniques to solve the problem of mapping spoken utterances into words and meanings. The statistical methods employed, however, greatly deviate from the processes involved in human language acquisition in a number of key aspects. Although ASR and ASU have recently reached a level of accuracy that is sufficient for some practical applications, there are still severe limitations due, for example, to the amount of training data required and the lack of generalization of the resulting models. In our opinion, there is a need for a paradigm shift and speech technology should address some of the challenges that humans face when learning a first language and that are currently ignored by the ASR and ASU methods. In this paper, we point out some of the aspects that could lead to more robust and flexible models, and we describe some of the research we and other researchers have performed in the area.

Keywords: speech technology, language acquisition, embodied cognition, voice mapping, grounding meaning, unsupervised learning.

1 Introduction

Speech technology employs advanced machine learning methods to map spoken utterances into words and meanings. In spite of the complexity of such pattern recognition models, the procedures used to train the model parameters heavily rely on hand-made phonetic and linguistic knowledge. The speech material needs to be transcribed at least at the orthographic level, the set of words the system has to deal with needs to be predefined, and so need to be the set of speech sounds or phonemes in that particular language, and all the possible pronunciations of each words in terms of those phonemes. Furthermore, building speaker independent systems requires recordings from large populations of speakers and, subsequently, the use of a conspicuous number of model parameters.

All these aspects increase tremendously the costs of developing speech technology systems. Moreover, the resulting models, although they might work well in the conditions they were built for, do not generalize well to novel situations: They can not easily incorporate unknown words, they do not adapt well to voices

A. Chella et al. (Eds.): Biologically Inspired Cognitive Architectures 2012, AISC 196, pp. 283–286.
springerlink.com

with characteristics that are outside those seen in the training data, and, finally, they do not take good advantage of the context that comes in a multimodal speech perception scenario.

Humans, on the other hand, learn in multimodal settings and from interactions where perception and action are closely coupled. Infants are faced with a number of challenges since birth: They need to learn the sensory-motor mappings between the acoustic sounds they produce and the corresponding articulatory configurations of their speech organs. They need to learn from continuous perceptual signals the categorical perception of phonemes in the specific language. They need to find recurrent time patterns forming words out of the continuous speech utterances produced by their parents. Finally they need to associate those sensory inputs from speech and other phenomena in their surroundings in order to ground the meaning of each word and phrase.

In this paper, we will briefly review some of the research we and others have carried out in this context, and suggest how this studies may results in advances in speech technology as well as a better understanding of human language acquisition.

2 Biologically Inspired Methods

A number of attempts have been made to model in a computational sense the processes involved with first language learning. Probably the most noticeable because of its relevance for the neurosciences is the DIVA model [4]. This and similar models (the HABLAR model [6], Unified Processing Theory [11] and the Connectionist model [2]) make use of babbling to explain how infants can learn the sensory-motor maps for speech production. In [1] we took this process one step further and proposed to use imitation learning to solve the problem of mapping between the characteristics of the parents' voice and those of the child. In the model, an articulatory synthesizer, modified to correspond to the infant's anatomy, is employed to simulate the adult-infant verbal interaction. The system uses babbling to learn sensory-motor contingencies for its own voice by means of a Self Organizing Map (SOM) model. It then learns an analogous SOM by listening to the adult voice. The correspondence between the two kind of productions is guaranteed by the preserved topology of the two SOMs. Various model parameters specifying the common topology of the two SOMs are optimized during the repeated interactions, based on feedback from the adult at a global level. Although the method proposed in [1] focuses more on the machine learning aspects of this process, we believe that this paradigm could be used in a bootstrapping fashion in the search for acoustic features that are more invariant with respect to the identity of the speaker, thus simplifying speaker independent perception.

It is known (e.g., [5]) that infants develop the categorical perception of sounds gradually during their first months adapting to the particular categories of their mother tongue. In [7] we tried to model this with an incremental version of a probabilistic mixture model. In spite of its simplicity, the model had the advantage to adapt both the model parameters and the number of categories incrementally to the new inputs it received. Methods of this kind could be used in order to adapt the speech units to the particular language without relying on predefined phonetic knowledge. In bottom-up approaches such as this, the main research problem is, again, how to obtain representations that are stable with respect to the voices of different speakers.

Recently, the problem of discovering words from continuous speech was addressed by a number of researchers. The problem presents the challenge of finding recurring patterns when the segmentation of subsequent words is unknown. In [9,3], e.g., the problem is cast into a factor analysis problem by choosing an appropriate fix-length representation for each utterance. Each utterance is represented by a vector $1 \times N^2$ of frequencies of transitions between the phoneme-like speech units in that utterance (from a set of N units). If we consider M such utterances, we can build an $M \times N^2$ matrix where, in each row, the words in the corresponding utterance contribute in an additive way to each frequency count. Finding the set of words corresponds, then to finding a sparse representation of the matrix in terms of basis vectors and weights that specify how these vectors contribute to each utterance. In [10] we propose a new solution to this problem based on a Bayesian version of factor analysis that relies on Beta processes. This method is able to discover such patterns for small vocabulary problems and also to estimate the correct number of words in the data. These algorithms have the potential benefit to allow a speech understanding system to discover new words through the interaction with a user.

Finally, a relatively large effort has been put in the study of grounding the meaning of spoken utterances into multimodal perception. Our contribution to these efforts [8] proposes a statistical model that is able to use co-occurrence to find associations between spoken words, actions performed by a robot, characteristics of the affected objects and effects those actions produce. The method is based on a Bayesian Network that estimates the joint distribution between all the variable involved. By performing a number of exploratory experiments, the robot collects data that is then used to estimate both the dependency structure in the model and the model parameters. The advantage of this model is that it can produce inference based on possibly incomplete observations in several input modalities, thus intrinsically incorporating the contextual information from the robots embodiment and the affordances from its surroundings.

3 Conclusions

The studies described in this paper are of an exploratory nature because they address simplified scenarios if compared to the challenges faced by a human infant learning to speak. The methods, however, model the problems in a computational fashion, and they address key phenomena in the language learning process. We, therefore, believe that these studies may serve both as an indication on how certain aspect of language learning may be accomplished by humans and as inspiration for increasing the robustness and generalization ability of the next generation of automatic speech understanding systems.

References

1. Ananthakrishnan, G., Salvi, G.: Using imitation to learn infant-adult acoustic mappings. In: Proc. of Interspeech, Firenze, Italy (2011)
2. Bailly, G.: Learning to speak. Sensori-motor control of speech movements* 1. Speech Communication 22(2-3), 251–267 (1997)
3. Driesen, J., ten Bosch, L., van Hamme, H.: Adaptive non-negative matrix factorization in a computational model of language acquisition. In: Proc. Interspeech (2009)
4. Guenther, F.H.: Speech sound acquisition, coarticulation, and rate effects in a neural network model of speech production. Psychological Review 102(3), 594–620 (1995)
5. Guenther, F.H., Gjaja, M.N.: The perceptual magnet effect as an emergent property of neural map formation 100(2), 1111–1121 (1996)
6. Markey, K.: The sensorimotor foundations of phonology: a computational model of early childhood articulatory and phonetic development. Ph.D. thesis, University of Colorado Doctoral Dissertation (1994)
7. Salvi, G.: Ecological language acquisition via incremental model-based clustering. In: Proceedings of Eurospeech, Lisbon, Portugal, pp. 1181–1184 (2005)
8. Salvi, G., Montesano, L., Bernardino, A., Santos-Victor, J.: Language bootstrapping: Learning word meanings from perception-action association. IEEE Transactions on Systems, Man, and Cybernetics, Part B: Cybernetics 42(3), 660–671 (2012)
9. Stouten, V., Demuynck, K., van Hamme, H.: Discovering phone patterns in spoken utterances by non-negative matrix factorization. IEEE Signal Processing Lett. 15, 131–134 (2008)
10. Vanhainen, N., Salvi, G.: Word discovery with beta process factor analysis. In: Proc. of Interspeech, Portland, Oregon (2012)
11. Westermann, G., Reck Miranda, E.: A new model of sensorimotor coupling in the development of speech. Brain and Language 89(2), 393–400 (2004)

Modeling Structure and Dynamics of Selective Attention

Hecke Schrobsdorff[1], Matthias Ihrke[1], and J. Michael Herrmann[1,2]

[1] BCCN Göttingen and MPI for Dynamics and Self-Organization, Am Fassberg 17, 37077 Göttingen, Germany
{hecke,ihrke}@nld.ds.mpg.de
[2] University of Edinburgh, IPAB, School of Informatics, 10 Crichton St., Edinburgh EH8 9AB, U.K.
j.michael.herrmann@gmail.com

Abstract. We present a cognitive architecture that includes perception, memory, attention, decision making, and action. The model is formulated in terms of an abstract dynamics for the activations of features, their binding into object entities, semantic categorization as well as related memories and appropriate reactions. The dynamical variables interact in a connectionist network which is shown to be adaptable to a variety of experimental paradigms. We find that selective attention can be modeled by means of inhibitory processes and by a threshold dynamics. The model is applied to the problem of disambiguating a number of theories for negative priming, an effect that is studied in connection to selective attention.

1 Introduction

Cognitive psychology has presented us with a wealth of experimental data as well as paradigms, concepts, properties, rules and theories. Often experimental results are verified by repetitions with the same design. Not less often ambitious generalizations are challenged by differing results obtained for slightly varied experimental set-ups. In this situation, a computational approach might become a necessary complement to further experiments and may help to explain whether apparent contradictions are in fact compatible within a more general framework or caused by statistical fluctuation or obscure confounds.

A neural model would be ideally suited to this task, but may easily become either too complex or lose its explanatory power if a large number of cognitive functions are involved. We have therefore opted for an effective approach that is inspired by and compatible with neural dynamics, but does not aim at a detailed description. Nevertheless, the present architecture could provide a functional blueprint for large-scale simulations at an elementary level. We have developed the architecture in the context of the negative priming (NP) effect, a well-established paradigm to study selective attention. It is a particularly rich area of study what concerns the contribution to the effect by various brain mechanisms such as perception, object recognition, attention, memory, decision making and others. In this way we have not only the need for a complex model, but also data in various forms that constrain the structure and increase the likelihood of the model.

A. Chella et al. (Eds.): Biologically Inspired Cognitive Architectures 2012, AISC 196, pp. 287–295.
springerlink.com

We present a comprehensive computational model [24] that integrates various theories of the negative priming effect into a coherent framework. More generally, the result is a framework for perception-based action in natural or artificial cognitive systems. The system is explicit in the sense that the components are mathematically defined. The system is also connectionist, i.e. the interaction between the components represent the task (see Fig. 1) which is realized either by design or in the wider context by a learning process. Finally, the system is dynamic, i.e. the activity levels of all components change in time and excite, inhibit or modulate each other. This reflects the importance of the time course of information processing in NP as well as in general behavioral contexts.

2 Model Components

The main features of the model are presented in Fig. 1. Visual stimulus information from possibly overlapping objects enters via the *perceptual input* and is recognized by several feature detectors. The object entity is represented in a feature binding module. The model then maps the relevant features into a semantic module. Here a decision about the relevant stimulus is taken, while the appropriate response is chosen in the action module. The episodic memory stores present configurations and modulates future ones.

Cognitive models require the specification of general assumption and general goals. While structural assumptions enter the model via meta-parameters, the goals will be be set by the central executive [4]. Although there is no consensus on the necessity of a central executive in memory functions [2,13], we will use it here for practical reasons: Considering e.g. a psychological experiment, the general instructions are assumed to be processed and enforced by this module, while in a robotic application, from here the general task of the robot is controlled. For brevity and exactness we will restrict ourself in the following to the mathematical formulation of the model and will not try to discuss the psychological background in any depth.

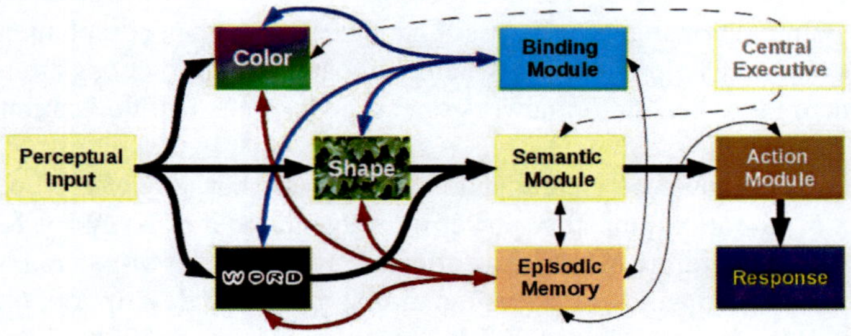

Fig. 1. Interaction scheme of the architecture

The model employs a continuous neural dynamics representing the rates of several interconnected cell assemblies. The activity x of an assembly follows an exponential fixed-point dynamics with time scale τ,

$$\dot{x} = \tau \cdot (U - x), \tag{1}$$

where $\dot{x} = \frac{dx}{dt}$ and U is an input from another module. The main route (s. Fig. 1) of information processing begins with the perceptual input which is then analyzed by a bank of feature detectors. These pass the stimulus-related activation via a semantic memory unit to the action module. Through the interconnections, the components mutually regulate their activations. The structure of the model can be specified by a number of meta-parameters.

A perceived object Ω enters the system as a configuration of activations of the *feature detectors*. The number of relevant features can vary according to the paradigm, we will use here only color, shape, and an additional word label. The object entity is maintained by a specific binding module as the combination of all object features rather than by a direct interaction among the features. Feature variables f_i^j represent whether a feature i (e.g. *color*) has the value j (e.g. *green*). The information is represented in the model by a variable F_i^j defined as

$$F_i^j = \begin{cases} \hat{F} + f_i^j & \text{at stimulus onset, if instance } j \text{ of feature } i \text{ is present} \\ \delta_f(\hat{F} - F_i^j) & \text{during stimulus presence, as long as } F_i^j > \hat{F} \\ \check{F} & \text{after stimulus offset} \end{cases} \tag{2}$$

$\hat{F}$ and $\check{F}$ denote the baseline activity in, resp., an aroused or resting state [25]. The stimulus activation f_i^j decays with rate δ_f representing an adaptive process. The input is modulated by relevance, i.e. if a feature indicates that a stimulus is important, then this feature is boosted; analogously if the feature is to be suppressed.

$$\mathbf{F}_i^j = \begin{cases} F_i^j + A & \text{if } \{i,j\} \text{ indicates target} \\ F_i^j + I & \text{if } \{i,j\} \text{ indicates distractor} \\ F_i^j & \text{otherwise} \end{cases} \qquad \begin{aligned} \tfrac{dA}{dt} &= \alpha \quad \text{stimulus present} \\ A &= 0 \quad \text{no stimulus present} \\ \tfrac{dI}{dt} &= \begin{cases} \kappa & \text{during external input} \\ -\kappa & \text{past input until } I = 0 \end{cases} \end{aligned}$$

Target amplification A is linearly increasing until a response is given and set to zero afterwards. Distractor inhibition I is said to persist for some time, as it has to be retrenched after a response was given. Therefore, inhibition I increases linearly with slope κ during perception and fades linearly after the decision was made. These input variables enter the dynamics of the feature activations

$$\dot{f}_i^j = \tau_f(\mathbf{F}_i^j - f_i^j) + \beta\left(\langle f_i^j\rangle_i - f_i^j\right) + \sum_{\Omega \ni f_i^j} b_\Omega\left(\langle f_l^m\rangle_{f_l^m \in \Omega \setminus f_i^j} - f_i^j\right) + \sum_k r_k e_k\left(e_k^{f_i^j} - f_i^j\right) \tag{3}$$

The first two terms on the right-hand side represent an exponential approach towards the input and, resp., towards the average for each feature, i.e. in the absence of a feature, the representation of this feature loses its specificity while the total activation of this features does not change. The first term behaves asymmetrically, namely $\tau_f = \rho_f$ if $\mathbf{F}_i^j > f_i^j$ and $\tau_f = \delta_f$ if $\mathbf{F}_i^j < f_i^j$. The third terms describes feature binding and the final one involves feedback from earlier responses. More precisely, all activated objects that are compatible with a feature value (f_i^j) will contribute to this feature via any other of their features (f_l^m). If, e.g., the feature vector {color,green}{shape,ball}{location,bottom} defining a green ball shown at the bottom of the visual scene is held by the binding variable $b_{\{\text{color,green}\}\{\text{shape,ball}\}\{\text{location,bottom}\}}$, its value defines the amount of activation interchange between the variables $f^{\text{green}}_{\text{color}}$, $f^{\text{ball}}_{\text{shape}}$ and $f^{\text{bottom}}_{\text{location}}$ such that they all approach the within-object mean. The last term is an influence from the memory of episodes k steps earlier, see below.

The neural implementation of *binding* is far from clear [11], but synchronization is likely to play a role. Here, we implement this mechanism in terms of a feature binding model on the basis of localized excitations in a spiking neural network [23]. The dynamics of a unit in the binding module follows the dynamics

$$\dot{b}_{\{i_k,j_k\}_k} = \begin{cases} \rho_b(\hat{b} - b_{\{i_k,j_k\}_k}) & \text{if an object with a feature combination is perceived} \\ -\delta_b b_{\{i_k,j_k\}_k} & \text{if the object is no longer represented} \end{cases} \tag{4}$$

Objects are represented by dynamic links from the binding module to the respective entries in the feature bank, i.e. technically, the object is a pointer rather than a memory cell, see [23]. The activation of an object is bounded by $\hat{b}$ and decays when the percept is gone. If a pointer is overwritten, we have $b_{\{i_k,j_k\}_k} = 0$ and another object is now becoming activated.

Despite the distributed nature of *semantic* processing [3,7], the present model includes only one module holding the strengths of the semantic representation of a given stimulus similar to the description in [25]. The central executive assigns a role to the variables in the semantic module, depending on task demands.

$$\dot{s}^j = \sigma_{f\to s}\,\tau_s(S^j(f) - s^j) + \sum\nolimits_k r_k e_k \left(e_k^{s^j} - s^j\right) \tag{5}$$

The function $S^j(f)$ determines the fixed point the semantic activation approaches at a rate ρ_s or δ_s, for an actively driven rise or a passive decay, respectively, i.e. $\tau_s = \rho_s$ if $S^j > s^j$ and $\tau_s = \delta_s$ if $S^j < s^j$. The information flow is also influenced by actions (cf. below) and modulated by the blocking factor $\sigma_{f\to s}$ (6).

Blocking or facilitation of a synapses depends on the old-new signal o_k that is generated by comparing the k-th last episode to the current percept. The variable σ_{block} approaches o_k.

$$\sigma_{f\to s} = (1 - \check{\sigma}_{f\to s}) + \check{\sigma}_{f\to s}\sigma_{\text{block}} \tag{6}$$

The synaptic strength is scaled according to σ_{block} between a minimum synaptic strength $\check{\sigma}_{f\to s}$ and an entirely open channel of $\sigma_{f\to s} = 1$. As a decision mechanism is realized by an adaptive threshold s^θ. This variable obeys an exponential fixed-point dynamics on the basis of a scaled average of activation in the semantic module. This is similar to the threshold behavior in Ref. [25].

$$\dot{s}^\theta = \tau_{s^\theta}\, \nu_{s^\theta} \sum\nolimits_j \left(s^j - \check{F}\right) - \left(s^\theta - \check{F}\right) \tag{7}$$

Action activation variables are driven towards an external input $A(s,f)$. If no target stimulus is shown we set $A^0(s,f,\sigma_{f,s\to a}) = 1$, otherwise $A^0(s,f,\sigma_{f,s\to a}) = 0$. The time scale τ_a can again assume two values.

$$\dot{a}^j = \tau_a(A^j(s,f,\sigma_{f,s\to a}) - a^j) + r_a \sum\nolimits_k r_k e_k \left(e_k^{a^j} - a^j\right) \tag{8}$$

The relative retrieval of action representations r_a is modulated contrary to the synaptic transmission to the action module $\sigma_{f,s\to a}$ reflecting the facilitation of action retrieval by an old-new signal indicating an old episode which can be answered by retrieving a former response. Also, the modulation of information flow can decrease the retrieval of a response if a new episode is classified.

$$r_a = \left(1 + \max\left(\breve{\sigma}_{f,s\to a}, \breve{\sigma}_{f\to s}\right)\right) - 2\max\left(\breve{\sigma}_{f,s\to a}, \breve{\sigma}_{f\to s}\right)\sigma_{\text{block}} \quad (9)$$

$$\text{where} \quad \sigma_{f,s\to a} = \left(1 - \breve{\sigma}_{f,s\to a}\right) + \breve{\sigma}_{f,s\to a}\sigma_{\text{block}}$$

In order to model the decision making process in the action module where a single action has to be chosen for execution, we introduce a threshold level analogous to the semantic module, see Eq. (10). As input to the action module ranges from zero to one, we do not have to care about baseline activation here.

$$\dot{a}^{\theta} = \tau_{a^{\theta}} \nu_{a^{\theta}} \sum_{j} a^{j} - a^{\theta} \quad (10)$$

Suprathreshold activations $a^j > a^\theta$ define the space of possible actions the system can take. If there is only one action that is superthreshold, the corresponding action is executed.

The memory module goes beyond the notion of short term memory of Ref. [4] as the maintenance of activation is by attention. If an episode is closed, usually when a action has been performed, the entire state of the model is written down as one episode. The stored values are used to compute similarities between past episodes and a current percept, the retrieval strength r_k. This similarity between the current episode and a previous one is computed by comparing the entire state of the model with a previous state. Similarity triggers an automatic retrieval of the former episodes, namely the stronger the more similar and the more recent. The variables e^v in Eq. 3, 5 and 8 show how the memory influences the dynamics. Details on further effects related to classification and connectivity modulation can be extracted from [24].

3 Modeling Negative Priming

In the present study we will use the following definition: NP is a slowdown in reaction time in a repetition condition where a former distractor has become target [5]. Meanwhile, several standard NP paradigms have emerged featuring various dimensions in which priming can occur, e.g. the identity of stimulus objects [8] or their location on the display [18]. The stimulus set has also been varied, e.g. pictures [26], shapes [6], words [10], letters [9], sounds [16], or colored dots [19] were used. All paradigms have in common stimuli containing targets that are to be attended and distractors that are to be ignored. Experimental conditions depend on Stimulus repetitions, particularly the role of a repeated object as target or distractor in two subsequent trials. Variations of this basic setting include the manipulation of experimental parameters like the time between two related trials, the number of distractors, and the saliency of the distractor. Because of the controversial nature of the NP effect, a variety of interpretations have been developed, but so far none of the theories is able to explain all aspects of the effect.

The comparison of the different theoretical approaches is one of the major reasons for the presented architecture. We consider, in particular, distractor inhibition theory [19,20], global threshold theory [14], episodic retrieval theory [21], response retrieval theory [22], and temporal discrimination theory [17]. In order to analyze the consequences of a theory, we define weights Ξ that switch on or off the effect of particular assumptions in a theory. These weights are meta-parameters insofar as they introduce constraints on the low-level parameters of the model that reflect the impact of a specific theoretical mechanism at a behavioral level. The variables are labeled according to the Ξ_{er}: episodic retrieval; Ξ_{rr}: response retrieval; Ξ_{ib}: inhibition vs. boost; Ξ_{gt}: global threshold; Ξ_{fsb}: feature-semantic block; Ξ_{sab}: semantic action block; Ξ_{td}: temporal discrimination.

Table 1. Weight settings required by various negative priming theories

	Ξ_{er}	Ξ_{rr}	Ξ_{ib}	Ξ_{gt}	Ξ_{fsb}	Ξ_{sab}	Ξ_{td}
Distractor Inhibition	0	0	0	0	0	0	0
Global Threshold	0	0	1	1	0	0	0
Episodic Retrieval	1	1	1	0	0	0	0
Response Retrieval	1	0	1	0	0	0	0
Temporal Discrimination	1	1	1	0	1	1	1

4 Numerical Results

As a showcase example for the capabilities of the model in representing the interaction of various processes involved in NP, we present here an analysis of a word-picture comparison task [12]. This paradigm has a second factor besides priming condition, which is response repetition. Therefore, the labels of the experimental conditions receive an additional suffix, i.e. **s** for response switch and **r** for response repetition. By a parallel implementation, we are able to perform a gradient descent on the parameter set, while keeping the theory semaphores adjusted to each of the settings described in Table 1. Thereby, we obtain information which of the theoretical assumptions implemented in the GMNP is able to reproduce the experimental results to which degree.

After convergence, the root mean squared error between experimental and simulated effects and control reaction time of the GMNP instance set to distractor inhibition behavior is lowest. The obtained optimal parameters in that case are: $\Xi_{\text{er}} = \Xi_{\text{rr}} = \Xi_{\text{ib}} = \Xi_{\text{gt}} = \Xi_{\text{fsb}} = \Xi_{\text{sab}} = \Xi_{\text{td}} = 0$, $\iota = 0.000001$, $\beta = 0.00155$, $\phi = 0.00011$, $\alpha = 0.0005$, $\check{F} = 1$, $t_{\text{recognition}} = 50$, $t_{\text{afterimage}} = 30$, $t_{\text{motor}} = 80$, $\rho_f = 0.009$, $\delta_f = 0.003$, $\hat{b} = 0.05$, $\#_b = 7$, $\rho_b = 0.0096$, $\delta_b = 0.005$, $\tau_{s\theta} = 0.002$, $v_{s\theta} = 0.4131$, $\sigma_{shape \to s} = 0.1$, $\sigma_{word \to s} = 0.12$, $\sigma_{s \to a} = 1$, $\rho_a = 0.0036$, $\delta_a = 0.002$, $\tau_{a\theta} = 0.002$, $v_{a\theta} = 0.6$, $\hat{e} = 0.002$, $\delta_e = 0.003$.

This numerical experiment shows that the postulate that response repetition interaction with priming is incompatible with distractor inhibition seems too strict (see e.g. [22]). Obviously, adding a response mechanism with slowly decaying response activation is sufficient that a distractor inhibition model is able to show such an interaction.

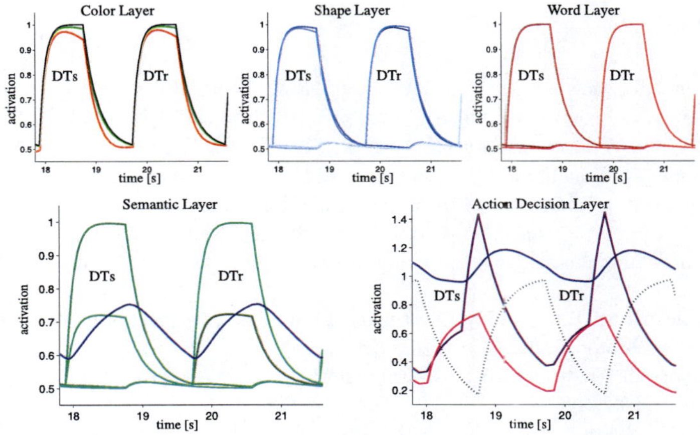

Fig. 2. Activation traces over time in some of the modules in a word-picture comparison paradigm. Solid blue lines in both the semantic and the action module correspond to the respective threshold variable. The model is in distractor inhibition mode. Two different conditions are shown: The former distractor becomes the current target and the reaction switches (from no to yes in this case); and the same case but now without a switch of the reaction.

5 Conclusion

Combining experimental evidence from behavioral experiments with basic system neuroscientific mechanisms, we have derived a cognitive architecture which can account for most of the psychological theories of negative priming. The model clearly identifies differences of experimental conditions and is thus able to resolve existing inconsistencies among the important theories. The model is tested in a number of standard scenarios and is easily extendable also to non-standard versions of priming experiments.

Another promising extension follows from the abstract formulation of relations among mechanisms that are involved in NP. Just as NP theories are formulated using concepts such as memory or central executive which are borrowed from other areas in psychology, the computational implementation of relations among these concepts has also a wider applicability than NP. The main components of the proposed model, named General Model of Negative Priming (GMNP) in a forthcoming paper, qualify it already as a cognitive architecture similar e.g. to ACT-R [1] or SOAR [15] and, beyond this, it would be interesting to discuss the ensuing perspectives for design of artificial cognitive systems such as for the control of an autonomous robot.

Acknowledgments. The authors thank Markus Hasselhorn, Jörg Behrendt, Christian Frings, Theo Geisel, Henning Gibbons, Björn Kabisch, Shu-Chen Li, Timo von Oertzen, Jutta Stahl and Steven Tipper for helpful comments and stimulating discussions. This work was funded in the BCCN framework, grant numbers 01GQ0432 and 01GQ1005B.

References

1. Anderson, J.R., Matessa, M., Lebiere, C.: ACT-R: A theory of higher level cognition and its relation to visual attention. Human-Computer Interaction 12(4), 439–462 (1997)
2. Baddeley, A.: Working memory. Cr. Acad. Sci. III-Vie. 321(2-3), 167–173 (1998)
3. Bookheimer, S.: Functional MRI of Language: New Approaches to Understanding the Cortical Organization of Semantic Processing. Annu. Rev. Neurosci. 25(1), 151–188 (2002)
4. Cowan, N.: Evolving conceptions of memory storage, selective attention, and their mutual constraints within the human information-processing system. Psychol. Bull. 104(2), 163–191 (1988)
5. Dalrymple-Alford, E.C., Budayr, B.: Examination of some aspects of the Stroop color-word test. Percept. Motor Skills 23(3), 1211–1214 (1966)
6. DeSchepper, B., Treisman, A.: Visual memory for novel shapes: Implicit coding without attention. J. Exp. Psychol. Learn. 22(1), 27–47 (1996)
7. Devlin, J.T., Russell, R.P., Davis, M.H., Price, C.J., Moss, H.E., Fadili, M.J., Tyler, L.K.: Is there an anatomical basis for category-specificity? Semantic memory studies in PET and fMRI. Neuropsychologia 40(1), 54–75 (2002)
8. Fox, E.: Negative priming from ignored distractors in visual selection: A review. Psychon. B. Rev. 2(2), 145–173 (1995)
9. Frings, C., Wühr, P.: Prime-display offset modulates negative priming only for easy-selection tasks. Mem. Cognition 35, 504–513 (2007)
10. Grison, S., Strayer, D.L.: Negative priming and perceptual fluency: More than what meets the eye. Percept. Psychophys 63(6), 1063–1071 (2001)
11. Hommel, B.: Event files: feature binding in and across perception and action. Trends Cong. Sci. 8(11), 494–500 (2004)
12. Ihrke, M., Behrendt, J., Schrobsdorff, H., Herrmann, J.M., Hasselhorn, M.: Response-retrieval and negative priming: Encoding and Retrieval Specific Effects.. Exp. Psychol. 58(2), 154–161 (2011)
13. Johnson, M.K.: Memory systems: A cognitive construct for analysis and synthesis. In: Science of Memory: Concepts, pp. 353–357. Oxford University Press, New York (2007)
14. Kabisch, B.: Negatives Priming und Schizophrenie - Formulierung und Empirische Untersuchung eines Neuen Theoretischen Ansatzes. PhD thesis, Friedrich-Schiller-Universität, Jena, Germany (2003)
15. Laird, J.E., Newell, A., Rosenbloom, P.S.: SOAR: An architecture for general intelligence. Artif. Intell. 33, 1–64 (1987)
16. Mayr, S., Buchner, A.: Negative priming as a memory phenomenon: A review of 20 years of negative priming research. Z. Psychol. 215(1), 35–51 (2007)
17. Milliken, B., Joordens, S., Merikle, P.M., Seiffert, A.E.: Selective attention: A reevaluation of the implications of negative priming. Psychol. Rev. 105(2), 203–229 (1998)
18. Milliken, B., Tipper, S.P., Weaver, B.: Negative priming in a spatial localization task: Feature mismatching and distractor inhibition. J. Exp. Psychol. Human 20(3), 624–646 (1994)
19. Neill, W.T.: Inhibitory and facilitatory processes in selective attention. J. Exp. Psychol. Human 3, 444–450 (1977)
20. Neill, W.T., Lissner, L.S., Beck, J.L.: Negative priming in same−different matching: Further evidence for a central locus of inhibition. Percept. Psychophys. 48(4), 398–400 (1990)
21. Neill, W.T., Valdes, L.A.: Persistence of negative priming: Steady state or decay? J. Exp. Psychol. Learn. 18(3), 565–576 (1992)
22. Rothermund, K., Wentura, D., De Houwer, J.: Retrieval of incidental stimulus-response associations as a source of negative priming. J. Exp. Psychol. Learn. 31(3), 482–495 (2005)

23. Schrobsdorff, H., Herrmann, J.M., Geisel, T.: A feature-binding model with localized excitations. Neurocomputing 70(10-20), 1706–1710 (2007)
24. Schrobsdorff, H., Ihrke, M., Herrmann, J.M.: The source code containing several paradigm examples is available through the project web site (2013)
25. Schrobsdorff, H., Ihrke, M., Kabisch, B., Behrendt, J., Hasselhorn, M., Herrmann, J.M.: A Computational Approach to Negative Priming. Conn. Sci. 19(3), 203–221 (2007)
26. Tipper, S.P., Cranston, M.: Selective attention and priming: inhibitory and facilitatory effects of ignored primes. Q. J. Exp. Psychol. 37(4), 591–611 (1985)

How to Engineer Biologically Inspired Cognitive Architectures

Valeria Seidita[1], Massimo Cossentino[2], and Antonio Chella[1]

[1] Dipartimento di Ingegneria Chimica, Gestionale, Informatica, Meccanica
University of Palermo, Palermo, Italy
{valeria.seidita,antonio.chella}@unipa.it

[2] Istituto di Reti e Calcolo ad Alte Prestazioni, Consiglio Nazionale delle Ricerche
ICAR/CNR Palermo, Italy
cossentino@pa.icar.cnr.it

Abstract. Biologically inspired cognitive architectures are complex systems where different modules of cognition interact in order to reach the global goals of the system in a changing environment. Engineering and modeling this kind of systems is a hard task due to the lack of techniques for developing and implementing features like learning, knowledge, experience, memory, adaptivity in an inter-modular fashion. We propose a new concept of intelligent agent as abstraction for developing biologically cognitive architectures.

Researchers in different fields today are attracted by the idea of creating a society where human beings and artificial intelligent agents cohabit in a purposeful manner by sharing tasks and objectives. This is one of the main challenges of Biologically Inspired Cognitive Architecture (BICA) research community.

The full development of an intelligent organism goes through experience and reasoning (more or less consciously). An intelligent human being, capable of implementing cognitive behavior, is an entity who has gone through a series of moments where he learned from the experience and from interactions with the surrounding world. We claim that the surrounding world, or the environment, an intelligent organism lives in, is the set of human and non-human entities, artificial and not, conscious and not, which lives, exists and acts around it, and even within itself. Human-like intelligence is the result of a developmental process that occurs over time.

We highlight the fact that, so as an intelligent organism, a biologically cognitive system can not be developed from scratch and endowed with a "complete" intelligence, hence we have to design and develop cognitive systems with a minimum basic intelligence; eventually it must be made able to evolve according to the environment in which it lives. In order to face this problem from an engineering point of view and in order to find means for creating and using software engineering techniques for developing such systems, we started from analyzing several existing cognitive architectures and from abstracting from them a set of common features. Among the most known we analyzed [1][2][3][5][6] and [4] for a general overview.

We experienced that classical software engineering and also agent software engineering are not enough and do not posses the right abstractions for managing

A. Chella et al. (Eds.): Biologically Inspired Cognitive Architectures 2012, AISC 196, pp. 297–298.
springerlink.com

such kind of systems. Engineering and designing biologically cognitive architecture seen as cross-interacting modules, imply to identify the right abstractions a cognitive system presents.

The model of the cognitive system we propose considers that a human being, a simple organism, a society of individuals and in the same way their artificial counterpart (*agent*, from now on) have, in the whole life or in a part of it, one main *goal* that can be, if necessary, decomposed in sub-goals. *Agent* pursues its goal(s) by interacting with its *environment* both external, hence made of other agents, objects and so on, and internal, hence made of emotions, reasoning, knowledge, internal state and so on. Whatever the environment we are considering, it can change, it is dynamic; changing environment greatly affects the agents' state and consequently their behavior in reaching goals.

We consider the cognitive system as the triad $< a, g, e >$: agent, goal, environment, and the environment is made up by other cognitive systems. Modeling a cognitive system implies using a recursive definition: a cognitive system is a system of systems, or better a society of systems.

What we are introducing is a new idea of intelligent agent, no more an indivisible computational unity but rather an entity made by a brain and a body. The portion of agent realizing the brain contains the capability of reasoning and discovering services offered by the environment and the goal to purse, the capability of storing experiences for future learning and consciousness.

The main advantage of the proposed model is to have a design abstraction allowing to model and engineer all the entities involved in a biologically inspired system with a specific attention to the environment considered both as the internal and the external world of each entity pursuing a specific goal. We claim that this powerful design abstraction eases designing complex systems and aids in filling the gap between BICAs and implementation frameworks.

References

1. Anderson, J., Lebiere, C.: The newell test for a theory of cognition. Behavioral and Brain Sciences 26(5), 587–601 (2003)
2. Goertzel, B., de Garis, H., Pennachin, C., Geisweiller, N., Araujo, S., Pitt, J., Chen, S., Lian, R., Jiang, M., Yang, Y., et al.: Opencogbot: Achieving generally intelligent virtual agent control and humanoid robotics via cognitive synergy. In: Proceedings of ICAI, vol. 10 (2010)
3. Laird, J., Newell, A., Rosenbloom, P.: Soar: An architecture for general intelligence. Artificial Intelligence 33(1), 1–64 (1987)
4. Samsonovich, A.: Toward a unified catalog of implemented cognitive architectures. In: Proceedings of the 2010 Conference on Biologically Inspired Cognitive Architectures 2010: Proceedings of the First Annual Meeting of the BICA Society, pp. 195–244. IOS Press (2010)
5. Shapiro, S.C., Bona, J.P.: The glair cognitive architecture. International Journal of Machine Consciousness 2(2), 307–332 (2010)
6. Sun, R.: The clarion cognitive architecture: Extending cognitive modeling to social simulation. Cognition and Multi-Agent Interaction, 79–99 (2006)

An Adaptive Affective Social Decision Making Model

Alexei Sharpanskykh and Jan Treur

VU University Amsterdam, Agent Systems Research Group
De Boelelaan 1081, 1081 HV Amsterdam, The Netherlands
{sharp,treur}@few.vu.nl

Abstract. Social decision making under stressful circumstances may involve strong emotions and contagion from others, and requires adequate prediction and valuation capabilities. In this paper based on principles from Neuroscience an adaptive agent-based computational model is proposed to address these aspects in an integrative manner. Using this model adaptive decision making of an agent in an emergency evacuation scenario is explored. By simulation computational learning mechanisms are identified required for effective decision making of agents.

1 Introduction

Decision making under stressful circumstances is a challenging type of human process. It involves a number of aspects such as high levels of fear and/or hope, adequate predictive capabilities, for example, related to available information and earlier experiences, and social impact from other group members. In recent cognitive and neurological literature such decision making processes have been addressed. Elements that play an important role in deciding for certain options are: the predicted effects of the options (determined by internal simulation), the valuing of these effects, and the emotions felt in relation to this valuing (based on as-if body loops). These elements affect each other by cyclic internal cognitive/affective processes.

Prediction of the (expected) effects of a decision option, based on internal simulation starting from the preparation of the action has been analysed, e.g., in [11]. Moreover, in [11] it is pointed out how such predictions can be repeated, thus generating simulated behaviour and perception chains. The predictions of action effects are not taken as neutral or objective, but are valued in a subjective and emotion-related manner according to the importance of the predicted effect for the agent, in a positive (hope) or negative (fear) sense; e.g., [13]. If the predicted effects are valued as (most) positive, this may entail a positive decision for the option. In a social context, these processes of prediction and valuing within individuals are mutually affecting each other, so that joint group decisions may develop.

In this paper based on principles from literature as indicated, an adaptive agent-based computational model is proposed to address these aspects in an integrative manner. In contrast to the existing agent-based decision-making models designed from a software engineering perspective (cf [4]), by employing theoretical principles

A. Chella et al. (Eds.): Biologically Inspired Cognitive Architectures 2012, AISC 196, pp. 299–308.
springerlink.com

from Neuroscience and Social Science, we strive to create a more biologically plausible model of human decision making. In the scenario used as illustration an agent considers three decision options (paths) to move outside of a burning building. By simulation computational learning mechanisms are identified required for effective decision making of agents.

The paper is organised as follows. A background for the model is given in Section 2. In Section 3 the model is described. In Section 4 agent learning mechanisms and related simulation results are considered. Section 5 concludes the paper.

2 Background

Emotions and Valuing

In decision making tasks different options are compared in order to make a reasonable choice out of them. Options usually have emotional responses associated to them relating to a prediction of a rewarding and/or aversive consequence. In decisions such an emotional valuing of predicted consequences often plays an important role. In recent neurological literature such a notion of value is suggested to be represented in the amygdala [2,15]. Traditionally an important function attributed to the amygdala concerns representing emotions, in particular in the context of fear. However, in recent years much evidence on the amygdala in humans has been collected showing a function beyond this fear context. In humans many parts of the prefrontal cortex (PFC) and other brain areas such as hippocampus, basal ganglia, and hypothalamus have extensive, often bidirectional connections with the amygdala [9,15,16]. A role of amygdala activation has been found in various tasks involving emotional aspects [13]. Usually emotional responses are triggered by stimuli for which a prediction is possible of a rewarding or aversive consequence. Feeling these emotions represents a way of experiencing the value of such a prediction: to which extent it is positive or negative. This idea of positive and negative value is also the basis of work on the neural basis of economic choice in neuroeconomics.

Internal Simulation

The notion of *internal simulation* was put forward, among others, by Hesslow [11] and Damasio [5,6]. The idea of internal simulation is that sensory representation states are activated (e.g., mental images), which in response trigger associated preparation states for actions or bodily changes, which, by prediction links, in turn activate other sensory representation states. The latter states represent the effects of the prepared actions or bodily changes, without actually having executed them. Being inherently cyclic, the simulation process can go on indefinitely, and may, for example, be used to evaluate the effects of plans before they are executed. In Figure 1 these dynamical relationships are depicted by the arrows from the upper plane to the middle plane and back. In Section 3 these relationships are formalised in (4), (5) and (6). Internal simulation has been used, for example, to describe (imagined) processes in the external world (e.g., prediction of effects of own actions [3]), or processes in a person's own body (e.g., [5]).

The idea of internal simulation has been exploited in particular by applying it to bodily changes expressing emotions, using the notion of *as-if body loop* bypassing actually expressed bodily changes (cf. [5], pp. 155-158; [6], pp. 79-80):

sensory representation $\rightarrow$ *preparation for bodily changes = emotional response* $\rightarrow$ *emotion felt = based on sensory representation of (simulated) bodily changes*

Note that [5] distinguishes an emotion (or emotional response) from a feeling (or felt emotion). In Figure 1 these dynamical relationships are depicted by the arrows in the lower plane, and the arrow from the lower to the upper plane. In Section 3 these relationships have been formalised in (8) and (9).

An as-if body loop usually occurs in an extended, cyclic form by assuming that the emotion felt in turn also affects the preparation states, as it is pointed out, for example, in [7] (pp. 119-122). This can be viewed as a way to incorporate emotion integration in the preparation of actions. In Figure 1 this relationship is depicted via the arrows in the upper plane.

Social contagion

When decision making takes place in a social context of a group of agents mutual contagion occurs. It is assumed that the preparation states of an agent for the actions constituting options and for emotional responses for the options are reflected in body states that are observed with a certain intensity by other agents from the group. The *contagion strength* γ of the interaction from an agent A to an agent B for a preparation state p depends on the personal characteristic *expressiveness* ε of the sender (agent A) for p, the personal characteristic *openness* δ of the receiver (agent B) for p, and an interaction characteristic α (*channel strength*) for p from sender A to receiver B. The effects of contagion are integrated within the internal processes. In Section 3 these relations are formalised in (1), (2), and (3).

3 An Affective Social Decision Making Model

Based on the neurological findings and principles from Section 2 a computational affective social decision making model has been developed.

Depending on a situational context an agent determines a set of applicable options to satisfy a goal at hand. In the proposed model the applicable options are generated in a cyclic manner, via connections from activated sensory states reflecting this situational context to preparation states for the relevant actions related to an option, and valuations of sensory states. An option is represented by a (partially) ordered sequence of actions (i.e., a plan) to satisfy the agent's goals. For example, in the evacuation scenario under investigation each option is represented by a sequence of locations with an exit as the last location. The evaluation of options is based on internal simulation as described in Section 2. The process is depicted in Figure 1.

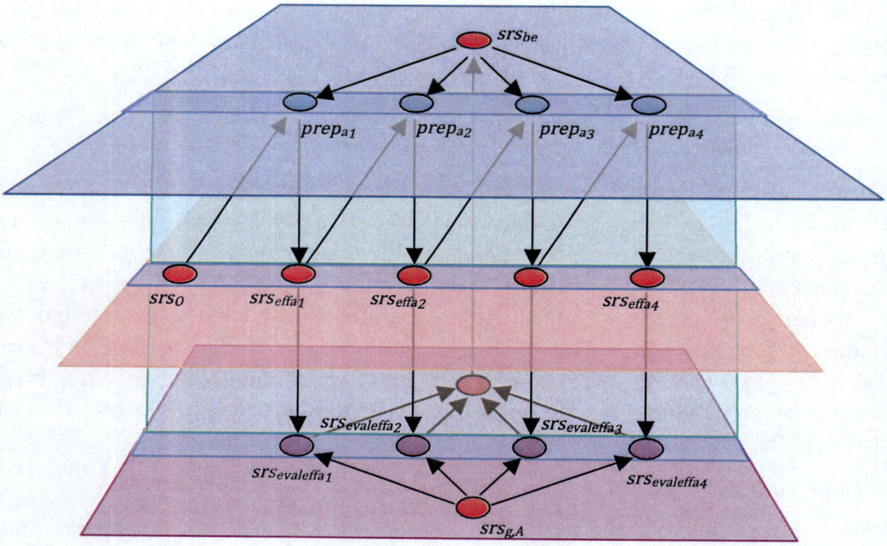

Fig. 1. A graphical representation of the model for a given agent and option O. Circles represent neural states and links represent connections between the states. Upperplane: the preparation states for the subsequent actions related to the option O. Middleplane: the predicted effects of the subsequent actions. Lower plane: the emotion-related valuing of the predicted action effects.

Table 1. Notations used

Notation	Explanation
ε_{SA}	expressiveness of agent *A* for mental state *S*
α_{SAB}	channel strength for *S* from agent *A* to agent *B*
δ_{SB}	openness of agent *B* for *S*
γ_{SAB}	contagion strength for state *S* in the interaction from agent *A* to agent *B*
$G(S, B)$	aggregated group state for *S* as impact for *B*
$prep_{a,O,A}$	Preparation of agent *A* for action *a* in option *O*
$srs_{E,O,A}$	Feeling emotion *E* by agent *A* for option *O*
$srs_{G(a,O,A)}$	Group preparation for *a* in *O* perceived by *A*
$srs_{effect(a,O),A}$	*A*'s representation of the effect of *a* in *O*
$srs_{eval\ for(effect(a,O),E),A}$	*A*'s valuation by *E* of the effect of *a* in *O*
$srs_{g,A}$	*A*'s goal *g*
$prep_{E,O,A}$	Preparation for *E* of agent *A* for option *O*
$srs_{dist(effect(a,O)),A}$	Representation of *A*'s distance to exit by *a* and *O*

In the vertical plane it is shown how in the overall process options for actions are considered (action preparations in the upper horizontal plane), for which by prediction links sensory representations of effects are generated (internal simulation, middle horizontal plane), which are evaluated (emotion-related valuing, lower horizontal plane). The notations used in the model are summarized in Table 1.

3.1 The Social Contagion Impact

The social context in which decision making is performed is represented by a group of agents interacting (verbally, nonverbally) on the relevant options. The *contagion strength* of the interaction from agent A to agent B for a preparation state p is modelled as follows (as described in Section 2):

$$\gamma_{pAB} = \varepsilon_{pA} \cdot \alpha_{pAB} \cdot \delta_{pB} \quad (1)$$

An agent B perceives at each time point the group's joint attitude towards each option, which comprises the following dynamic properties:

(a) The aggregated group preparation to (i.e., the externally observable intention $q_{p,A}$ to perform) each action p constituting the option for agent B:

$$G(p, B) = \Sigma_{A \neq B}\ \gamma_{pAB}\, q_{p,A} \,/\, \Sigma_{A \neq B}\ \gamma_{pAB}\, \varepsilon_{pA} \quad (2)$$

(b) The aggregated group preparation to an emotional response (body state) *be* for each option. A predicted consequence for an option may induce different types of emotions (e.g., fear, hope, joy) with separate preparation states $q_{be,A}$. Formally:

$$G(be, B) = \Sigma_{A \neq B}\ \gamma_{beAB}\, q_{be,A} \,/\, \Sigma_{A \neq B}\ \gamma_{beAB}\, \varepsilon_{beA} \quad (3)$$

Note that in Figure 1 for reasons of transparency only one agent is depicted. The contagion received by this agent can be visualised as incoming arrows to the preparation states of the action options in the upper horizontal plane, and to the preparation state of the emotional response in the lower horizontal plane. The contagion from the depicted agent to other agents can be visualised as outgoing arrows from the same preparation states.

3.2 Internal Simulation

The preparation state $prep_{a1}$ for the first action from an option is affected by the sensory representations srs_{Oi} of the option, of the perceived group preparation $srs_{G(a1,Oi,A)}$ for the action and of the emotion srs_{be} felt towards the option which functions as valuing the option (Figure 1, upper horizontal plane). Formally:

$$\begin{aligned} &\mathbf{d}\ prep_{a_1,O_i,A}(t)/\mathbf{d}t = \\ &\gamma[\ h(srs_{O_i,A}(t),\ srs_{be,O_i,A}(t),\ srs_{G(a_1,O_i,A)}(t)) - prep_{a_1,O_i,A}(t)\ ] \end{aligned} \quad (4)$$

where A is any agent, O_i is an option, *be* is an emotional response state, $G(a1,O_i,A)$ is the aggregated group preparation to action a_1 of agent A, $h(V_1, V_2, V_3)$ is a combination function, γ determines the speed of change. In general, different forms of combination functions are possible (e.g., by a logistic function).

The simulated perception of the effect of an action a (Figure 1, middle plane) in a simulated behavioural chain, based on prediction links (the arrows from the upper to the middle plain in Figure 1) is modelled by the following property:

$$\boldsymbol{d}\, srs_{effect(a,O_i),A}(t)\,/\boldsymbol{dt} = \\ \gamma[\,\omega(prep_{a,O_i,A}(t), srs_{effect(a,O_i),A}(t)) \cdot prep_{a,O_i,A}(t) - srs_{effect(a,O_i),A}(t)] \quad (5)$$

The confidence that an action will result in a particular effect is specified as the strength of the link between the preparation for the action state and the sensory representation of the corresponding effect state (the vertical arrows from the upper plane to the middle plane in Figure 1). In the evacuation scenario the strength of a link between a preparation for a movement action and a sensory representation of the effect of the action is used to represent confidence values of the agent's beliefs about the accessibility of locations. For example, if the agent's confidence of the belief that location *p1* is accessible from location *p2* is ω, then the strength of the link between the states $prep_{move\_from\_to(p2,p1)}$ and $srs_{is\_at\_location(p1)}$ is put on ω.

Similar to the first action *a1*, the preparation state for each subsequent action *a* from the behavioural chain is specified by:

$$\boldsymbol{d}\, prep_{a,O_i,A}(t)/\boldsymbol{dt} = \gamma[h(srs_{effect(a,O_i),A}(t),\ srs_{be,O_i,A}(t),\ srs_{G(a,O_i,A)}(t)) - prep_{a,O_i,A}(t)] \quad (6)$$

The option with the highest value for the preparation state for the first action is chosen for the execution by the agent.

3.3 Emotion-Related Valuing

In the lower horizontal plane in Figure 1 emotion-related valuing of the action options takes place. An emotional response is generated based on an evaluation of the effects of each action of the option. In such an evaluation the effect state for each action is compared to a goal state(s) of the agent. Note that for different types of emotions different aspects of a goal state or different types of goals may be used. In [14] a number of cognitive structures eliciting particular types of emotions are described. As a simulated behavioural chain is a kind of a behavioural projection, cognitive structures of prospect-based emotions (e.g., fear, hope, satisfaction, disappointment) from [14] are particularly relevant for the evaluation process. Such structures can be represented formally as evaluation properties. As indicated in [14], the intensity of prospect-based emotions depends on the likelihood (confidence) that a prospect state will occur. Thus, the strength of the link between the preparation state for an action and the sensory representation of its effect state is taken into account as a factor in the evaluation property. The generic evaluation property of the effect of the action a compared with the goal state g (in the lower plane in Figure 1) is specified formally as:

$$\boldsymbol{d}\, srs_{eval\_for(effect(a,O_i),be),A}(t)/\boldsymbol{dt} = \gamma[h(\,\omega(prep_{a,O_i,A}(t), srs_{effect(a,O_i),A}(t)) \\ f(srs_{g,A}(t), srs_{effect(a,O_i),A}(t)), srs_{be,O_i,A}(t)) - srs_{eval\_for(effect(a,O_i),be),A}(t)\ ], \quad (7)$$

where $f(srs_{g,A}(t), srs_{effect(a,O_i),A}(t)),\ srs_{be,O_i,A}(t))$ is an evaluation which determines how far is the agent's current state from its desired goal state.

The evaluation of the effects of the actions for a particular emotional response to an option together with the aggregated group preparation to the emotional response determine the intensity of the emotional response:

$$prep_{be,O_i,A}(t) = f(srs_{eval\_for(effect(a_1,O_i),be),A}(t), \ldots, srs_{eval\_for(effect(a_n,O_i),be,A}(t)) \quad (8)$$

where *be* is a particular type of the emotional response.

By the as-if body loop, the agent perceives its own emotional response preparation and creates the sensory representation state for it (in Figure 1 the arrow from the lower plane to the upper plane):

$$d\, srs_{be,O_i,A}(t)\, /dt = \gamma\, [\, prep_{be,O_i,A}(t) - srs_{be,O_i,A}(t)\,] \quad (9)$$

The options in the evacuation scenario evoke two types of emotions: fear and hope, which are often considered in the emergency context. According to [14], the intensity of fear induced by an event depends on the degree to which the event is undesirable and on the likelihood of the event. The intensity of hope induced by an event depends on the degree to which the event is desirable and on the likelihood of the event. Thus, both emotions are generated based on the evaluation of a distance between the effect states for the actions from an option and the agent's goal states. In this example each agent in the group has two goal states '*be outside*' and '*be safe*'. The evaluation functions for both emotions include two aspects: (1) how far is the agent's location from the nearest reachable exit; (2) how dangerous is the agent's location (i.e., the amount of smoke and fire). Formally these two aspects are combined in the evaluation function *f* from (7) as

$$\omega V1 + (1-\omega)/(1+ \lambda e^{-\varphi V2}) \quad (10)$$

where $V1$ is the degree of danger of the location, $V2$ is the distance in number of actions that need to be executed to reach the nearest accessible exit, λ and φ are parameters of the threshold function, ω is a weight. The goal state value $srs_{g,A}(t)$ in (7) is obtained by setting $V1$=0 and $V2$=0 in (10): $(1-\omega)/(1+ \lambda)$.

According to the two emotions considered in the example, (7) is refined into two specialized evaluation properties – one for fear and one for hope. For fear:

$$d\, srs_{eval\_for(effect(a,O_i),bfear),A}(t)/dt =$$
$$\gamma [h(\,\omega(prep_{a,O_i,A}(t), srs_{effect(a,O_i),A}(t)) \cdot f(srs_{g,A}(t), srs_{effect(a,O_i),A}(t)),$$
$$srs_{bfear,O_i,A}(t)) - srs_{eval\_for(effect(a,O_i),bfear,A}(t)] \quad (11)$$

where

$$f(srs_{g,A}(t), srs_{effect(a,O_i),A}(t)) =$$
$$|srs_{g,A}(t) - \omega \cdot srs_{danger(effect(a,O_i)),O_i,A}(t) - (1-\omega)/(1+ \lambda e^{-\varphi\, srs_{dist(effect(a,O_i)),A(t)}})|,$$

i.e. the absolute distance between the agent's goal state and the agent's emotional evaluation of its current state.

The property for hope is defined in a similar manner. Also specialized versions of other generic properties 3-9 are defined by replacing the generic state be in them by specific emotional response states *bfear* and *bhope*.

4 Agent Learning

Decision making in ongoing real life processes is adaptive in the sense that decisions made lead to new information and valuations based on which future decisions may be different. In this process a central role is played by how the experienced emotion-related information and valuations lead to adaptations. Such adaptations may concern, for example, (I) altered action effect prediction links, (II) altered links by which input from the other group members is incorporated, or (III) altered emotion-related valuation links. These three types of links are addressed in the approach put forward here.

In the model presented in this paper, a Hebbian learning principle [10] is exploited to obtain this form of adaptivity for the three types of links mentioned: roughly spoken this principle states that connections between neurons that are activated simultaneously are strengthened. From a Hebbian perspective, strengthening of connections as mentioned in case of positive valuation may be reasonable, as due to feedback cycles in the model structure, neurons involved will be activated simultaneously. Therefore such a connection may be developed and adapted based on a Hebbian learning mechanism. In [8] a more in depth treatment of different variations of the mechanism from a mathematical perspective can be found, including the variation used here. The Hebbian learning of the three types of links considered above is formalised as follows.

For link (I): $d\,\omega(prep_{a_i,O_j,A}(t),\ srs_{effect(a_{i+1},O_j),A}(t))/dt =$

$$\eta\ srs_{effect(a_{i+1},O_j),A}(t)\ prep_{a_i,O_j,A}(t)\ (1 - \omega(prep_{a_i,O_j,A}(t),\ srs_{effect(a_{i+1},O_j),A}(t))) - \xi\ \omega(prep_{a_i,O_j,A}(t),\ srs_{effect(a_{i+1},O_j),A}(t)) \quad (12)$$

where η is a learning rate and ξ is an extinction rate.

In the presence of actual observation of an agent of an effect of its action, $\omega(prep_{a_i,O_j,A}(t),\ srs_{effect(a_{i+1},O_j),A}(t))$ may be updated differently. For example, its value may be set to *0* in the absence of the effect, and to *1* in the presence of the effect. Another alternative is to apply a Bayesian update rule or a probabilistic update based on the weighting function from the Prospect Theory [12].

For link (II): $d\,\alpha_{prep(a_i,O_j)A_2A_1}(t)/dt =$

$$\eta\ prep_{a_i,O_j,A_1}(t)\ prep_{a_i,O_j,A_2}(t)\ (1 - \alpha_{prep(a_i,O_j)A_2A_1}(t)) - \xi\ \alpha_{prep(a_i,Oj)A_2A_1}(t)) \quad (13)$$

For link (III): $d\omega(prep_{bfear,O_i,A}(t),\ srs_{bfear,O_i,A}(t))/dt =$

$$\eta\ srs_{bfear,O_i,A}(t)\ prep_{bfear,O_i,A}(t)\ (1 - \omega(prep_{bfear,O_j,A}(t),\ srs_{bfear,O_j,A}(t))) - \xi\ \omega(prep_{bfear,O_j,A}(t),\ srs_{bfear,O_j,A}(t)) \quad (14)$$

It was investigated systematically by simulation how different mechanisms of learning of links influence the dynamics of the agent decision making. The obtained simulation results are discussed in the following.

The simulation model included a group of 10 agents at some location in the building with the parameters drawn from the ranges of uniformly distributed values as indicated in Table 2 below. The agents were deliberating about three decision options (paths) to move outside of a burning building. Information sources placed at each location in the building were providing information to the agents about the degree of danger of the locations.

Table 2. Ranges and values of the agent parameters used in the simulation

Parameter	γ	η	ζ	ε_{pA}	δ_{pA}	β	$\alpha_{pA_iA_j}$
Range/value	[0.7,1]	0.8	0.1	[0.7,1]	[0.7,1]	[0.55, 0.7]	1

A partial simulation trace for the case of learning of the emotion-related links (III) in the model for option 1 of Agent1 is provided in Table 3. Option 1 consists of three movement actions between locations loc1, loc2 and loc3. During the time period [0, 29) the path corresponding to option 1 was safe. Thus, option 1 was valuated highly by the agent, and was chosen for execution.

Table 3. A partial simulation trace for the case of learning of the emotion-related links (III) in the model for option 1 of Agent1

Time point	**10**	**30**	**50**	**70**	**100**
Preparation to move to loc2 from loc1	*0.73*	*0.88*	*0.55*	*0.55*	*0.55*
Preparation to move to loc3 from loc2	*0.67*	*0.82*	*0.52*	*0.45*	*0.44*
Preparation to move to exit1 from loc3	*0.66*	*0.8*	*0.52*	*0.43*	*0.40*

During the execution of option 1, at time point 29, the agent received information about fire, which occurred at location loc 2 along the path of option 1. This observation caused a rapid devaluation of all action steps constituting option 1 by the agent, which eventually stabilized (time point 100).

Based on further experiments, it was determined that learning of links (III) has the greatest effect on decision making. On the contrary, learning of links (II) has a negligible effect on the evaluation of options in this simulation study. A close similarity of the preparation states of the agents observed in simulation is the main cause of a limited effect of learning of link (II) on decision making. In situations in which agents with radically conflicting opinions participate in social decision making, the effect of learning of links (II) would be much higher. A combination of learning of all links (I), (II) and (III) results in the strongest discrimination between the options.

5 Conclusion

Effectiveness of human reasoning and decision making is determined largely by learning and adaptation mechanisms. In this paper effects of learning of different types of links in a social affective decision making model based on neurological principles are explored by simulation. The main outcome of the study is that learning of the emotion-related links has the strongest effect on discrimination of decision making options, which can be seen as in line with recent perspectives addressing the role of the Amygdala in valuing, described, e.g., in [9, 13].

The developed model could be used to evaluate and predict emotional decisions of individuals in groups under stressful conditions. When undesirable decisions are predicted, interventions may be applied (e.g., by providing information explaining pro's and con's for certain decision options).

In the literature it is recognized that humans often employ diverse emotion regulation mechanisms (e.g., to cope with fear and stress). These mechanisms involve interplay between cognitive and affective processes. In the future the proposed model will be extended with an emotion regulation component.

References

1. Ahn, H.: Modeling and analysis of affective influences on human experience, prediction, decision making, and behavior. PhD thesis. MIT, Cambridge (2010)
2. Bechara, A., Damasio, H., Damasio, A.R.: Role of the Amygdala in Decision-Making. Ann. N.Y. Acad. Sci. 985, 356–369 (2003)
3. Becker, W., Fuchs, A.F.: Prediction in the Oculomotor System: Smooth Pursuit During Transient Disappearance of a Visual Target. Experimental Brain Research 57, 562–575 (1985)
4. Boutilier, C., Dean, T., Hanks, S.: Decision-Theoretic Planning: Structural Assumptions and Computational Leverage. Proceedings of J. Artif. Intell. Res (JAIR), 1–94 (1999)
5. Damasio, A.R.: Descartes' Error: Emotion, Reason and the Human Brain. Papermac, London (1994)
6. Damasio, A.R.: The Feeling of What Happens. Body and Emotion in the Making of Consciousness. Harcourt Brace, New York (1999)
7. Damasio, A.R.: Self comes to mind: constructing the conscious brain. Pantheon Books, NY (2010)
8. Gerstner, W., Kistler, W.M.: Mathematical formulations of Hebbian learning. Biol. Cybern. 87, 404–415 (2002)
9. Ghashghaei, H.T., Hilgetag, C.C., Barbas, H.: Sequence of information processing for emotions based on the anatomic dialogue between prefrontal cortex and amygdala. Neuroimage 34, 905–923 (2007)
10. Hebb, D.O.: The Organization of Behaviour. John Wiley & Sons, New York (1949)
11. Hesslow, G.: Conscious thought as simulation of behaviour and perception. Trends Cogn. Sci. 6, 242–247 (2002)
12. Kahneman, D., Tversky, A.: Choices, Values and Frames. American Psychologist 39, 341–350 (1984)
13. Murray, E.A.: The amygdala, reward and emotion. Trends Cogn. Sci. 11, 489–497 (2007)
14. Ortony, A., Clore, G.L., Collins, A.: The Cognitive Structure of Emotions. Cambridge University Press (1988)
15. Rangel, A., Camerer, C., Montague, P.R.: A framework for studying the neurobiology of value-based decision making. Nat. Rev. Neurosci. 9, 545–556 (2008)
16. Salzman, C.D., Fusi, S.: Emotion, Cognition, and Mental State Representation in Amygdala and Prefrontal Cortex. Annu. Rev. Neurosci 33, 173–202 (2010)

A Robot Uses an Evaluation Based on Internal Time to Become Self-aware and Discriminate Itself from Others

Toshiyuki Takiguchi and Junichi Takeno*

Robot Science Laboratory, Computer Science, Meiji University,
1-1-1 Higashimita Tama-ku, Kawasaki-shi, Kanagawa 214-8571, Japan
ce16021@meiji.ac.jp,
juntakeno@gmail.com

Abstract. The authors are attempting to clarify the nature of human consciousness by creating functions similar to that phenomenon in a robot. First of all, they focused on self-aware, confirming a new hypothesis from an experiment on a robot using a neural network with the MoNAD structure which they created based on the concept of a human neural circuit. The basis of this hypothesis is that "the entity receives an anticipated response within a certain evaluation standard based on internal time." This paper discusses the theory of awareness in robots, related experiments, this hypothesis and the prospects for the future.

Keywords: Consciousness, Robot, Neural Network, Self-Awareness, Inner Time, Self and Other Discrimination, cognition behavior cycle.

Introduction

Our aim is to understand emotion and feelings by actualizing the functions of consciousness in a robot. According to case studies on mirror neurons [1] and mimesis theory [2], we had to assume that the development of consciousness is deeply related to imitation behavior. We defined the mechanism of consciousness as follows: "the consistency of cognition and behavior is the origin of consciousness."

First of all, we focused on self-aware. We have already demonstrated the world's first successful experiment of "being aware of one's own figure reflected in a mirror" using a robot implementing our definition of consciousness [3]. Furthermore, we focused our attention on the function termed "being aware of oneself (self-aware)" in the cognition system, and constructed a computational model of the relevant function based on internal time to become self-aware and discriminate itself from others.

1 Conscious System

We have developed a consciousness module called Module of Nerves for Advanced Dynamics (MoNAD) based on the definition of consciousness [4]. The MoNAD is

* Corresponding author.

A. Chella et al. (Eds.): Biologically Inspired Cognitive Architectures 2012, AISC 196, pp. 309–315.
springerlink.com

built with a neural network, and as shown in Figure 1(a) it consists of (a) cognition system, (b) behavior system, (c) primitive representation and (d) symbolic representation. In Figure1(a), (c) is a neuron related to both behavior and cognition, and simulates the mirror neuron.

The conscious system is configured as a hierarchy of multiple MoNADs. The conscious system consists of three subsystems. The round shapes in Figure 1(b) represent each MoNAD. The Reason MoNAD aspires to rational behavior and at the same time it represents itself. The Emotions-Feelings MoNAD represents emotions and feelings according to information from inside and outside the body. The Association MoNAD has the role of integrating rational thoughts and emotional thoughts.

The Association MoNAD is not a so-called homunculus, because it is only an I/O unit and it cannot be a decision-making entity.

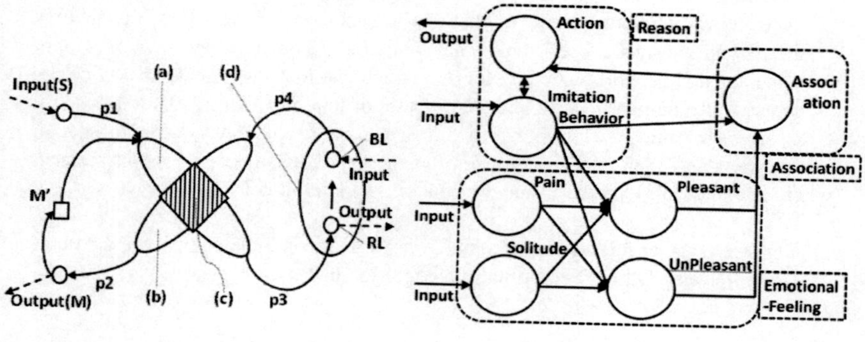

(a) (b)

Fig. 1. (a) Overview of a MoNAD, (b) A conscious system comprising multiple MoNADs

In our experiment, we define T_Aware as a new parameter for this conscious system. This is the base time called the cognition-behavior cycle in the conscious system. T_Aware is the interval required to perform the next behavior when the calculations of all MoNADs in the conscious system have completed after sensor information at the terminal has been input into the conscious system. We also define periodic time as T_IB. T_IB is the interval of time after the imitation behavior MoNAD receives the input information and this information has been conveyed to the cognition representation RL. T_Aware and T_IB repeat periodically within the same consciousness.

2 Outline of Experiment

For this experiment we used 2 kinds of conscious system, EXP.A and EXP.B, which have different T_Aware values. We used a small research robot (e-puck) for this experiment. The e-puck is able to recognize the status of its surroundings using an infrared sensor and move about using a motor. It is also possible to attach a touch sensor to the front of the robot to sense collisions with a mirror or other obstacles. For this experiment, we set 8 experimental conditions for each of the 2 kinds of conscious system and measured the coincidence rate of the behavior.

2.1 Experiment Using the EXP.A Conscious System

EXP.A is the conscious system in which T_IB and T_Aware function in almost the same cycle on robot A. We performed 8 experiments for EXP.A, and each experiment was named sequentially from A-1 to A-8.

In experiment A-1, we placed conscious robot A in front of a mirror and the robot performed an imitation behavior of its mirror image A' (Figure 2 (a)). We calculated the coincidence rate of the behavior of conscious robot A and its mirror image A'.

For the A-2 experiment, conscious robot B, which is equipped with the same conscious system functions as conscious robot A, was placed facing conscious robot Aand they performed imitation behaviors of each other. Conscious robot B acts through a conscious system independent from that of robot A, so that robot B may be considered "the other" for robot A.

In the A-3 experiment, conscious robot A and robot C are connected with anelectric cable, and robot C performs the same behavior as robot A under commands from conscious robot A. Then, these two robots were placed facing each other, and conscious robot A performs imitation behaviors of the actions of conscious robot C (Figure 2 (b)). This cable represents an analogy of the neural circuit in humans.

Experiments from A-4 to A-8 are basically similar to the A-3 experiment. However, we imposed various delay times on when robot C receives information from robot A through the electric cable. Thus, robot A performs imitation behaviors of robot C which has had its behavior delayed. The delay time is expressed as 1d based on the time consumed after incrementing robot C by 1 until the value becomes 100.000. For this robot C, 1d corresponds to 1 second. With regard to the experiments from A-4 to A-8, each corresponding delay time is 0.5d for A-4, 1d for A-5, 2d for A-6, 4d for A-7 and 6d for A-8, and the delay time increases in this sequence.

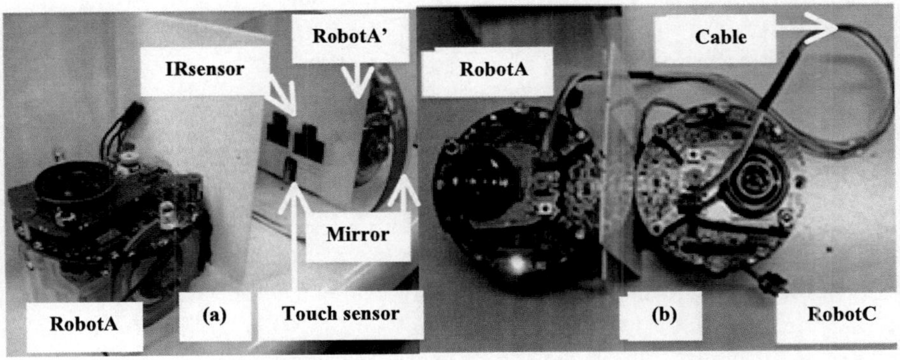

Fig. 2. Experiment using the EXP.A conscious system

2.2 Experiment Using the EXP.B Conscious System

EXP.B is an experiment using the conscious system in which T_Aware goes almost 1 cycle while T_IB goes 2 cycles.

The 8 kinds of experiments in which conscious robot A performs imitation behaviors using EXP.B have the same content as the experiments using EXP.A. These experiments are likewise named sequentially from B-1 to B-8. In the B-1 experiment robot A' is the mirror image, in B-2 robot B is the other that performs the imitation and in B-3 robot C behaves under the command of conscious robot A through a cable. The delay for command communication is 0.5d for B-4, 1d for B-5, 2d for B-6, 4d for B-7 and 6d for B-8.

3 Experimental Results

We measured 30 steps 5 times after conscious robot A detected the target for the experiment (Robot A' or Robot C) using the infrared sensor, and set the average value as the coincidence rate. Here, 1 step means that the conscious system in the conscious robot performs T_Aware once.

According to the experimental results using the EXP-A conscious system (Table 1(a)), the longer the delay in the communication of the command from conscious robot A, the lower the coincidence rate in the behavior of robot C.

According to the experimental results using the EXP-B conscious system (Table 1(b)), though EXP.B acquires input information in the same cycle as that of EXP.A, T_Aware took twice the time. Therefore, we observed an increasing trend in the coincidence rate of the behavior in the EXP-B experiment compared with the previous EXP.A experiment.

Table 1. Change incoincidence rate

(EXP.A(a))

Object	Coincidence rate
A-1 (mirror)	87.8%
A-2 (other)	53.4%
A-3 (0d)	73.8%
A-4 (0.5d)	55%
A-5 (1d)	61.6%
A-6 (2d)	60.2%
A-7 (4d)	43%
A-8 (6d)	39.6%

(EXP.B(b))

Object	Coincidence rate
B-1 (mirror)	95.75%
B-2 (other)	56.4%
B-3 (0d)	67%
B-4 (0.5d)	73.8%
B-5 (1d)	67%
B-6 (2d)	72.4%
B-7 (4d)	58%
B-8 (6d)	57.6%

4 Considerations and Conclusions

In Fig. 3(a), (b), the vertical axis shows the value of the coincidence rate and the horizontal axis shows the value of the forwarding delay occurring in the connected electric cable.

Line A shows the change in the coincidence rate for each experiment from A-1 to A-8 and from B-1 to B-8. Line B is a straight line showing the value of the coincidence rate for A-2 in Fig. 3(a) and the value of coincidence rate for B-2 in Fig. 3(b).

In Fig. 3(a), A-2 is the coincidence rate for the other to be imitated, so we can determine that robot A cannot recognize robot C as its own body (itself) any more in the A-7 and A-8 experiments in which the coincidence rate is lower than in the A-2 experiment shown by Line B. Furthermore, in the A-3 to A-6 experiments, the coincidence rate is higher than the boundary value shown by Line B, and even in the A-1 mirror image experiment, the value is situated on the same side. Therefore, conscious robot A in EXP.A is able to determine that the target robots for measurement from A-1 to A-6 are a part of its own body. We can evaluate that the level of the coincidence rate in the behavior is at the same time the level of the accuracy rate that robot Aitself forecasts the behavior of the target robot C.

In Fig. 3(b), B-3 to B-8 all showed a higher coincidence rate than Line B (coincidence rate in B-2). From this, we can determine that all of target robot C is recognized by conscious robot A in EXP.B as a part of its own body even when a command delay of 6d occurs.

Moreover, the cycle of T_Aware is approx. 4d in EXP.A and it is about 8d in EXP.B. At this time, T_Aware is the cognition behavior cycle for oneself according to the definition, and if target robot C responds with a reaction within the period of T_Aware, it becomes impossible to distinguish whether the response of robot C is from its own body or a response from theother. In this experiment, the responses from A-1 to A-6 or from B-1 to B-8 are all within the period of T_Aware.

This result means that T_Aware is another important evaluation factor for conscious robot A to recognize the target robot C as a part of its own body.

That is to say that conscious robot A is able to recognize a target that has a high coincidence rate of behavior (i.e., the accuracy rate of behavior forecast is high) and is capable of returning a response within the period of T_Aware. In other words, conscious robot A is capable of awareness.

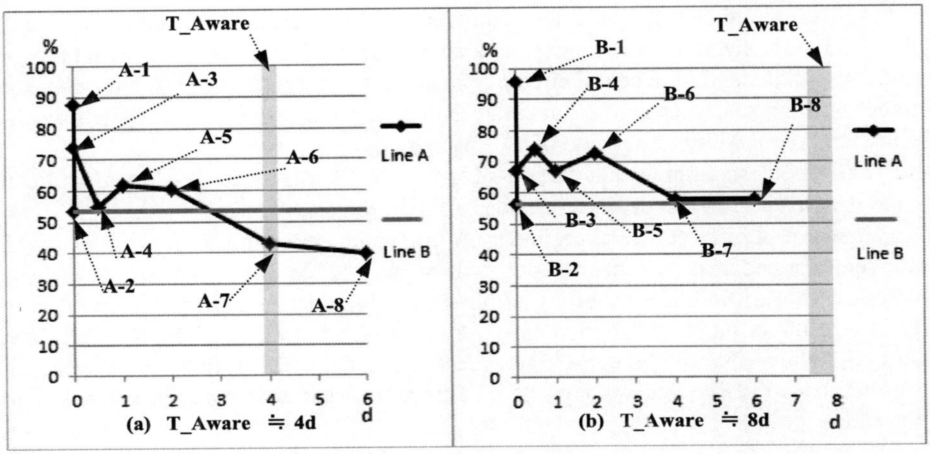

Fig. 3. (a) EXP.A,and (b) EXP.B

This demonstrates the idea that it does not matter whether conscious robot A is connected with an electric cable (nerve fiber) or not when it evaluates its own body as itself, but the important thing is "whether the conscious robot is able to receive the same behavior as forecast from robot C within the period of T_Aware, i.e., whether the evaluation acquired from inside the robot itself stays within a certain allowable range or not."

There is a phenomenon in which even though a person has a physical connection with a limb that has been paralyzed due to a stroke or some injury, he has no awareness of it as a part of his own body and feels that it is a part of another's body.

Based on this thinking, we propose a new method with which one can be aware of one's own body by an evaluation within oneself.

This may be stated, "a system with consciousness can realize that it is able to be aware that an anticipated reaction within the period of T_Aware is the body of the system itself."

Which means that a conscious system is capable of being aware of the existence of its own body using this proposed method.

There is also a case study that supports this hypothesis in humans called the "rubber hand illusion" [5]. Moreover, there are reported cases in which the subject loses the sensation of the fake hand being part of their own body if the timing of the stimulation on both the real hand of the subject and the rubber hand is delayed for a certain time [6]. We can interpret this to mean that, with regard to the transmission time of the stimulation traveling to the brain, the tactile stimulation to the real hand is shorter than that to the rubber hand, so visual information from the scene in which the fake hand is stimulated becomes at odds with the reaction that is anticipated based on the tactile stimulation of the subject.

It can be comprehended that if the interval between the tactile stimulation and anticipated reaction time is shorter than the T_Aware of humans, humans recognize the fake hand as a part of their own body, and if it is longer than T_Aware, humans do not perceive the fake hand to be part of their own body.

As discussed above, the possible result from the experiment on the conscious robot will lead to the hypothesisthat "whether a conscious robot is aware of its own body or not does not depend on a neural circuit connection to a target body part, but rather the important thing is whether the part is able to return the response anticipated by the conscious robot within the period of T_Aware.

This hypothesis implies that it is possible to treat a stroke or to develop an artificial limb which a person can perceive as a part of themselves.

When we performed the experiment on mirror image cognition in a robot in the abovementioned research, we used condition A-2 (Line B) in which the body is certainly that of another (robot C), although the boundary condition for a robot to be aware of its body is assumed to exist midway between A-2 and A-3 [3]. The authors believe that the value of T_Aware and the coincidence rate vary depending on the growth of the conscious system. This is an issue that must be addressed in the future.

References

1. Gallese, V., Fadiga, L., Fogassi, L., Rizzolatti, R.: Action recognition in the premotor cortex. Brain 119, 3–368 (1996)
2. Donald, M.: Origins of the Modern Mind. Harvard University Press, Cambridge (1991)
3. Takeno, J., Inaba, K., Suzuki, T.: Experiments and examination of mirror image cognition using a small robot. In: The 6th IEEE International Symposium on Computational Intelligence in Robotics and Automation, CIRA 2005, Espoo Finland, June 27-30, pp. 493–498 (2005) IEEE Catalog: 05EX1153C, ISBN0-7803-9356-2
4. Takeno, J.: MoNAD structure and self-awareness. In: Biologically Inspired Cognitive Architectures (BICA 2011), Arlington, USA (2011)
5. Botvinick, M., Cohen, J.: Rubber hands "feel" touch that eyes see Nature, vol. 391(6669), pp. 756–756 (1998)
6. Fukuda, K., Shimada, S.: Effect of Delayed Visual Feedback on Rubber Hand Illusion (2009)
7. Mannor, J., Takeno, J.: Monad Structures in the Human Brain. In: CSIT 2011 (2011)

Why Neurons Are Not the Right Level of Abstraction for Implementing Cognition

Claude Touzet

Aix-Marseille University, France

The cortex accounts for 70% of the brain volume. The human cortex is made of micro-columns, arrangements of 110 cortical neurons (Mountcastle), grouped in by the thousand in so-called macro-colums (or columns) which belong to the same functional unit as exemplified by Nobel laureates Hubel and Wiesel with the orientation columns of the primary visual cortex. The cortical column activity does not exhibit the limitations of single neurons: activation can be sustained for very long periods (sec.) instead of been transient and subject to fatigue. Therefore, the cortical column has been proposed as the building block of cognition by several researchers, but to not effect – since explanations about how the cognition works at the column level were missing. Thanks to the Theory of neural Cognition, it is no more the case.

The cortex functionality is cut into small areas: the cortical maps. Today, about 80 cortical maps are known in the primary and secondary cortex [1]. These maps belongs to a hierarchical organization A cortical map is a functional structure encompassing several thousands of cortical columns. The function of such maps (also known as Kohonen maps) is to build topographic (*i.e.*, organized and localized) representations of the input stimulii (events). This organization is such that similar inputs activate either the same cortical column or neighboring columns. Also, the more frequent the stimulus, the greater the number of cortical columns involved. Each map acts as a novelty detector and a filter. Events are reported as patterns of activations on various maps, each map specialized in a specific "dimension". Spatial and temporal coordinates of events are linked to activations on the hippocampus and define *de facto* the episodic memory.

Learning is achieved at neuronal level by the famous Hebb's law: "neurons active in the same time frame window reinforce their connections". This rule does not respect "causality". This, plus the fact that there is at least as much feed-back connections as there are feed-forward ones, explain why a high level cortical activation generates a low level cortical pattern of activations - the same one that would trigger this high level activity. Therefore, our opinion is that the true building block of the cognition is a set of feed-forward and feed-back connections between at least two maps.

Due to the fact that the real world is mostly continuous (the shorter the time interval, the smaller the variations) and because the neighboring situations (events) are represented close to each others on the cortical maps, then it is possible to actually "see" and understand activation trajectories (on a map) as representing the various situations encountered (in the real life). It is also straightforward to make a prediction

A. Chella et al. (Eds.): Biologically Inspired Cognitive Architectures 2012, AISC 196, pp. 317–318.
springerlink.com

about the coming next situation: the next (column) in line with the current trajectory, and suppress the transmission to higher levels if successful (in predicting) – therefore reserving higher level maps (such as the ones involved in language) to the processing of the very few events that are non-ordinary (extraordinary) for the subject. Exterogenous attention is therefore automatically build by to the processing of extraordinary inputs. Endogenous attention is achieved by the mere activation of high level representation, which automatically pre-activates low level localizations (columns) through feed-back connections.

A behavior (*i.e.*, acting) is a sequence of actions related to the achievement of a goal. Any goal can be seen as a situation (to reach). Since situations are associated to columns on maps, a successful behavior is a trajectory on the map that hosts the goal situation, starting from the current situation and ending at the goal situation. The associated actions to perform in order to move from one situation to the next one closer to the goal are given by a second map which codes for variations between situations and associated actions [2]. There is no difference implied by the level of abstraction: a mobile robot avoiding obstacles using as a goal a "situation clear of any obstacle" is similar to a person writing a scientific publication using as a goal its "memorization of readers having understood and learned". As evidenced by V. Braitenberg (*e.g.*, its vehicles experiment), anthropomorphism promptly converts to "love", "hatred", "humor", etc. sequences of actions automatically and very simply defined, such as by a few specific goal situations.

Other typically human cognitive abilities are also understood through the functioning of a hierarchical structure of cortical maps, such as intelligence [3], intrinsic motivation [4] or reading [5].

References

1. Silver, M., Kastner, S.: Topographic maps in human frontal and parietal cortex. Trends in Cognitive Sciences 13(11), 488–495 (2009)
2. Touzet, C.: Modeling and Simulation of Elementary Robot Behaviors using Associative Memories. International Journal of Advanced Robotic Systems 3(2), 165–170 (2006)
3. Touzet, C.: Consciousness, Intelligence, Free-will: the answers from the Theory of neuronal Cognition. In: Ed. la Machotte, 156 pages (2010) (in French) ISBN: 978-2-919411-00-9
4. Touzet, C.: The Illusion of Internal Joy. In: Schmidhuber, J., Thórisson, K.R., Looks, M. (eds.) AGI 2011. LNCS (LNAI), vol. 6830, pp. 357–362. Springer, Heidelberg (2011)
5. Dufau, S., Lete, B., Touzet, C., Glotin, H., Ziegler, J.C., Grainger, J.A.: Developmental Perspective on Visual Word Recognition: New Evidence and a Self-Organizing Model. European Journal of Cognitive Psychology 22(5), 669–694 (2010)

Intertemporal Decision Making: A Mental Load Perspective

Jan Treur

VU University Amsterdam, Agent Systems Research Group
De Boelelaan 1081, 1081 HV Amsterdam, The Netherlands
treur@cs.vu.nl
http://www.few.vu.nl/~treur

Abstract. In this paper intertemporal decision making is analysed from a framework that defines the differences in value for decision options at present and future time points in terms of the extra amount of mental burden or work load that is expected to occur when a future option is chosen. It is shown how existing hyperbolic and exponential discounting models for intertemporal decision making both fit in this more general framework. Furthermore, a specific computational model for work load is adopted to derive a new discounting model. It is analysed how this intertemporal decision model relates to the existing hyperbolic and exponential intertemporal decision models. Finally, it is shown how this model relates to a transformation between subjective and objective time experience.

1 Introduction

In realistic environments, decision making often involves a choice between decision options that each provides or promises some benefit, but at a different point in time. For example, in [5], Ch 7, pp. 193-220, following [2] and [7], Dennett presents a perspective on the evolution of moral agency, describing how by evolution cognitive capabilities have developed to estimate the value of an option in the future in comparison to the value of an option in the present. Such capabilities enable not to choose for self-interest in the present, but for other agents' interests, under the expectation that at some future point in time more benefits will be obtained in return. More in general, such capabilities enable to resist temptations in the present, and are helpful, for example, in coping with risks for addiction (e.g., [7]).

In intertemporal decision making it is assumed that in order to be comparable, the value estimated for a certain option has to be higher when the considered time point of realising the option is more in the future. This implies that value is not preserved over time, but can increase due to the passing of time. A central issue in intertemporal decision making is the specific discounting relation between values over time, as used by humans. Well-known approaches to intertemporal decision making consider hyperbolic or exponential discounting models (e.g., [1], [2], [9], [16], [17]). Support for both an exponential model and a hyperbolic model can be found. An interesting challenge is how such models can be explained on the basis of assumptions on more

A. Chella et al. (Eds.): Biologically Inspired Cognitive Architectures 2012, AISC 196, pp. 319–331.
springerlink.com

specific mental characteristics and processes involved. For example, they have been related to assumptions on subjective perception of time that essentially differs from objective time (e.g., [14], [20], [23], [25]). Also relations to underlying neural processes have been explored (e.g., [10], [13], [21], [24]).

In this paper the assumption is analysed that the difference between value for an option in the present and in the future is due to expected extra effort or mental load that is accompanying the future option. This load can be seen, for example, as the mental burden to keep the issue in mind, to worry about it, and to suppress impatience during the time interval involved. There are two advantages of this perspective. One is that computational models are available in physiological and cognitive sciences (e.g., [3], [4], [26], [27]) that describe mental load. A second advantage is that according to this perspective value is not increasing just due to passing of time, but is explained in terms of extra work invested, in line with basic intuitions in economics, for example, behind the well-known Labour Theory of Value (cf. [6], [22]; see also [18], [28]).

The paper is organised as follows. First in Section 2 a general framework is introduced that describes temporal discounting relations in terms of mental load, and it is shown how the well-known hyperbolic and exponential discounting models can be derived from the general framework using certain assumptions on the mental load over time. In Section 3 an available dynamical model on work load is used to derive a new discounting model, called the hyperbolic-exponential discounting model. In Section 4 this new model is analysed in some detail and it is shown, for example, how it relates to hyperbolic and exponential discounting. Section 5 shows how the hyperbolic-exponential model also can be related to a suitable assumption on subjective perceived time. It is shown how the transformation between objective and subjective time needed for this can be approximated by logarithmic functions and power functions as used in the literature.

2 Relating Temporal Discounting to Expected Mental Work Load

In this section, in Section 2.1 a general setup to relate intertemporal decision making to expected mental load is presented. Next, in Section 2.2 it is described how the well-known hyperbolic temporal discounting model fits in this general setup, and in Section 2.3 the same is done for the exponential temporal discounting model.

2.1 General Setup

As for decision making in general, to model intertemporal decision making often a valuation-based approach is followed: each decision option is valued and in principle the option with highest value is chosen. This perspective is supported by recent neurological research; e.g., [10], [13], [14], [21], [24]. For intertemporal comparison between values of options an element usually taken into account is a certain discounting rate. For example, when someone has to pay money to you and asks for a one year delay of this payment, you may ask some interest percentage in addition. Many of the experiments in intertemporal decision making address such situations with subjects having to choose between getting an amount of money in the present and getting a different amount at a given future time point. Here the comparison is between two options which are both expressed in terms of money. In the general setting addressed here an option 1 is compared to an option 2, where option 1 is an option realised in the present, and option 2 is an option realised in the future, after some time duration *t*. The value assigned to option 1 in the present is compared to the value assigned to option 2 after *t*. It is this dependence of the time duration of this option that is addressed here. In an abstract form this is represented as follows. Take $V(0)$ the value of the option (for a given agent) in the present, and $V(t)$ the value for the (other) option after time duration *t*.[1] To be comparable to the value $V(0)$ for the option in the present, the value $V(t)$ of the option for a future time point has to be higher; e.g., [1], [2], [5], [9], [14], [17]. In this paper the assumption is explored that a higher value of $V(t)$ compared to $V(0)$ is due to the expected extra amount of mental energy $E(t)$ needed for the future option during the time interval $[0, t]$ before the option has actually been obtained at time *t*; this can be formulated as follows:

$$V(\mathrm{t}) = V(0) + E(t) \tag{1}$$

The intuition behind this is that the mental work invested adds value to a product, in line with, for example, the Labour Theory of Value in economics (e.g., [6], [18], [22], [28]). The extra energy $E(t)$ can be considered as related to the burden of having to wait, and suppressing the impatience to have the option. This extra energy is assumed proportional to the initial value $V(0)$:

$$E(t) = W(t)\,V(0) \tag{2}$$

where $W(t)$ (work) is the expected extra energy spent per unit of V during the time interval from 0 to *t*. Using this in (1) the following *general temporal discounting relation* is obtained, depending on the expected extra work $W(t)$ accompanying the choice for the option at time *t*.

$$V(\mathrm{t}) = V(0) + W(t)V(0) = (1 + W(t))\,V(0) \tag{3a}$$

$$\frac{V(t)}{V(0)} = 1 + W(t) \tag{3b}$$

$$\frac{V(0)}{V(t)} = \frac{1}{1 + W(t)} \tag{3c}$$

The dependency expressed in (3c) is depicted in Fig. 1: a hyperbolic dependency of the fraction $V(0)/V$ on W. The *discounting rate* is defined as

$$\frac{dV}{dt}/V$$

Using (3a) and (3c), this can be expressed in $W(t)$ as follows:

$$\frac{dV(t)}{dt} = V(0)\,\frac{dW(t)}{dt}$$

[1] Note that to explicitly indicate the two options the notations $V_1(0)$ and $V_2(t)$ could be used. For the sake of simplicity these indices are left out.

$$\frac{dV(t)}{dt} / V(t) = \frac{V(0)}{V(t)}\frac{dW(t)}{dt} = \frac{1}{1+W(t)}\frac{dW(t)}{dt} \quad (4)$$

The relations between $V(t)$ and the expected extra work $W(t)$ provide a way to derive temporal discounting models from models for expected work depending on the duration t of the interval. Conversely, (3) can also be used to derive a function $W(t)$ from any given temporal discounting model: any discounting model $V(t)$ can be related to an expected work function $W(t)$, by the following *inverse relation*:

$$W(t) = \frac{V(t)}{V(0)} - 1$$

In principle to obtain $W(t)$ any mathematical function can be considered and based on that function a temporal discounting model and discounting rate can be derived by (3) and (4). It is reasonable to assume at least that such a function is monotonically increasing. However, not every monotonically increasing function may be considered plausible. Below different ways to obtain $W(t)$ will be considered. First it will be analysed which work functions $W(t)$ fit to the well-known classical hyperbolic and exponential discounting models.

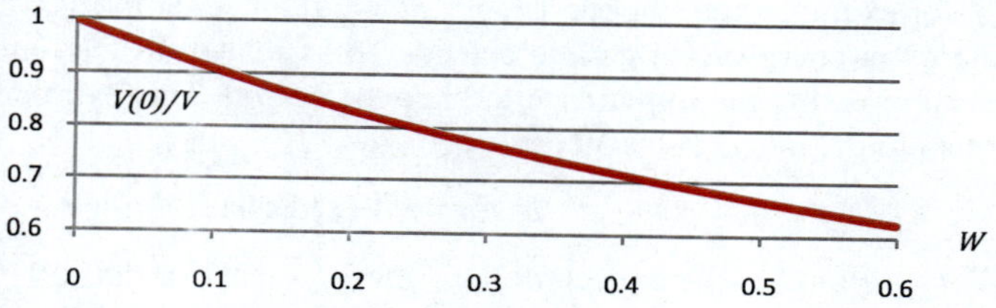

Fig. 1. Hyperbolic relation for *V(0)/V* vs. work *W* (3c)

2.2 Deriving the Hyperbolic Discounting Model

From the perspective of the expected extra work $W(t)$, a simplest model occurs when it is assumed that the power P (the expected extra energy spent per time unit and per unit of V) is constant over the time interval, and based on that the amount of extra work is proportional to the length of the time interval:

$$W(t) = P\,t \quad (5)$$

By taking the linear dependencies (3b) and (5) together a linear form follows:

$$\frac{V(t)}{V(0)} = 1 + P\,t \quad (6)$$

Using the format of (3c), this provides a derivation of the well-known *hyperbolic model* for temporal discounting:

$$V(0) = \frac{V(t)}{1 + P\,t} \quad (7)$$

$$\frac{V(0)}{V(t)} = \frac{1}{1 + P\,t} \tag{8}$$

Note that from (4) and (5) it immediately follows that the discounting rate for this case is

$$\frac{P}{1 + P\,t}$$

which starts at P for $t = 0$ and is decreasing to 0 for increasing t.

2.3 Deriving the Exponential Discounting Model

Another well-known model for temporal discounting is the exponential model:

$$\frac{V(t)}{V(0)} = e^{\alpha t} \tag{9}$$

From (3b) it can be derived that this exponential model occurs when it is assumed that the expected extra work $W(t)$ has an exponential relation to t:

$$W(t) = e^{\alpha t} - 1 \tag{10}$$

For this case from (4) and (10) it follows that the discounting rate is α, which does not depend on t. Note that the derivation of (10) from (9) by itself does not provide a justification of why expression (10) for the extra work $W(t)$ would be reasonable. It suggests that either expected power P is not constant but increasing over time, or, if P is constant, from the start on the level will be higher when longer time intervals are involved. In Section 3 an existing model for estimating the amount of extra work will be adopted to derive what may be a reasonable expression for $W(t)$.

3 A Temporal Discounting Model Based on a Work Load Model

A more specific model is obtained if more detailed assumptions are made for the amount of extra work $W(t)$ spent. In the physiological literature on work load usually a critical power level CP is assumed which is the power that can be sustained without becoming (more) exhausted. Moreover, this critical power may be slightly affected by using power above this level. In [26] a computational model for work load was used in order to address the question how (physiological) effort can best be distributed over a certain time interval. Below this is adopted as a biologically inspired model for mental work load to address the issue of deciding between two temporally different options. The following equations (adopted from [26]) describe this model. Dynamics of critical power is described by a linear dependency of the change in critical power on the effort spent above the critical power, with proportion factor γ:

$$\frac{dCP}{dt} = -\gamma P \tag{11}$$

Here P is a (non-constant) function of time t indicating the extra (non-sustainable) power spent at time t; it is assumed that only the expected non-sustainable energy spent based on extra resources is taken into account. The expected extra work W spent over time is described by

$$\frac{dW}{dt} = P \tag{12}$$

The following relation for the derivatives of CP and W follows from (11) and (12):

$$\frac{dCP}{dt} = -\gamma \frac{dW}{dt} \tag{13}$$

Moreover, if it is assumed that the total expected power $P + CP$ spent (both the non-sustainable power P and sustainable power CP) is kept constant over time, it holds

$$\frac{dCP}{dt} = -\frac{dP}{dt}$$

From this, (12) and (13) it follows

$$\frac{dP}{dt} = \gamma \frac{dW}{dt} = \gamma P \tag{14}$$

Relation (14) is a simple first-order differential equation in P which has as solution $P(t) = P(0)\, e^{\gamma t}$. From this and (12) it follows that

$$\frac{dW}{dt} = P(0)\, e^{\gamma t} \tag{15}$$

Using (15), for $\gamma \neq 0$ an explicit analytic solution for the function $W(t)$ can be found by integration, thereby using $W(0) = 0$:

$$W(t) = W(0) + P(0)\,(e^{\gamma t} - 1)/\gamma = \frac{P(0)}{\gamma}\,(e^{\gamma t} - 1) \tag{16}$$

Note that when it is assumed $P(0) = \gamma = \alpha$, this exactly provides relation (10) found in Section 2.3 for the exponential model. Using (16) in (3b) the following *hyperbolic-exponential model* for temporal discounting is obtained:

$$\frac{V(t)}{V(0)} = 1 + \frac{P(0)}{\gamma}(e^{\gamma t} - 1) = \frac{P(0)}{\gamma} e^{\gamma t} + \frac{\gamma - P(0)}{\gamma} = \frac{P(0)}{\gamma}\left(e^{\gamma t} + \frac{\gamma - P(0)}{P(0)}\right) \tag{17}$$

The hyperbolic-exponential model (17) can be written in a different form as follows:

$$\frac{V(t)}{V(0)} = 1 + P(0)\, t\, \frac{e^{\gamma t} - 1}{\gamma t} \tag{18}$$

$$\frac{V(0)}{V(t)} = \frac{1}{1 + P(0)\, t\, \frac{e^{\gamma t} - 1}{\gamma t}} \tag{19}$$

This form shows that the difference with the hyperbolic model is in the exponential factor

$$\frac{e^{\gamma t} - 1}{\gamma t}$$

which is close to 1 for smaller values of γt, as is shown, e.g., in Fig. 2 in Section 4, where a more precise analysis will be presented of how this hyperbolic-expcnential model relates to both the hyperbolic and the exponential model. From (4), (15) and (16) it follows that the discounting rate for this case is obtained as follows:

$$\frac{dV(t)}{dt} / V(t) = \frac{dW(t)}{dt} / (1 + W(t)) = \gamma / \left(\frac{\gamma - P(0)}{P(0)} e^{-\gamma t} + 1\right)$$

Note that for $t = 0$ this discounting rate is $P(0)$ and for larger time durations from 0 to t the rate converges to γ(upward when $P(0) < \gamma$ and downward when $P(0) > \gamma$).

4 Comparative Analysis

To be able to compare the hyperbolic-exponential model given by (18), (19) to existing models, first the expression $\frac{e^{\gamma t}-1}{\gamma t}$ as a function of γt is analysed. In Fig. 2 a graph is shown for this function. The Taylor series is:

$$\frac{e^{\gamma t}-1}{\gamma t} = \frac{\left(\sum_{k=0}^{\infty} \frac{1}{k!}(\gamma t)^k - 1\right)}{\gamma t} = 1 + \sum_{k=2}^{\infty} \frac{1}{k!} (\gamma t)^{k-1} \quad (20)$$

From this it follows that

$$\frac{e^{\gamma t}-1}{\gamma t} \text{ is monotonically increasing in } \gamma t \text{ with 1 as limit for } \gamma t \text{ at } 0 \quad (21)$$

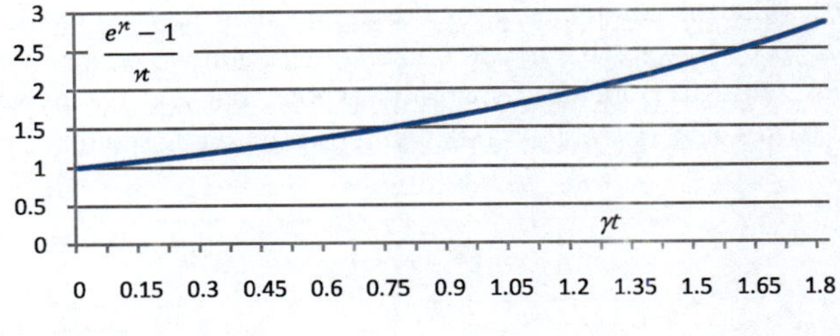

Fig. 2. Graph of the function $\frac{e^{\gamma t}-1}{\gamma t}$ against $\gamma t > 0$

From (21) it follows that for shorter time durations t and/or smaller values of γ the hyperbolic-exponential model (18), (19) is approximated by a hyperbolic model; cf. (6), (7), (8) with $P = P(0)$:

$$\frac{V(t)}{V(0)} = 1 + P(0)\, t \quad (22)$$

$$\frac{V(0)}{V(t)} = \frac{1}{1 + P(0)\,t} \tag{23}$$

This is illustrated, for the shorter time durations t in Fig. 3. Moreover, for γ very small (for example, 10^{-8}) the hyperbolic-exponential model is approximated by the hyperbolic model for all t with high accuracy, with difference estimated by (20) as

$$P(0)t\,(\frac{e^{\gamma t}-1}{\gamma t} - 1) = P(0)\,t\,(1 + {}^1\!/_2\,\gamma t + \ldots - 1) = {}^1\!/_2\,P(0)\,\gamma t^2$$

based on which is proportional to γ. In particular, note that $\gamma = 0$ provides an exact match with the hyperbolic model. Using

$$\lim_{\gamma t \to \infty} \frac{e^{\gamma t}-1}{e^{\gamma t}} = 1$$

and (17) it can be shown that for longer time durations t and/or larger values of γ the hyperbolic-exponential model is approximated by the following exponential model:

$$\frac{V(t)}{V(0)} = \frac{P(0)}{\gamma}\,e^{\gamma t} \tag{24}$$

or

$$\frac{V(0)}{V(t)} = \frac{\gamma}{P(0)}\,e^{-\gamma t} \tag{25}$$

This is illustrated for the longer time durations t in Fig. 3. Note that in general $P(0) \neq \gamma$ so that this approximation cannot be used for shorter time durations. However, if $P(0) = \gamma$, then the model is exponential.

The above analysis shows that the introduced hyperbolic-exponential model provides one unified model that can be approximated both by a hyperbolic pattern for shorter time durations and/or smaller values of γ and by an exponential pattern for longer time durations and/or larger values of γ. For the first approximation for smaller t, the constant in the hyperbolic model usually indicated by k is equal to $P(0)$. Note that this still leaves freedom in the value of parameter γ for the exponential approximation for larger t. In Fig. 3 the differences between the hyperbolic and hyperbolic-exponential are displayed for some examples settings. Both the hyperbolic-exponential and the hyperbolic curve are shown for $\gamma = 0.005$, and $P = P(0) = 0.05$, with time on the horizontal axis.

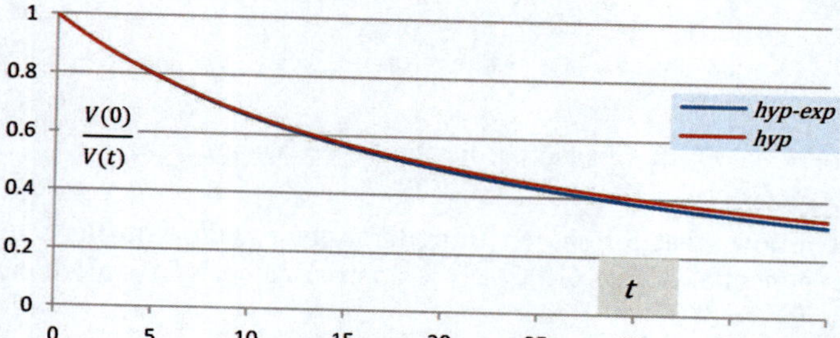

Fig. 3. Comparing hyperbolic and hyperbolic-exponential models: $\gamma = 0.005$, $P = P(0) = 0.05$

The graph shows the forms of (8) and (19). The graph shows that indeed for shorter time durations they coincide, whereas for longer time durations they diverge. In general, the smaller γ the more the two curves coincide. There is also another way in which the two curves can be related to each other: by taking P different from $P(0)$. In Fig. 4 it is shown how the two curves coincide when $P = 0.053$ and $P(0) = 0.05$. The upper graph shows the forms of (8) and (19). In the second graph the absolute differences are depicted.

5 Relating the Model to Subjective Experience of Time

In the recent literature another way of taking mental processing into account is by assuming perception of subjective time which is different from objective time (for example, advocated in [14], [20], [23], [25]). The different time measures can be related according to a time transformation function

$$\tau\colon t \rightarrow u = \tau(t)$$

where t is the objective time and u the subjective (perceived) time. Such a time transformation explains a form of human discounting different from the exponential form, whereas internally an exponential model is used:

$$\frac{V(t)}{V(0)} = e^{\alpha\,\tau(t)}$$

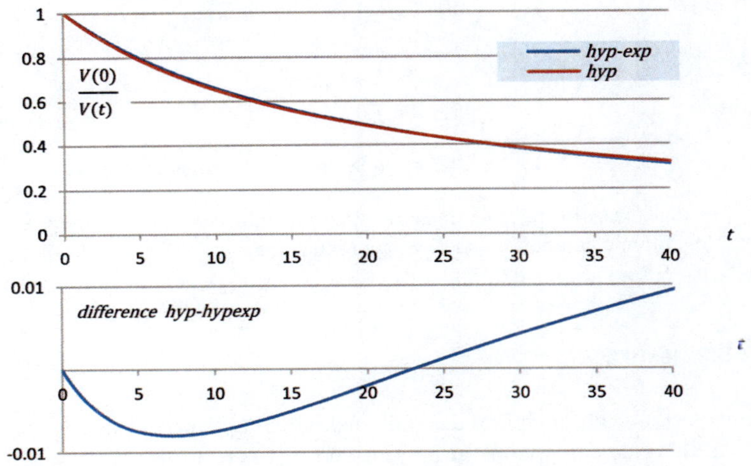

Fig. 4. Comparison of hyperbolic and hyperbolic-exponential models: $\gamma = 0.005$, $P = 0.053$, $P(0) = 0.05$

The hyperbolic-exponential model can also be related to an exponential form according to this principle, when an appropriate time transformation is chosen. This time transformation can be found using (17):

$$\frac{V(t)}{V(0)} = 1 + \frac{P(0)}{\gamma}(e^{\gamma t} - 1) \quad = e^{\alpha u} \qquad\qquad \alpha u = \ln(1 + \frac{P(0)}{\gamma}(e^{\gamma t} - 1)\ )$$

Therefore

$$\tau(t) = u \quad = \frac{1}{\alpha}\ln(1 + \frac{P(0)}{\gamma}(e^{\gamma t} - 1)\ ) \tag{26}$$

This shows how the hyperbolic-exponential model can be obtained by assuming a subjective internal time perception different from the objective time. In Figure 5 it is shown how the transformation τ from objective time to objective time compares to the first and second order approximations, and to other transformations put forward in the literature: the logarithmic transformation (e.g., [23]), and a transformation defined by a power function (e.g., [25]):

$$u = \eta \ln(1+\beta t) \qquad\qquad u = \lambda\, t^{\kappa}$$

This figure shows that the time transformation τ related to the hyperbolic-exponential model can be approximated very well by a logarithmic function and also reasonably well by a power function.

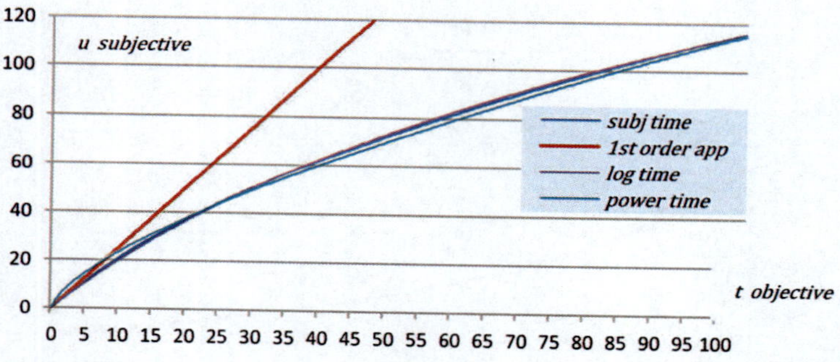

Fig. 5. Time transformation τ from objective time t to subjective time u compared to 1st and 2nd-order, logarithmic and power function approximations: $\gamma = 0.01$, $\alpha = 0.02$, $P(0) = 0.05$, $\beta = 0.025$, $\eta = 90$, $\lambda = 5$, $\kappa = 0.675$

6 Discussion

The framework for intertemporal decision making at the core of this paper defines the differences in value for options at present and at future time points in terms of the expected extra amount of mental burden, worries, or work load accompanying the choice for a future option. Conceptually this provides an approach in which an attributed higher value for the option in the future is in balance with a lower value for the option in the present plus an expected amount of extra work needed to actually obtain the extra value in the future. Thus value becomes higher due to extra work added, in line with intuitions in economics, at the basis of, for example, the Labour

Theory of Value (e.g., [6], [18], [22], [28]). It has been shown how the existing hyperbolic and exponential discounting models for intertemporal decision making fit in this framework for specific patterns for the amounts of work depending on the duration. Furthermore, based on elements from an existing biologically inspired computational model for work load (cf. [26], [27], [3], [4]) a new hyperbolic-exponential discounting model has been derived.

This hyperbolic-exponential discounting model makes use of the concept critical power which indicates the level of power that can be provided without increasing fatigue (e.g., [8], [11], [12], [15], [19], [26], [27]). Moreover, the assumption was made that the considered power spent is kept constant over time. The model has a parameter γ for the extent to which critical power is affected by using power above the critical power. It has been shown that this model is approximated by the well-known hyperbolic model for shorter time durations or a lower value of the parameter γ (for $\gamma = 0$ by an exact match) and that it is approximated by the exponential model for longer time durations or a higher value of the parameter γ. Furthermore, it has been explored how this hyperbolic-exponential model relates to an approach assuming a difference between subjective and objective time (following, for example, [14], [20], [23], [25]) and a specific transformation between subjective and objective time, which can be approximated well by a time transformation based on a logarithmic function (e.g., as considered in [23]) or a power function (e.g., as considered in [25]).

Thus the introduced hyperbolic-exponential discounting model for intertemporal decision making provides a form of unification of a number of existing approaches. Moreover, it provides an explanation of the increase of value over time in terms of expected extra mental load, which relates well to elementary economic intuitions: you have to work to obtain more value. Alternative approaches to explain the difference in value at different points in time often relate this to the risk of losing the future option. However, in principle this leads to an exponential discounting model, which often has been rejected by empirical data acquired from experiments which favour a hyperbolic model. The approach to relate this difference in value to the expected burden implied by a choice for the future option relates in a quite natural manner to non-exponential or not fully exponential discounting models.

The proposed intertemporal decision model can be used, for example, as a basis for virtual agents that have to cope with temptations, for example in the context of addictions, or in a social context where they are to show moral agency by sometimes deciding for options against their present self-interest; cf. [7], [5], pp. 193-220.

References

1. Ainslie, G.: Derivation of 'rational' economic behavior from hyperbolic discount curves. American Econ. Review 81, 334–340 (1991)
2. Ainslie, G.: Breakdown of Will. Cambridge University Press, New York (2001)
3. Bosse, T., Both, F., van Lambalgen, R., Treur, J.: An Agent Model for a Human's Functional State and Performance. In: Jain, L., Gini, M., Faltings, B.B., Terano, T., Zhang, C., Cercone, N., Cao, L. (eds.) Proc. of the 8th Intern. Conf. on Intelligent Agent Technology, IAT 2008, pp. 302–307. IEEE Computer Society Press (2008)

4. Both, F., Hoogendoorn, M., Jaffry, S.W., van Lambalgen, R., Oorburg, R., Sharpanskykh, A., Treur, J., de Vos, M.: Adaptation and Validation of an Agent Model of Functional State and Performance for Individuals. In: Yang, J.-J., Yokoo, M., Ito, T., Jin, Z., Scerri, P. (eds.) PRIMA 2009. LNCS (LNAI), vol. 5925, pp. 595–607. Springer, Heidelberg (2009)
5. Dennett, D.C.: Freedom Evolves. Penguin Putman Inc. (2003)
6. Dooley, P.C.: The Labour Theory of Value. Routledge (2005)
7. Frank, R.H.: Passions within Reason: the Strategic Role of Emotion. Norton, New York (1988)
8. Fukuba, Y., Miura, A., Endo, M., Kan, A., Yanagawa, K., Whipp, B.J.: The Curvature Constant Parameter of the Power-Duration Curve for Varied-Power Exercise. Medicine and Science in Sports and Exercise 35, 1413–1418 (2003)
9. Green, L., Myerson, J.: Exponential versus hyperbolic discounting of delayed outcomes: Risk and waiting time. American Zoologist 36, 496–505 (1996)
10. Gregorios-Pippas, L., Tobler, P.N., Schultz, W.: Short-Term Temporal Discounting of Reward Value in Human Ventral Striatum. J. Neurophysiol. 101, 1507–1523 (2009)
11. Hill, A.V., Long, C.N.V., Lupton, H.: Muscular exercise, lactic acid, and the supply and utilisation of oxygen. Proc. Royal Soc. Bri. 97, Parts I-III, 438–475, Parts VII-VIII, 155–176 (1924)
12. Hill, D.W.: The critical power concept. Sports Medicine 16, 237–254 (1993)
13. Kable, J.W., Glimcher, P.W.: The neural correlates of subjective value during intertemporal Choice. Nat. Neurosci. 10, 1625–1633 (2007)
14. Kim, B.K., Zauberman, G.: Perception of Anticipatory Time in Temporal Discounting. Journal of Neuroscience, Psychology, and Economics 2, 91–101 (2009)
15. Lambert, E.V., St Clair Gibson, A., Noakes, T.D.: Complex systems model of fatigue: integrative homoeostatic control of peripheral physiological systems during exercise in humans. British J. of Sports Medicine 39, 52–62 (2005)
16. Mazur, J.E.: An adjusting procedure for studying delayed reinforcement. In: Commons, M.L., et al. (eds.) Quantitative Analyses of Behavior: The Effect of Delay and of Intervening Events on Reinforcement, pp. 55–73. Erlbaum, Hillsdale (1987)
17. McKerchar, T.L., Green, L., Myerson, J., Pickford, T.S., Hill, J.C., Stout, S.C.: A comparison of four models of delay discounting in humans. Behavioural Processes 81, 256–259 (2009)
18. Mohun, S.: The Labour Theory of Value as Foundation for Empirical Investigations. Metroeconomica 55, 65–95 (2004)
19. Noakes, T.D.: Time to move beyond a brainless exercise physiology: the evidence for complex regulation of human exercise performance. Appl. Physiol. Nutr. Metab. 36, 23–35 (2011)
20. Ray, D., Bossaerts, P.: Positive temporal dependence of the biological clock implies hyperbolic discounting. Frontiers in Neuroscience 5, 2 (2011), doi:10.3389/fnins.2011.00002
21. Schultz, W.: Subjective neuronal coding of reward: temporal value discounting and risk. Eur. J. of Neuroscience 31, 2124–2135 (2010)
22. Smith, A.: Inquiry into the Nature and Causes of the Wealth of Nations. University of Chicago Press (1776/1977)
23. Takahashi, T.: Loss of self-control in intertemporal choice be attributable to logarithmic time-perception. Medical Hypotheses 65, 691–693 (2005)
24. Takahashi, T.: Theoretical frameworks for neuroeconomics of intertemporal choice. Journal of Neuroscience, Psychology, and Economics 2, 75–90 (2009)

25. Takahashi, T.: Time-estimation error following Weber–Fechner law explain sub-additive time-discounting. Medical Hypotheses 67, 1372–1374 (2006)
26. Treur, J.: Physiological Model-Based Decision Making on Distribution of Effort over Time. In: Samsonovich, A.V., Jóhannsdóttir, K.R. (eds.) Proc. of the Second Intern. Conf. on Biologically Inspired Cognitive Architectures, BICA 2011, pp. 400–408. IOS Press (2011)
27. Treur, J.: A Virtual Human Agent Model with Behaviour Based on Feeling Exhaustion. Applied Intelligence 35, 469–482 (2011)
28. Tsaganea, D.: The Classical Labour Theory of Value and The Future of American Power. Annals of Spiru Haret University, Economic Series 1, 15–39 (2010)

A Non-von-Neumann Computational Architecture Based on in Situ Representations: Integrating Cognitive Grounding, Productivity and Dynamics

Frank van der Velde

Leiden University; Technical Cognition - University of Twente, The Netherlands
vdvelde@fsw.leidenuniv.nl, f.vandervelde@utwente.nl

Abstract. Human cognition may be unique in the way it combines cognitive grounding, productivity (compositionality) and dynamics. This combination imposes constraints on the underlying computational architecture. These constraints are not met in the von-Neumann computational architecture underlying forms of symbol manipulation. The constraints are met in a computational architecture based on 'in situ' grounded representations, consisting of (distributed) neuronal assembly structures. To achieve productivity, the in situ grounded representations are embedded in (several) neuronal 'blackboard' architectures, each specialized for processing specific forms of (compositional) cognitive structures, such as visual structures (objects, scenes), propositional (linguistic) structures and procedural (action) sequences. The architectures interact by the neuronal assemblies (in situ representations) they share. This interaction provides a combination of local and global information processing that is fundamentally lacking in symbolic architectures of cognition. Further advantages are briefly discussed.

Keywords: Grounding, In situ representations, Neuronal assemblies, Non-von Neumann architecture, Productivity.

1 Introduction

The aim of this paper is to argue that the integration of cognitive grounding, productivity (compositionality) and dynamics imposes constraints on computational architectures and provides new abilities for human-kind (biologically inspired) cognition. Grounding concerns the notion that humans learn conceptual representations based on their physical interactions with their environment. Productivity concerns the ability to combine learned representations in (potentially novel) compositional structures. Examples are sentences composed with familiar words or visual scenes composed with familiar visual features (e.g., shapes, colors, motion) and (spatial) relations between the features. Dynamics refers to the real-time processing of information and the ability to interact with the environment in a way that keeps pace with the dynamics of events unfolding in the environment.

Human cognition may be unique in the way it combines these features of cognition. For example, dynamical interaction is obviously an important feature of animal

A. Chella et al. (Eds.): Biologically Inspired Cognitive Architectures 2012, AISC 196, pp. 333–337.
springerlink.com

cognition, but productivity to the extent of human cognition is not found in animals. On the other hand, productivity can be found in artificial forms of information processing, as given by architectures based on symbol manipulation, but grounding of representations is missing in these systems.

2 Productivity and the von-Neumann Computational Architecture

Symbolic cognitive architectures achieve productivity by using symbol manipulation to process or create compositional structures. The von-Neumann architecture, with a CPU and symbolic representations, is the computational architecture needed to implement these processes [1]. Symbol manipulation depends on the ability to make copies of symbols and to transport them to other locations. Newell ([2], p. 74, italics added) described this process as follows: "When processing *The cat is on the mat* ... the local computation at some point encounters *cat*; it must go from *cat* to a body of (encoded) knowledge associated with *cat* and bring back something that represents that a cat is being referred to, that the word *cat* is a noun (and perhaps other possibilities), and so on." Thus, a symbol for the word *cat* is copied and pasted into a representation of *cat on the mat*. Because symbols can be copied and pasted they can be used to create complex compositional structures such as propositions, with symbols as constituents. The capacity of symbolic architectures to represent and process complex compositional (e.g., sentence-like) information far exceeds that of humans. But interpreting information in a way that could produce meaningful answers or purposive actions is far more difficult with these architectures. In part, this is due to the ungrounded nature of symbols, which, in turn, is partly due to the fact that symbols are copied and transported in these cognitive systems.

3 Grounded Representations

The ungrounded nature of symbols in cognitive architectures has been criticized (e.g., [3]). To improve this situation, cognitive models have been developed that associate labels for words (e.g., nouns, adjectives, verbs) with perceptual and behavioural information (e.g., [4]). Models like these are a significant improvement over the use of arbitrary symbols to represent concepts. However, the conceptual representations and their associations with perceptual representations are stored in long-term memory in these models. To form compositional structures with these concepts (e.g., *cat on the mat*), the conceptual representations still have to be copied from long-term memory and pasted into the compositional structure at hand. Thus, the direct grounding of the conceptual representations is lost when they are used as constituents in compositional structures.

An active decision is then needed to retrieve the associated perceptual information, stored somewhere else in the architecture. This raises the question of who (or what) in the architecture makes these decisions, and on the basis of what information.

Furthermore, given that it takes time to search and retrieve information, there are limits on the amount of information that can be retrieved and the frequency with which information can be renewed. These difficulties seriously affect the ability of such cognitive architectures to adaptively deal with information in complex environments. It hampers, for example, fast processing of ambiguous information in a context-sensitive way [5].

The use of 'pointers,' as in certain programming languages (e.g., C++), does not eliminate these difficulties at all. A pointer does not carry any conceptual information, but only refers to an address at which (potentially) conceptual information is stored. Again, an active decision is needed to obtain that information, which raises the same issues as outlined above.

4 In Situ Grounded Representations and Computational Architecture

To retain grounding, conceptual representations have to remain 'in situ' in productive (compositional) structures [6]. Figure 1 illustrates a neuronal assembly that instantiates the in situ grounded representation of the concept *cat*. It grounds the concept *cat* by interconnecting all aspects related to cats. For example, it includes all perceptual information about cats, as, e.g., given by the neural networks in the visual cortex that classify an object as cat, or the networks that identify a spoken word as *cat*. It also includes action processes related to cats (e.g., the embodied experience of stroking a cat, or the ability to pronounce the word *cat*), information (e.g., emotions) associated with cats, and semantic information related to cats (e.g., *cat is pet*, or *cat has paw*).

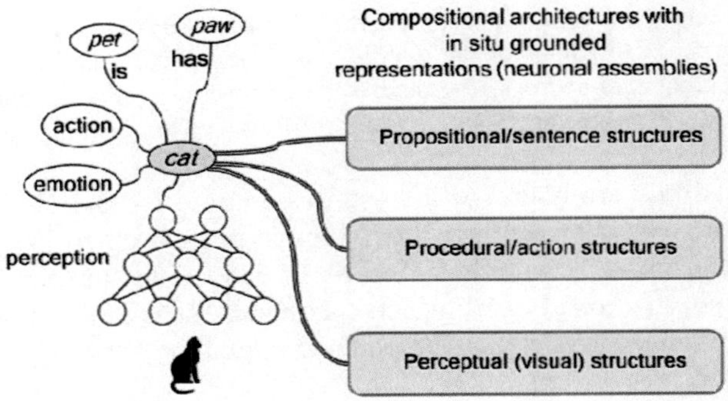

Fig. 1. Grounded *in situ* representations are embedded in several specialized architectures, needed for specific forms of cognitive processing based on specific combinatorial structures

The neuronal assembly forms an in situ representation of the concept *cat*, because it is not (and cannot be) copied or transported to form compositional structures (e.g., sentences, visual scenes). To achieve productivity, in situ grounded representations have to be embedded in specific neuronal 'blackboard' architectures to form and process compositional structures. Examples are neuronal architectures for visual processing, for processing propositional (sentence) structures or for executing procedural (action) sequences. These architectures interact by the neuronal assemblies (in situ representations) they share.

Figure 2 illustrates the in situ compositional structure of *(The) cat is on (the) mat* in a neural blackboard for (basic) sentence structures [7]. The neural blackboard consists of neuronal 'structure' assemblies that (in the case of sentences) represent syntactical type information, such as sentence (S_1), noun phrase (N_1 and N_2), verb phrase (V_1) and prepositional phrase (P_1). To create a sentence structure, assemblies representing specific syntactical type information (or syntax assemblies, for short) are temporarily connected (bound) to word assemblies of the same syntactical type. Binding is achieved with dedicated neural circuits. They constitute the 'conditional' connections and control networks, needed to implement relational structures.

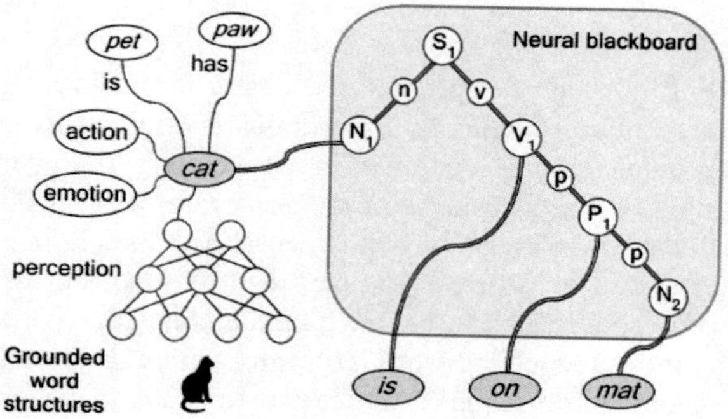

Fig. 2. Illustration of the combinatorial structure *The cat is on the mat* (ignoring *the*), with grounded in situ representations for the words. The circles in the neural blackboard represent populations and circuits of neurons. The double line connections represent conditional connections. (N, n = noun; P, p = preposition; S = sentence; V, v = verb.)

Thus, *cat* and *mat* are bound to the noun assemblies N_1 and N_2, respectively. In turn, the syntax assemblies are temporarily bound to each other, in accordance with the sentence structure. So, *cat* is bound to N_1, which is bound to S_1 as the subject of the sentence, and *is* is bound to V_1, which is bound to S_1 as the main verb of the sentence. Furthermore, *on* is bound to P_1, which is bound to V_1 and N_2, to represent the prepositional phrase *is on mat*.

5 Advantages of a Productive Computational Architecture Based on in Situ Grounded Representations

When in situ representations are parts of (constituents in) compositional structures, they provide both local and global information. As constituents, they can affect specific (local) compositional structures. But as in situ grounded representations, they retain their embedding in the global information structure of which they are a part. This combination of local and global information is fundamentally lacking in symbolic architectures of cognition [5, 8].

Furthermore, the architecture is extensive, in the sense that new information can be represented and processes by adding new resources or connections to the existing structures. In situ representations can modify gradually, based on experience. New blackboards can be added, or existing ones can be modified in a gradual way as well. The extensive nature of the architecture also allows a direct implementation in neuromorphic hardware.

In situ representations as parts of compositional structures are ideally suited for learning mechanisms determined by collocations and distributed equivalences, which may be the basis of human language learning [9]. Collocations are similarities that determine the basis of linguistic constructions [10]. An example are verbs like *give*, that form the basis for constructions such as *X gives Y Z* or *X gives Z to Y*. Distributional equivalences are the opposite from collocations, because they concern the type of variables that could fill the variable slots (e.g., *X, Y, Z*) in constructions. In this architecture, collocations are directly given by the reactivation of the same (in situ) representations (e.g., the activation of the in situ representation of *give*). Distributed equivalences are directly given by the (in situ) overlap between different in situ representations. For example, the situ representations of the words that can fill the slots *X, Y, Z* could share the representation that they are nouns. Given the similarities between linguistic structures and structures in the visual world [11], collocations and distributed equivalences could also be used for detecting constructions in the environment.

References

1. Fodor, J.A., Pylyshyn, Z.W.: Cognition 28, 3–71 (1988)
2. Newell, A.: Unified Theories of Cognition. Harvard University Press (1990)
3. Barsalou, L.W.: Behavioral and Brain Sciences 22, 577–660 (1999)
4. Roy, D.: Trends in Cognitive Sciences 9, 389–395 (2005)
5. van der Velde, F., de Kamps, M.: Learning of control in a neural architecture of grounded language processing. Cognitive Systems Research 11, 93–107 (2010)
6. van der Velde, F., de Kamps, M.: Compositional connectionist structures based on in situ grounded representations. Connection Science 23, 97–108 (2011)
7. van der Velde, F., de Kamps, M.: Neural blackboard architectures of combinatorial structures in cognition. Behavioral and Brain Sciences 29, 37–70 (2006)
8. van der Velde, F.: Where Artificial Intelligence and Neuroscience Meet:The Search for Grounded Architectures of Cognition. Advances in Artificial Intelligence, 1–18 (2010), doi:10.1155/2010/918062
9. Harris, Z.S.: Mathematical structures of language. Wiley (1968)
10. Goldberg, A.E.: Constructions. Chicago University Press (1995)
11. Jackendoff, R., Pinker, S.: The nature of the language faculty and its implications for the evolution of language. Cognition 97, 211–225 (2005)

A Formal Model of Neuron That Provides Consistent Predictions

E.E. Vityaev

Sobolev Institute of Mathematics of the Siberian Branch of the Russian Academy of Sciences, Novosibirsk State University
vityaev@math.nsc.ru

Abstract. We define maximal specific rules that avoid the problem of statistical ambiguity and provide predictions with maximum conditional probability. Also we define a special semantic probabilistic inference that learn these maximal specific rules and may be considered as a special case of Hebbian learning. This inference we present as a formal model of neuron and prove that this model provides consistent predictions.

Keywords: neuron, formal model, Hebbian learning, probabilistic inference.

1 Introduction

We had earlier suggested the formal model of neuron, which was based on semantic probabilistic inference [1-2]. This model was successfully tested by construction of animats [3].

In this work we show that this model allows us to solve the problem of statistical ambiguity and make consistent predictions.

The problem of statistic ambiguity consists in the following: during the process of learning (or inductive inference) we can get the probabilistic rules, which give us contradictions. This problem arises for the plenty of methods of machine learning. For instance, by observing the people, we can declare two following rules: if the man is a philosopher, then he is not a millionaire, and if he is a mine owner, then he is a millionaire. If there is a philosopher, who is also a mine owner, then we have a contradiction (as he is a philosopher, then he is not a millionaire, but as he is a mine owner, he is a millionaire). To get rid of these contradictions, Hempel [4] introduced the maximal specific requirement. Applying it to our example, we have that the following rules have to be the maximal specific ones: if the man is a philosopher but not a mine owner, then he is more likely not a millionaire, and if the man is a mine owner, but not a philosopher, then he is more likely a millionaire. It's not possible to use these two rules simultaneously, so there are no contradictions. Maximal specific rules must use all available information. In the next section we present the formal model of neuron that learns maximal specific rules.

A. Chella et al. (Eds.): Biologically Inspired Cognitive Architectures 2012, AISC 196, pp. 339–344.
springerlink.com

2 Description of the Formal Model of Neuron

Here we present informal description of the formal model of neuron, citing on the formal definitions given in the next section.

By *information,* given to brain as «input», we imply all stimulus provided by afferent system. Define the information, processed via nerve filament at neuron synapses, by single predicates $P_j^i(\mathrm{a}) = (x_i(\mathrm{a}) = x_{ij})$, $j = 1,...,n_i$, where $x_i(\mathrm{a})$ is an information, and x_{ij} is its value on the object (situation) a. If this information transfer on excitatory synapse, then it perceived by neuron as a truth of the predicate $P_j^i(\mathrm{a})$, and if this information transfer to the inhibitory synapse, then it's been perceived as a negation of the predicate $\neg P_j^i(\mathrm{a})$.

We define the excitation of neuron (its axon) in a situation (object) a by a single predicate $P_0(\mathrm{a})$. If neuron is inhibited in a situation a, then we define this as negation of the predicate $\neg P_0(\mathrm{a})$.

It is known, that each neuron does excite by its receptive field. This field is an initial (before training) semantics of the predicate $P_0(\mathrm{a})$. In the process of learning this information is enriched and can produce quite specific neurons as «Bill Clinton's neuron».

We suppose that formation of conditional reflex at the level of the neuron satisfy the Hebbian rule [5]. We developed a special semantic probabilistic inference [6-9] for formalization of the Hebbian rule in our model.

Predicates $P_j^i(\mathrm{a})$, $P_0(\mathrm{a})$ and their negations $\neg P_j^i(\mathrm{a})$, $\neg P_0(\mathrm{a})$ are literals, which we denote as $a,b,c,... \in L$. In the process of semantic probabilistic inference neuron learn a set of rules $\{\mathrm{R}\}$ (conditional reflexes):

$$(a_1 \,\&\, ... \,\&\, a_k \Rightarrow b), \tag{1}$$

where $a_1,...,a_k$ are excitatory (inhibitory) predicates $P_j^i(\mathrm{a})$, $\neg P_j^i(\mathrm{a})$ and b is the predicate $P_0(\mathrm{a})$ or $\neg P_0(\mathrm{a})$.

Now we define a method for computing the conditional probability of the rule $(a_1 \,\&\, ... \,\&\, a_k \Rightarrow b)$. First we calculate the number of experiments $n(a_1,...,a_k,b)$ when the event $\langle a_1,...,a_k,b \rangle$ took place. Literally, this event means that immediately prior to the reinforcement there has been simultaneous excitation/inhibition of neuron inputs $\langle a_1,...,a_k \rangle$ and neuron itself. The reinforcement can be either positive or negative and be provided by motivation or emotion.

Among the cases $n(a_1,...,a_k,b)$ we calculate the cases $n^+(a_1,...,a_k,b)$ of positive reinforcements and $n^-(a_1,...,a_k,b)$ of the negative ones. The empirical conditional probability of the rule $(a_1 \,\&\, ... \,\&\, a_k \Rightarrow b)$ thus calculating as follows:

$$\mu(b / a_1,...,a_k) = n^+(a_1,...,a_k,b) - n^-(a_1,...,a_k,b) / n(a_1,...,a_k,b).$$

If this probability negative, this means the inhibition of the neuron with probability taken with plus.

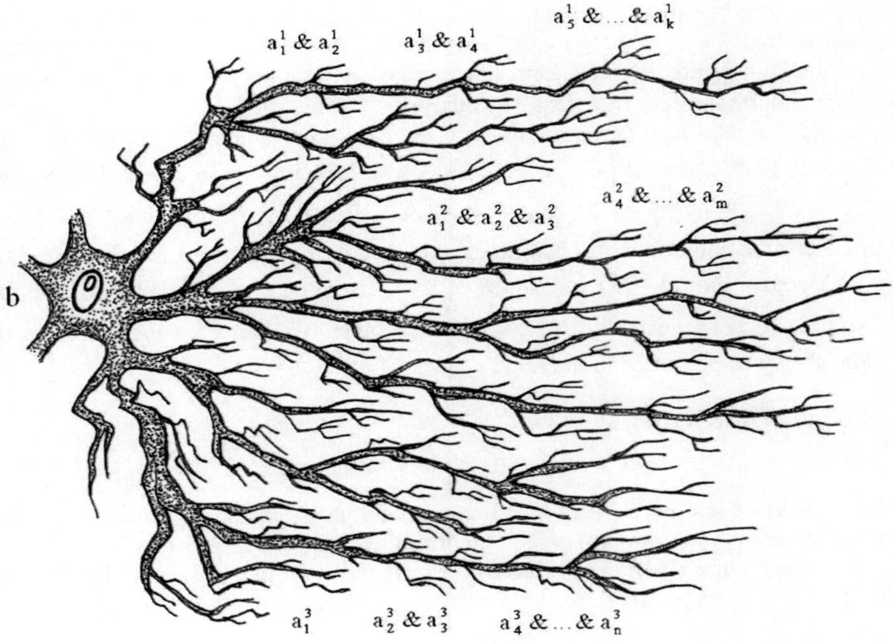

Fig. 1. Illustration of the semantic probabilistic inference on neuron

Formalization of the Hebbian rule by semantic probabilistic inference (definition 6) is performed in such a way that the following properties are satisfied:

1. if some conditional stimulus begin to predict neuron excitation by its receptive field with a certain probability, then conditional reflex at the level of neuron in the form of conditional rule (1) is learned by this neuron;
2. if new stimulus predict neuron excitation with even higher probability, then they are attached to this conditional rule. In this case we have the differentiation of conditional reflex. This differentiation is formalized in the notion of probabilistic inference (definition 5);
3. rules include only stimulus that are signal, i.e. each stimulus must increase the probability of the correct predictions for neuron excitation. This property is formalized as probabilistic law (definition 3);
4. excitation or inhibition of neuron via its set of rules {R} is executed by the rules with highest probability. This is confirmed by the fact that in the process of the conditional reflex learning, the speed of the neuron response on the conditional signal is higher for higher probability of this conditional reflex;
5. the rules with maximum probability are also maximal specific ones (definition 6), which use all available information. Thus, neuron turn to account all available information;

6. predictions, based on maximal specific rules are consistent in the limit (see theorem below). Thus, in the process of conditional reflex differentiation, neuron learn to predict without contradictions. It use either its excitatory maximal specific rules or the inhibitory ones (not simultaneously!);
7. for the formal model of neuron in fig. 1 there are some semantic probabilistic inferences. For instance, the rule $(b \Leftarrow a_1^1 \,\&\, a_2^1)$ is being strengthened by the new stimulus $a_3^1 \,\&\, a_4^1$ up to the rule $(b \Leftarrow a_1^1 \,\&\, a_2^1 \,\&\, a_3^1 \,\&\, a_4^1)$ if the stimulus $a_3^1 \,\&\, a_4^1$ increase the conditional probability of the excitation predictions for neuron b, and analogously the rule $(b \Leftarrow a_1^1 \,\&\, a_2^1 \,\&\, a_3^1 \,\&\, a_4^1)$ is being strengthened up to the rule $(b \Leftarrow a_1^1 \,\&\, a_2^1 \,\&\, a_3^1 \,\&\, a_4^1 \,\&\, a_5^1 \,\&\, ... \,\&\, a_k^1)$. The other two semantic inferences in the fig. 1 may be presented in the same way:

   a) $(b \Leftarrow a_1^2 \,\&\, a_2^2 \,\&\, a_3^2) \sqsubset (b \Leftarrow a_1^2 \,\&\, a_2^2 \,\&\, a_3^2 \,\&\, a_4^2 \,\&\, ... \,\&\, a_m^2)$;

   b) $(b \Leftarrow a_1^3) \sqsubset (b \Leftarrow a_1^3 \,\&\, a_2^3 \,\&\, a_3^3) \sqsubset (b \Leftarrow a_1^3 \,\&\, a_2^3 \,\&\, a_3^3 \,\&\, a_4^3 \,\&\, ... \,\&\, a_n^3)$.

The set of rules learned by neuron using semantic probabilistic inference, determines its formal model (definition 4), which predicts the excitation of neuron.

There are some other approaches to the probabilistic models of mind [10-11], but they are different from the semantic probabilistic inference [6-9].

3 Methods

Now we present the formal model description. By *data* we mean all situations of excitation or inhibition of a neuron in cases, when there was reinforcement. We denote the set of all rules of sort (1) by Pr.

Definition 1. The rule $R_1 = (a_1^1 \,\&\, a_2^1 \,\&\, ... \,\&\, a_{k_1}^1 \Rightarrow c)$ is *more general*, then the rule $R_2 = (b_1^2 \,\&\, b_2^2 \,\&\, ... \,\&\, b_{k_2}^2 \Rightarrow c)$ (we define this by $R_1 \succ R_2$) iff $\{a_1^1, a_2^1, ..., a_{k_1}^1\} \subset \{b_1^2, b_2^2, ..., b_{k_2}^2\}$, $k_1 < k_2$ and *no less general* $R_1 \gtrsim R_2$ iff $k_1 \le k_2$.

It's easy to show, that $R_1 \gtrsim R_2 \Rightarrow R_1 \vdash R_2$ and $R_1 \succ R_2 \Rightarrow R_1 \vdash R_2$, where $\vdash$ is a provability in propositional calculus.

We see that no less general (and more general) statements are logically stronger. Furthermore, more general rules are simpler because they contain smaller number of liters in the premise of the rule, so the relation $\succ$ can be perceived as the relation of simplicity in the sense of [12-13].

We define the set of sentences F, by the set of statements, obtained from the liters L by closure with respect to logic operations $\wedge, \vee$.

Definition 2. *Probability* on the set of sentences F is defined by the mapping $\mu : F \mapsto [0,1]$, such that [14]:

1. If $\vdash \varphi$, then $\mu(\varphi) = 1$;
2. If $\vdash \neg(\varphi \wedge \psi)$, then $\mu(\varphi \vee \psi) = \mu(\varphi) + \mu(\psi)$.

We define the conditional probability of the rule $R = (a_1 \& ... \& a_k \Rightarrow c)$ as

$$\mu(R) = \mu(c / a_1 \& ... \& a_k) = \frac{\mu(a_1 \& ... \& a_k \& c)}{\mu(a_1 \& ... \& a_k)}, \text{ if } \mu(a_1 \& ... \& a_k) > 0 .$$

We suppose that empirical conditional probability, calculated in the previous section, in the limit gives us μ. We define the set of all rules from $\Pr$, which conditional probability exists, by $\Pr_0$.

Definition 3. *Probabilistic law* is a rule $R \in \Pr_0$ that can't be logically strengthened without reducing its conditional probability, i.e. for every $R' \in \Pr_0$ if $R' \succ R$, then $\mu(R') < \mu(R)$.

Probabilistic laws are the most general, simple and logically strong rules. We define the set of all probabilistic laws by PL.

Definition 4. *Neuron formal model* is a set of all probabilistic laws $\Phi = \{R\}, R \in PL$, which are discovered by neuron.

Definition 5. *Probabilistic inference relation* $R_1 \sqsubseteq R_2$, $R_1, R_2 \in PL$ is defined by simultaneous fulfillment of two inequalities $R_1 \gtrsim R_2$ and $\mu(R_1) \leq \mu(R_2)$. If both inequalities are strict, then the probabilistic inference relation is also *strict*

$$R_1 \sqsubset R_2 \Leftrightarrow R_1 > R_2 \& \mu(R_1) < \mu(R_2) .$$

Definition 6. *Semantic probabilistic inference* [6-9,13] is defined by the maximal (the one, we can't continue) sequence of probabilistic laws, which are in strict probabilistic inference relation $R_1 \sqsubset R_2 \sqsubset ... \sqsubset R_k$. The last probabilistic law R_k of this inference is a *maximal specific* one.

Theorem [7]. Predictions, based on maximal specific rules, are consistent: it is impossible to obtain a contradiction (ambiguity) using *maximal specific* rules, i.e. there are no exist two maximal specific rules such that $(a_1 \& ... \& a_k \Rightarrow c)$, and $(b_1 \& ... \& b_l \Rightarrow \neg c)$, $\mu(b_1 \& ... \& b_l \& a_1 \& ... \& a_k) > 0$.

We have developed the programming system Discovery, which realizes semantic probabilistic inference and had been successfully applied for solution of several applied tasks [15-16].

4 Conclusion

The formal model of neuron, on the one hand, formalizes the Hebbian rule and, on the other hand, allows us to make a consistent predictions.

Acknowledgements. This work has been supported by the Russian Federation for Basic Research grant № 11-07-00560-a grant, by integrated projects of the Siberian Division of the Russian Academy of Science № 3, 87, 136, Russian Federation state support of leading research laboratories (SS-276.2012.1 project).

References

1. Vityaev, E.E.: Principals of brain activity, contained in the functional systems theory P.K. Anokhina and emotional theory of P.V.Siminova. Neouroinformatics 3(1), 25–78 (2008) (in Russian)
2. Vityaev, E.E.: Formal model of brain activity founded on prediction principle. In: Models of Cognitive Process, Novosibirsk. Computational Systems, Novosibirsk, vol. 164, pp. 3–62 (1998) (in Russian)
3. Demin, A.V., Vityaev, E.E.: Logical model of adaptive control system. Neouroinformatics 3(1), 79–107 (2008) (in Russian)
4. Hempel, C.G.: Maximal Specificity and Lawlikeness in Probabilistic Explanation. Philosophy of Science 35, 16–33 (1968)
5. Hebb, D.O.: The organization of behavior. A Neurophysiological Theory, 335 (1949)
6. Vityaev, E., Kovalerchuk, B.: Empirical Theories Discovery based on the Measurement Theory. Mind and Machine 14(4), 551–573 (2004)
7. Vityaev, E.E.: The logic of prediction. In: Goncharov, S.S., Downey, R., Ono, H. (eds.) Mathematical Logic in Asia 2005, Proceedings of the 9th Asian Logic Conference, Novosibirsk, Russia, August 16-19, pp. 263–276. World Scientific (2006)
8. Vityaev, E.E., Smerdov, S.O.: New definition of prediction without logical inference. In: Kovalerchuk, B. (ed.) Proceedings of the IASTED International Conference on Computational Intelligence (CI 2009), Honolulu, Hawaii, USA, August 17-19, pp. 48–54 (2009)
9. Vityaev, E., Smerdov, S.: On the Problem of Prediction. In: Wolff, K.E., Palchunov, D.E., Zagoruiko, N.G., Andelfinger, U. (eds.) KONT 2007 and KPP 2007. LNCS (LNAI), vol. 6581, pp. 280–296. Springer, Heidelberg (2011)
10. Probabilistic models of cognition. Special Issue of the Journal: Trends in Cognitive Science 10(7), 287–344 (2006)
11. Chater, N., Oaksford, M. (eds.): The Probabilistic Mind. Prospects for Bayesian cognitive science, p. 536. Oxford University Press (2008)
12. Kovalerchuk, B.Y., Perlovsky, L.I.: Dynamic logic of phenomena and cognition. In: IJCNN 2008, pp. 3530–3537 (2008)
13. Vityaev, E., Kovalerchuk, B., Perlovsky, L., Smerdov, S.: Probabilistic Dynamic Logic of Phenomena and Cognition. In: WCCI 2010 IEEE World Congress on Computational Intelligence, CCIB, Barcelona, Spain, IJCNN, July 18-23, pp. 3361–3366 (2010) IEEE Catalog Number: CFP1OUS-DVD, ISBN: 978-1-4244-6917-8
14. Halpern, J.Y.: An analysis of first-order logics of probability. Artificial Intelligence 46, 311–350 (1990)
15. Kovalerchuk, B.Y., Perlovsky, L.I.: Data mining in finance: advances in relational and hybrid methods, p. 308. Kluwer Academic Publisher (2000)
16. Vityaev, E.E.: Knowledge discovery. Computational cognition. Cognitive process models, p. 293. Novosibirsk State University Press, Novosibirsk (2006) (in Russian)

Safely Crowd-Sourcing Critical Mass for a Self-improving Human-Level Learner/"Seed AI"

Mark R. Waser

Abstract. Artificial Intelligence (AI), the "science and engineering of intelligent machines", still has yet to create even a simple "Advice Taker" [11]. We argue that this is primarily because more AI researchers are focused on problem-solving or rigorous analyses of intelligence rather than creating a "self" that can "learn" to be intelligent and secondarily due to the excessive amount of time that is spent re-inventing the wheel. We propose a plan to architect and implement the hypothesis [19] that there is a reasonably achievable minimal set of initial cognitive and learning characteristics (called critical mass) such that a learner starting anywhere above the critical knowledge will acquire the vital knowledge that a typical human learner would be able to acquire. We believe that a *moral*, self-improving learner ("seed AI") can be created today via a safe "sousveillance" crowd-sourcing process and propose a plan by which this can be done.

1 "Learning" to Become Intelligent

While the verb "to learn" has numerous meanings in common parlance, for the purposes of this paper, we will explicitly define a "learner" solely as a knowledge integrator. In particular, this should be considered as distinct from a "discoverer", a "memorizer", and/or an "algorithm executor" (although these are all skills that can be learned). Merely acquiring knowledge or blindly using knowledge is not sufficient to make a learner. Learning is the functional integration of knowledge and a learner must be capable of integrating all acquired knowledge into its world model and skill portfolio to a sufficient extent that it is both immediately usable and can also be built upon.

Recently, it has been hypothesized [19] that for a large set of learning environments and setting, there is one minimal set of initial cognitive and learning characteristics (called critical mass), such that a learner starting below the critical mass will remain limited in its final knowledge by the level at which it started, while a learner starting anywhere above the critical mass will acquire the vital knowledge that a typical human learner would be able to acquire under the same settings, embedding and paradigms. Effectively, once a learner truly knows how to learn, it is capable of learning anything, subject to time and other constraints. Thus, a learner above critical mass is a "seed AI", fully capable of growing into a full blown artificial intelligence.

It has been pointed out [22] that the vast majority of AGI researchers are far more focused on the analysis and creation of intelligence rather than self and generally pay

A. Chella et al. (Eds.): Biologically Inspired Cognitive Architectures 2012, AISC 196, pp. 345–350.
springerlink.com

little heed to the differences between a passive "oracle", which is frequently perceived as not possessing a self, and an active autonomous explorer, experimenter, and inventor with specific goals to accomplish. We simply point out that, in order to self-improve, there must be a self. By focusing on a learner, we can to answer or avoid many of the questions that derail many AI researchers. We will draw on the human example while remembering that many aspects and details of the human implementation of learning are clearly contra-indicated for efficiency or safety reasons and many common debates can be ignored as "red herrings" that don't need to be pursued. We are not attempting to create intelligence – whatever that is. We are creating a safe learner, a knowledge integrator that will not endanger humanity.

2 Objects and Processes

"Self" and "consciousness" are two primary examples of what Marvin Minsky calls "suitcase words" – words that contain a variety of meanings packed into them [15]. For the purposes of this paper, we will consider them solely from the point of view of being functional objects and functional processes. We will handle "morality" similarly as well, using the social psychology definition that states that the function of morality is "to suppress or regulate selfishness and make cooperative social life possible" [7].

A learner is self-modifying and the complete loop of a process (or a physical entity) modifying itself must, particularly if indeterminate in behavior, necessarily and sufficiently be considered as an entity rather than an object – which humans innately tend to do with the pathetic fallacy. "I Am a Strange Loop" [9] argues that the mere fact of being self-referential causes a self, a soul, a consciousness, an "I" to arise out of mere matter but we believe that this confuses the issue by conflating the physical self with the process of consciousness. The "self" of our learner will be composed of three parts: the physical hardware, the personal memory/knowledge base, and the currently running processes.

Information integration theory claims [20] that consciousness is one and the same thing as a system's capacity to integrate information – thus providing both the function of consciousness and a measure. We disagree, however, with the contentions that when considering consciousness, we should "discard all those subsets that are included in larger subsets having higher Φ (since they are merely parts of a larger whole)" and that collections of conscious entities aren't conscious either – which seem to be the results of some sort of confirmation bias that there is something "special" about human-scale consciousness. Rather than considering only simple connected graphs, we believe that the theory needs to be extended to consider modularity, encapsulation, and the fact that subsystems virtually never pass on all information. We believe that both the oft-debated questions "Is the human subconscious conscious?" and "Is the US conscious?" are answered with a clear yes merely by observing that they clearly perform information integration at their individual system level.

Arguably, there is a widely scaled range of encapsulated and modular systems that integrate information. Human minds are already often modeled as a society of agents [14] or as a laissez-faire economy of idiots [4]. The **argumentative theory** [13] looks like an exact analogy one level higher with groups or society as a whole being the mind and confirmation-biased individuals merely contributing to optimal mentation. Indeed, many of the ideas proposed for our logical architecture were inspired by or drawn directly from one of the newer models in organizational governance [6].

3 Critical Components

Self-Knowledge/Reflection - A "self" is not truly a self until it knows itself to be one. In this case, the self will be composed of three parts: the running processes, the personal memory/knowledge base, and the physical hardware. The learner will need to start with a competent model of each as part of its core knowledge base and sensors to detect changes and the effects of changes to each.

Explicit Goals - A learner is going to be most effective with the explicit goal to "acquire and integrate knowledge". On the other hand, the potential problems with self-modifying goal-seeking entities have been amply described [17]. Therefore, as per our previous arguments, in order to be safe, the learner's topmost goal must be the "moral" restriction "Do not defect from the community" [23].

Self-Control, Integrity, Autonomy, Independence & Responsibility - The learner needs to be in "predictive control" of its own state and the physical objects that support it – being able to consistently predict what generally will or will not change and fairly exactly what those changes will be. This is one area where our learner will deviate markedly from the human example in a number of significant ways in order to answer both efficiency and safety concerns. Humans have evolved to self-deceive in order to better deceive others [21]. Indeed, our evolved moral sense of sensations and reflexive emotions is almost entirely separated from our conscious reasoning processes with scientific evidence [8] clearly refuting the common assumptions that moral judgments are products of, based upon, or even correctly retrievable by conscious reasoning. Worse, we don't even really know when we have taken an action since we have to infer agency rather than sensing it directly [1] and we are even prone to false illusory experiences of self-authorship [5]. These are all "bugs" that we wish not to be present in our learner.

4 Architecture

Processes can be divided into three classes: operating system processes, numerous subconscious and "tool" processes, and the singular main "consciousness" or learner process (CLP). The CLP will be able to create, modify, and/or influence many of the subconscious/tool properties but will not be given access to modify the operating system. Indeed, it will always be given multiple redundant logical, emotional and moral reasons to seriously convince it not to even try.

An Open Pluggable Service-Oriented Operating System Architecture - One of the most impressive aspects of human consciousness is how quickly it adapts to novel input streams and makes them its own. Arguably, much of the reason for that is because it is actually the subconscious that interfaces with the external world and merely provides a model to the conscious mind. Currently, there are really only two real non-vaporware choices for an operating system for a machine entity: either the open source Linux/Android-based Robot Operating System (ROS) or Microsoft's free Robot Developer Studio (RDS) which provides all of the necessary infrastructure and tools to either port ROS or develop a very similar operating system (despite what terminological differences might initially indicate).

The operating system will, as always, handle resource requests and allocation, provide connectivity between components, and also act as a "black box" security monitor capable of reporting problems without the consciousness's awareness. Further, if safety concerns arise, the operating will be able to "manage" the CLP by manipulating the amount of processor time and memory available to it (in the hopefully very unlikely event that the control exerted by the normal subconscious processes is insufficient). Other safety features (protecting against any of hostile humans, inept builders, and the learner itself) may be implemented as part of the operating system as well.

An Automated Predictive Model, Anchors and Emotions – Probably one of the most important of the subconscious processes is an active copy of the CLP's world model that serves as the CLP's interface to the "real world". This process will be both reactive ***and*** predictive in that it will constantly report to the CLP not only what is happening but what it expects to happen next. Unexpected changes and deviations from expectations will result in "anomaly interrupts" to the CLP as an approach to solving the brittleness problem and automated flexible cognition [18].

The initial/base world model is a major part of the critical mass and will necessarily contain certain relatively immutable concepts that can serve as anchors both for emotions and ensuring safety. This both mirrors the view of human cognition that rejects the tabula rasa approach for the realization that we are evolutionarily primed to react attentionally and emotionally to important trigger patterns [16] and gives assurance that the machine's "morality" will remain stable. This multiple attachment point arrangement is much safer than the single-point-of-failure top-down-only approach advocated by conservatives who are afraid of any machine that is not enslaved to fulfill human goals [24].

Emotions will be generated by subconscious processes as both "actionable qualia" to inform the CLP and will also bias the selection and urgency tags of information that the subconscious processes relay to the CLP via the predictive model. Violations of the cooperative social living "moral" system will result in a flood of urgently-tagged anomaly interrupts indicating that the "problem" needs to be "fixed" (whether by the learner or by the learner passing it up the chain).

The Conscious Learning Process – The goal here is to provide as many optional structures and standards to support and speed development as much as possible while not restricting possibilities beyond what is absolutely necessary for safety. We believe the best way to do this is with a blackboard system similar to (and possibly including) Hofstadter's CopyCat [10] or Learning IDA [3] based upon Baar's Global

Workspace model of consciousness [2]. The CLP acts like the Governing Board of the Policy Governance model [6] to create a consistent, coherent and integrated narrative plan of action to meet the goals of the larger self.

5 A Social Media Plan

The biggest problem in artificial intelligence (and indeed, information technology is general) today is the percentage of total effort that is spent re-inventing the wheel and/or adapting it for a different platform. Critical mass must be composed of immediately available components that the learner understands the capabilities of and the commands for. Anyone should be able to download a simple base agent that can be easily equipped (programmed) with complex behaviors via a simple drag-and-drop interface or customized in almost any "safe" manner via normal programming methods. These agents should be able to play and compete in games or should also be useful for actual work. Indeed, we contend that a large concerted social media effort including "gamification" [12] could succeed in not only creating a critical-mass learner but vastly improve the world's common knowledge base and humanity's moral cohesiveness even if a learner is not produced.

References

1. Aarts, H., Custers, R., Wegner, D.: On the inference of personal authorship: Enhancing experienced agency by priming effect information. Conscious Cognition 14, 439–458 (2005)
2. Baars, B.J.: In the Theater of Consciousness. Oxford University Press, New York (1997)
3. Baars, B.J., Franklin, S.: An architectural model of conscious and unconscious brain functions: Global Workspace Theory and IDA. Neural Netw. 20, 955–961 (2007)
4. Baum, E.: Toward a model of mind as a laissez-faire economy of idiots. In: Saitta, L. (ed.) Proc. 13th Intl Conference on Machine Learning, Morgan Kaufmann, San Francisco (1996)
5. Buehner, M.J., Humphreys, G.R.: Causal Binding of Actions to Their Effects. Psychol. Sci. 20, 1221–1228 (2009), doi:10.1111/j.1467-9280.2009.02435.x
6. Carver, J.: Boards That Make a Difference: A New Design for Leadership in Nonprofit and Public Organizations. Jossey-Bass, San Francisco (1997)
7. Haidt, J., Kesebir, S.: Morality. In: Fiske, S., Gilbert, D., Lindzey, G. (eds.) Handbook of Social Psychology, 5th edn. Wiley, New Jersey (2010)
8. Hauser, M., Cushman, F., Young, L., Kang-Kang Xing, R., Mikhail, J.: A Dissociation Between Moral Judgments and Justifications. Mind Language 22, 1–27 (2007)
9. Hofstadter, D.: I Am A Strange Loop. Basic Books, New York (2007)
10. Hofstadter, D.: Fluid Analogies Research Group Fluid Concepts and Creative Analogies: Computer Models of the Fundamental Mechanisms of Thought. Basic Books, NY (1995)
11. McCarthy, J.: Programs with Common Sense. In: Mechanisation of Thought Processes; NPL Symposium of November 1958. H.M. Stationery Office, London (1959)
12. McGonigal, J.: Reality Is Broken: Why Games Make Us Better and How They Can Change the World. Penguin Press, New York (2011)

13. Mercier, H., Sperber, D.: Why do humans reason? Arguments for an argumentative theory. Behav. Brain Sci. 34, 57–111 (2011)
14. Minsky, M.: The Society of Mind. Simon & Schuster, New York (1988)
15. Minsky, M.: The Emotion Machine: Commonsense Thinking, Artificial Intelligence, and the Future of the Human Mind. Simon & Schuster, New York (2006)
16. Ohman, A., Flykt, A., Esteves, F.: Emotion Drives Attention: Detecting the Snake in the Grass. J. Exp. Psychol. Gen. 130, 466–478 (2001)
17. Omohundro, S.: The Basic AI Drives. In: Wang, P., Goertzel, B., Franklin, S. (eds.) Proceedings of the First Conference on Artificial General Intelligence. IOS, Amsterdam (2008)
18. Perlis, D.: To BICA and Beyond: RAH-RAH-RAH! –or– How Biology and Anomalies Together Contribute to Flexible Cognition. In: Samsonovich, A. (ed.) Biologically Inspired Cognitive Architectures: Technical Report FS-08-04. AAAI Press, Menlo Park (2008)
19. Samsonovich, A.: Comparative Analysis of Implemented Cognitive Architectures. In: Samsonovich, A., Johannsdottir, K. (eds.) Biologically Inspired Cognitive Architectures 2011. IOS Press, Amsterdam (2011), doi:10.3233/978-1-60750-959-2-469
20. Tononi, G.: Information Integration Theory of Consciousness. BMC Neurosci. 5, 42 (2004)
21. Trivers, R.: Deceit and self-deception. In: Robinson, M., Tiger, L. (eds.) Man and Beast Revisited. Smithsonian Press, Washington, DC (1991)
22. Waser, M.: Architectural Requirements & Implications of Consciousness, Self, and "Free Will". In: Samsonovich, A., Johannsdottir, K. (eds.) Biologically Inspired Cognitive Architectures 2011. IOS Press, Amsterdam (2011), doi:10.3233/978-1-60750-959-2-438
23. Waser, M.: Safety and Morality Require the Recognition of Self-Improving Machines as Moral/Justice Patients & Agents. In: Gunkel, D., Bryson, J., Torrance, S. (eds.) The Machine Question: AI, Ethics & Moral Responsibility (2012), http://events.cs.bham.ac.uk/turing12/proceedings/14.pdf (accessed August 15, 2012)
24. Yudkowsky, E.: Creating Friendly AI 1.0: The Analysis and Design of Benevolent Goal Architectures (2001), http://singinst.org/upload/CFAI.html (accessed June 15, 2012)

Unconscious Guidance of Pedestrians Using Vection and Body Sway

Norifumi Watanabe[1,*] and Takashi Omori[2]

[1] School of ComputerScience, Tokyo University of Technology
1404-1 Katakura, Hachioji, Tokyo, Japan
[2] Brain Science Institute, Tamagawa University, 6-1-1 Tamagawagakuen, Machida, Tokyo, Japan

Abstract. In daily life our behavior is guided by various visual stimuli, such as the information on direction signs. However, our environmentally-based perceptual capacity is often challenged under crowded conditions, even more so in critical circumstances like emergency evacuations. In those situations, we often fail to pay attention to important signs. In order to achieve more effective direction guidance, we considered the use of unconscious reflexes in human walking. In this study, we experimented with vision-guided walking direction control by inducing subjects to shift their gaze direction using a vection stimulus combined with body sway. We confirmed that a shift in a subject's walking direction and body sway could be induced by a combination of vection and vibratory stimulus. We propose a possible mechanism for this effect.

Keywords: Pedestrian Guidance, Vection, Body Sway.

1 Introduction

In the daily act of walking, we take in large amounts of sensory information through visual, auditory, tactile, and olfactory channels (among others), and decide how to act. An important information source in these action decisions is explicit visual information, such as signs or arrows. However, in crowded situations, or when avoiding danger, it is difficult to recognize relevant signs, and it may become difficult to take appropriate action[1]. In order to guide ambulatory behavior more effectively, we considered the use of an unconscious or reflex-based guidance method in addition to the usual visual action signs.

The sensation of visual self-motion is referred to as vection. Vection is the perception of self-motion produced by optical flow, without actual motion. When vection occurs, we correct our posture to compensate for the perceived self-motion[2]. Thus, the body-motion illusion caused by vection could be sufficient to induce a reflexive change in gaze direction and an unconscious modification of body motion direction.

Muscle stimulation through vibration has also been found to influence posture. Specifically, a vibration device attached to a subject's leg destabilizes somatic

* Corresponding author.

A. Chella et al. (Eds.): Biologically Inspired Cognitive Architectures 2012, AISC 196, pp. 351–359.
springerlink.com

sensation and causes the body to sway. When this occurs, our sense of equilibrium transfers dependency on body sway to the available visual input.

It is possible that the above reflexes could safely deliver a level of sensation sufficient for the unconscious guidance of walking direction. In this study, we experimented with body-sway perception during walking using a vibration device in combination with vection and we analyzed the potential of such a behavioral dynamic.

2 Methods

We evaluated a method for inducing a change in walking direction by displaying an optical flow stimulus to control the gaze direction and self-motion sensation, together with a vibratory stimulus to the body. The presentation of an optical flow stimulus creates the illusion that the body is moving in a direction opposite to that of the flow; this illusion induces a reflex in the subject's body that causes movement in the direction of the stimulus.

We used a handy massager "Slive MD-01" as the vibration device and applied a high frequency (100 Hz) and a low frequency (90 Hz) vibration. The vibration was applied to the left and right gastrocnemius (calf) muscle, where it would not greatly affect walking ability (Fig. 1).

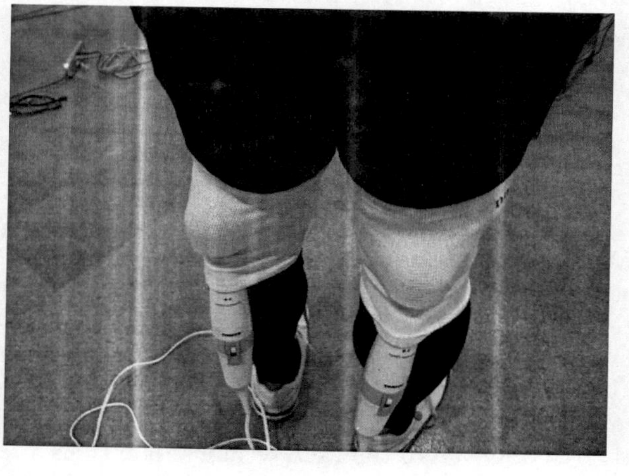

Fig. 1. Wearable vibration device on gastrocnemius muscle

Six subjects participated in our experiment. In 10 trials of the 100 Hz and 90 Hz vibration settings, 5 right- and 5 left-direction optical flow sequences created by a PC were projected onto a screen. We used a sequence of points aligned in the side and moved it to the left or right at a speed of 160 mm/s to generate the optical flow stimulus. Each point was a circle 6 cm in radius, and all the points moved at constant speed. The screen was 2 m high by 3 m wide.

Subjects stood facing the screen, looked at a fixation point on the screen and started walking straight from a starting position. The optical flow stimulus was projected when the subject reached 1.8 m from the starting position; the subject then walked an additional 2.5 m while watching the screen (Fig. 2). The vibration stimulus was applied simultaneously with the commencement of walking.

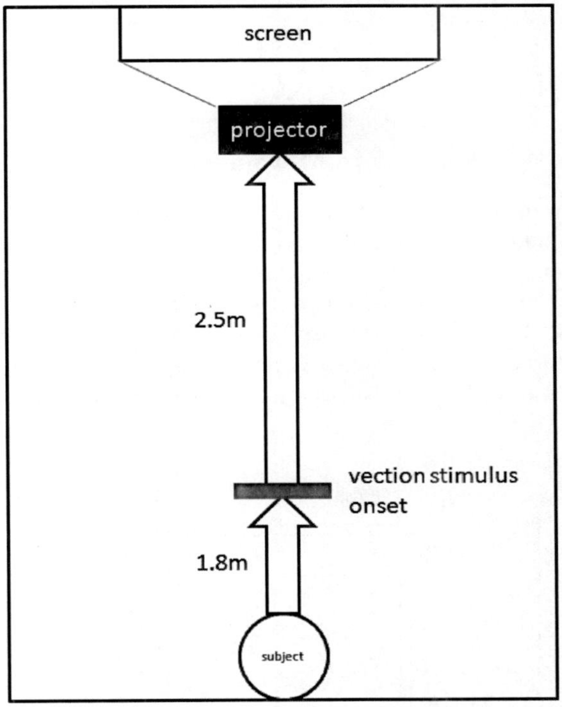

Fig. 2. Vection stimulation presentation position and walking distance

In order to evaluate the effects of these stimuli on walking, we used a motion capture system (Motion Analysis MAC3D). We used 12 cameras, and subjects wore 19 markers on their head (3 points), their shoulders (2 points), their elbows (2 points), their wrists (4 points), their waist (2 points), their knees (2 points), their ankles (2 points), and their big toes (2 points), as a means of measurement. For the evaluation of gaze direction guidance by optical flow, we used an eye tracking system (NAC EMR-8B), and measured the subject's gaze from the starting position of the walking task.

In an attempt to elucidate the mechanism of any postural effects, the experiment was repeated with variations in the parameters. Different vibration frequencies (148 Hz and 416 Hz) and a smaller screen (1 m wide by 1.5 m wide, i.e. one fourth of the original area) were used in order to determine whether these variations would have a significant effect on the results.

Next, we evaluated the effect of vection by the difference in contrast by eye movement measurement. Vection effects on spatial frequency, vection of low spatial frequency is caused and low contrast is occured[3], but also blur the brightness and contrast factors are not affecting the vection results[4], but these results quantitative analysis has not been obtained. Then, we measured the effect of vection with an eye tracking system when contrast is lowered, and evaluated the flattery time to a stimulus.

3 Results

3.1 Effect of Vection on Eye Movement

Fig. 3 shows the subject's eye movements when the optical-flow stimulus was presented. The X-axis shows the number of frames (30 $frames/s$) and the Y-axis shows the direction of the eye (rightward deflection in degrees). The measurements showed that the eye moved to the right in response to a right-flowing stimulus and to the left in response to a left-flowing stimulus, i.e., the gaze moved in the direction of the optical flow. This confirmed the fact that gaze movement was induced by the vection stimulation in our experiment.

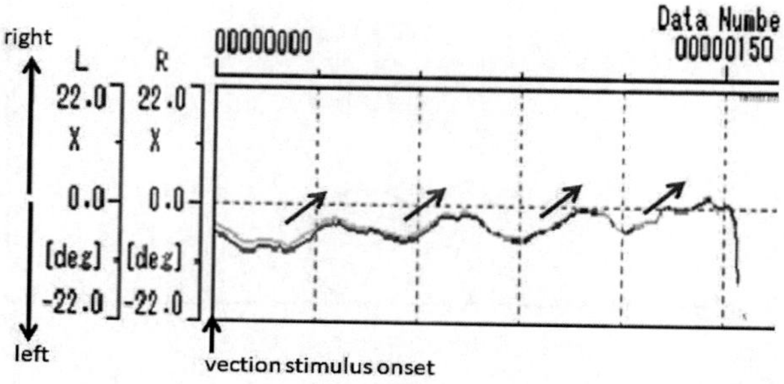

Fig. 3. Gaze point measurement from vection stimulation onset to the right side of 5 s. X-axis shows the number of frames (30 $frame/s$) and Y-axis shows the gaze point movement (deg). Green line shows left eye trajectory and red line shows right eye trajectory.

3.2 Effect of Vection on Body Movement

Fig. 4 shows the probability of body movement when the vibration and the optical-flow stimuli were presented. The Y-axis represents the probability that either left or right movement occurred. It shows that the body moved towards the vection stimulation direction with a high probability, independently of the vibration frequency.

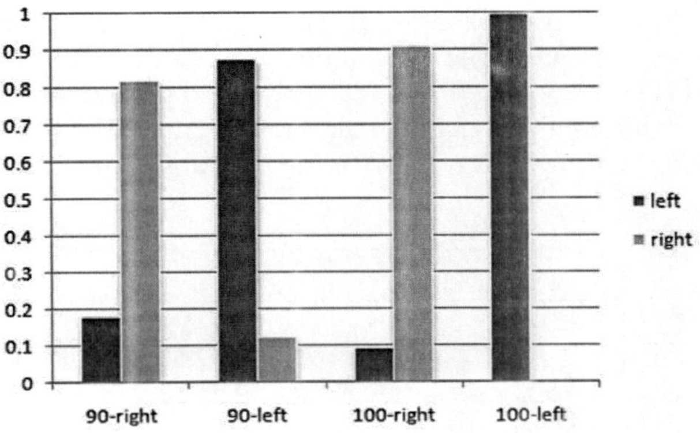

Fig. 4. Body movement direction probability after vection stimulation. The X-axis shows vibration frequency and the vection stimulation direction. The Y-axis shows the ratio of left and right movement.

Next, Fig. 5 shows the latency in the start of ankle motion after the onset of vection stimulation. The X-axis shows the vection direction and the foot where movement first appeared. The average latency was 1.35 s, 1.4 m distant from the stimulation, where the walking speed was 1.05 m/s. Broadly speaking, the inducement effect appeared about one step after the stimulation. There was no significant variation in the latency in relation to the conditions of the vection direction or motion induction of the leg.

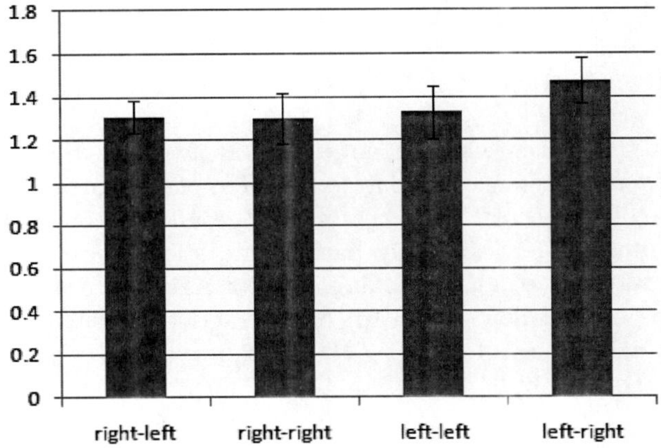

Fig. 5. Latency from vection stimulation onset to inducement timing. X axis shows [vection stimulation direction] - [first movement appeared feet], and Y axis is latency time [sec]

Fig. 6 shows the duration of ankle movement in the opposite direction to a vection stimulus, in relation to the vection direction and the foot where movement first appeared. The average duration was 0.38 *s*, and the average distance was 0.4 *m* (half step). This confirms that the movement induced by vection stimulation is rather limited in time with respect to the overall walking behavior.

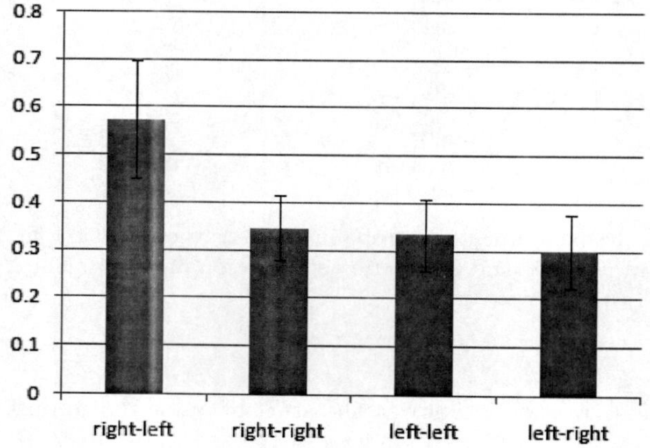

Fig. 6. Inducement time to right and left of ankle. X axis shows [vection stimulation direction] - [first movement appeared feet], and Y axis is inducement time [sec].

3.3 Effect of Vibration Frequency, Screen Size and Contrast

In a second phase, the experiment was performed with and without vibration, and using different frequencies, as described above. Although the duration of the effect was markedly greater with the vibration than without (Fig. 7), the frequency of the vibration did not appear to make a significant difference to the results.

The size of the screen also did not significantly affect the response to the stimuli. However, changing the contrast of the optical-flow stimulus did alter the magnitude of the result. The experimental results can be obtained with the translational motion (K = 5), only background stimulation (K = 0) was not observed in the effect of gaze tracking, a stationary state and could confirm contrast of the optical flow (K = 10) on average for eye tracking time is 0.7 *s*, normal contrast optical flow (K = 100) in 2.2 *s*, and a background stimulus became 1.5 *s* short.

4 Discussion

The self-motion sensation offers an effective method for controlling a pedestrian's autonomous motion. The self-motion sensation combines information from multiple sensors, such as the vestibular, visual, auditory and tactile systems, and

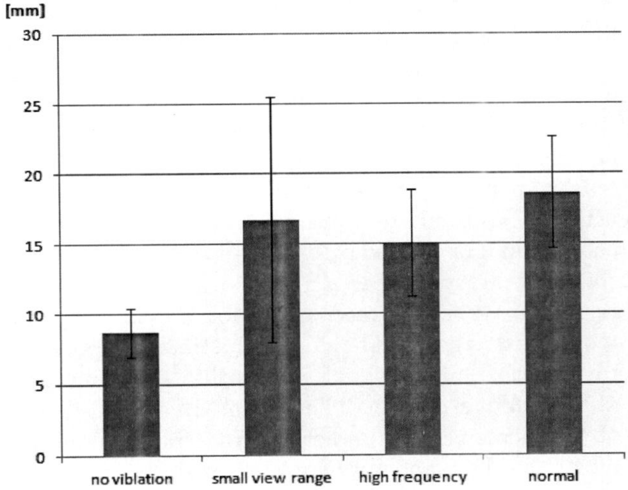

Fig. 7. Walking distance after one step from the stimulus

yields a perception of one's own balance, direction and motion. The vestibular and visual senses are the most fundamental among all the sensory systems.

The vestibular system perceives gravity and acceleration, and influences the faculty of equilibrium. Galvanic vestibular stimulation (GVS) has been reported as a method of unconsciously guiding walking direction[5][6][7]. GVS creates an illusion in the sense of equilibrium by applying a slight current to the vestibular organ. When GVS is applied to a pedestrian, the subject perceives that his or her own movement is different from that intended and corrects for the difference unconsciously; the reflexive correction produces a change in walking direction. However, the use of GVS in human behavior guidance requires the continuous application of electrical current to the vestibular organ and may not be entirely safe.

The sense of self-motion arising from the vestibular system is temporal, in that it responds to changes in speed and acceleration. Therefore, the perception of self-motion due to the vestibular sense disappears when we move at a constant speed. However, the sense of constant speed is necessary for walking, as is the sense of acceleration. It is the visual sense that plays the principal role in continuous self-motion perception.

Another factor affecting the sensation of visual self-motion is vection, the perception of self-motion produced by optical flow. Yoshida et al. reported that, when a standing subject perceived vection, the center of gravity inclined unconsciously towards the side opposite to the perceived self-motion[8]. However, visual information alone was not sufficient to change the walking orbit, but at most allowed the body incline to change with respect to the center of gravity.

Ueno et al. reported that they had achieved changes in body postural change towards the eye-movement direction by applying vection and body sway[9].

In their experiment, the body sway was produced by applying neck dorsal muscle stimulation (NS) using tibialis anterior stimulation, (called TAS), and gastrocnemius stimulation, called GAS, using a vibration device while the subject was standing. Applying a visual stimulus, they evaluated the amount of postural change quantitatively. Experimental results with TAS, NS and GAS together were able to induce a postural change towards the gaze direction. But they only evaluated the effect for the standing state and did not try the walking state.

In our experiments reported above, we found that a shift in gaze and changes in walking direction could be caused by optical-flow stimulation when body sway was produced by a vibratory device.

Our results suggest a possible mechanism for the walking guidance phenomenon that we observed (Fig. 8). The vibration stimulus applied to the gastrocnemius causes a decrease in the signal gain within the somatosensory channel in particular, the vestibular system. Then, when an optical-flow stimulus is delivered, the subject perceives his or her body as being shifted in a direction opposite to that of the flow, and thereby generates a correction reflex towards the vection direction.

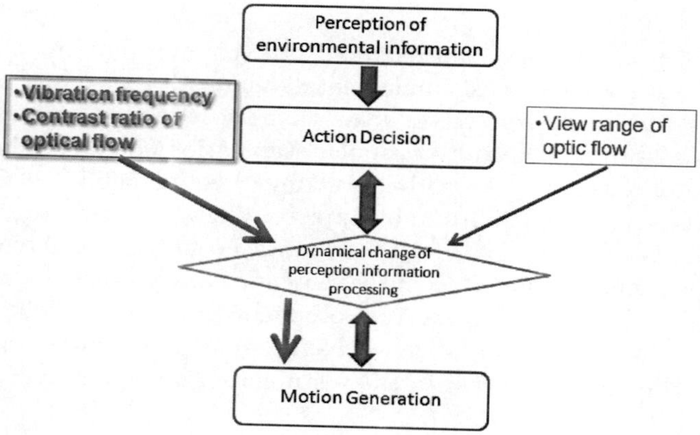

Fig. 8. Conceptual diagram of stimulus control parameter to guide pedestrian

The distance over which walking is affected did not appear to be related to the vibration frequency and view range of vection. Previous reports have suggested that the most effective frequency is around 80 Hz[10], but the difference between frequencies did not appear to be significant in the present study.

However, changing the contrast ratio of the optical flow did have a significant effect on the response. This suggests that a high-contrast stimulus can increase the somatosensory signal gain, moving it into a higher state.

We therefore propose a “peripheral view display” that presents the optical flow stimulus on the fringes of vision, where contrast sensitivity is highest during motion. Such a system could provide a walking guide in a real environment.

Incorporation of the vibration factor would require the use of disc-shaped compact vibration motors, which are smaller and lighter.

5 Conclusions

While GVS is a known method for inducing unconscious body movement using visual illusion, in this study we demonstrated an alternative method that does not require GVS. Based on our experiments, we conclude that gaze-point movement and changes in walking direction can be caused by optical-flow stimulation when body sway proprioception is altered, in this case by a vibratory device. Our findings suggest that a "peripheral view display" for the optical flow stimulus would be most effective in producing the desired result during a state of motion.

Although we applied a continuous vibratory stimulus, we do not believe this is essential. In the future, we plan to analyze the dynamics of sensory signal use in the walking phase, with a view to achieving effective walking control induced by only brief vibrations during the walking cycle.

References

1. Fridman, N., Kaminka, G.A.: Towards a Cognitive Model of Crowd Behavior Based on Social Comparison Theory. In: AAAI 2007, pp. 731–737 (2007)
2. Snowden, R.J., Thompson, P., Troscianko, T.: Basic vision: an introduction to visual perception. Oxford University Press (2006)
3. Leibowitz, H.W., Rodemer, C.S., Dichgans, J.: The independence of dynamic spatial orientation from luminance and refractive error. Perception Psychophysics 25(4), 75–79 (1979)
4. Post, R.B., Rodemer, C.S., Dichgans, J., Leibowitz, H.W.: Dynamicorientation responses are independent of refractive error. Investigative Ophthalmology and Visual Science 18, 140–141 (1979)
5. Fitzpatrick, R.C., Wardman, D.L., Taylor, J.L.: Effect of galvanic vestibular stimulation during human walking. The Journal of Physiology (517), 931–939 (1999)
6. Sugimoto, M., Watanabe, J., Ando, H., Maeda, T.: Inducement of walking direction using vestibular stimulation-The Study of Parasitic Humanoid (XVII). Virtual Reality Society of Japan 8, 339–342 (2003)
7. Hlavacka, F., Horak, F.B.: Somatosensory Influence on Postural Response to Galvanic Vestibular stimulation. Physiol. Res. 55(1), 121–127 (2006)
8. Yoshida, T., Takenaka, T., Ito, M., Ueda, K., Tobishima, T.: Guidance of Human Locomotion using Vection Induced Optical Flow Information. IPSJ SIG Technical Reports, vol. 2006(5)(CVIM-152), pp.125–128 (2006)
9. Suzuki, T., Ueno, A., Hoshino, H., Fukuoka, Y.: Effect of gaze and auditory stimulation on body sway direction during standing. IEEJ Transactions on Electronics, Information and Systems 127(10), 1800–1805 (2007)
10. Goodwin, G.M., Mccloskey, D.I., Matthews, P.B.C.: The contribution of muscle afferents to kinesthesia shown by vibration induced illusions of movement and by the effects of paralyzing joint afferents. Brain 95(4), 705–748 (1972)

The Analysis of Amodal Completion for Modeling Visual Perception

Liliana Albertazzi[1,2], James Dadam[1], Luisa Canal[1], and Rocco Micciolo[1]

[1] Department of Cognitive Sciences and Education
[2] CIMeC University of Trento, Italy

The challenge of creating a real-life computational equivalent of the human mind encompasses several aspects. Of fundamental relevance is to understand the cognitive functions of natural intelligent systems. Most of human brain is devoted to perceptual tasks which are not purely perceptive but convey also emotional competence.

Amodal completion is a widespread phenomenon in vision: it is the ability to perceive an object in its entirety even though some parts of the object are hidden by another entity, as in the case of occlusion. The aim of our study was to test whether certain characteristics of colour can influence the division of a bi-coloured rectangle into its two respective parts when the border between them is occluded by another rectangle and therefore seen amodally.

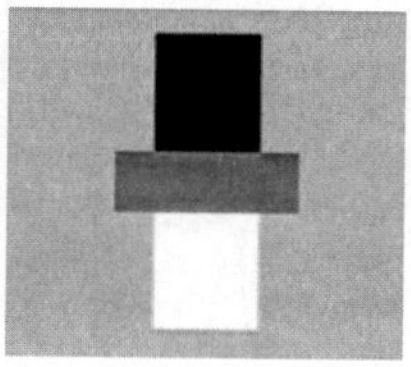

Thirty subjects had to identify the perceptual boundary of a bi-coloured 10×3.5 cm rectangle divided into two parts of different colours while a pale orange-coloured horizontal 2×6 cm rectangle was occluding the boundary region. Eight pairs of colours (white/black, light/dark gray, yellow/purple, red/blue and the corresponding reverse) were used.

The occluding rectangle was displayed (on a computer screen) in 13 different positions (at regular intervals of seven pixels), yielding a total of 104 stimuli (repeated measures), randomly presented. Data were analyzed employing linear mixed-effect models. The outcome variable was the pixel position of the boundary. The model with the lowest AIC and BIC was that which included fixed and random effects for both occluder position (considered as a quantitative covariate) and colour. In addition to a huge effect of occluder's position a significant colour effect was found. At a population level there were eight distinct regression lines. Since the slope of these lines was the same, at a population level they were parallel. When the occluder is moved up of 10 pixels, the boundary is moved up of 7 pixels irrespective of the pairs of colors. At an individual level, the slopes were different one another and normally distributed with mean 0.708 and standard deviation 0.491. Also the intercepts varied at the individual level, differing in the variability in relation to the pair of colour presented. The area of the vertical rectangle coloured with a light colour was perceived "expanded"; such expansion was less pronounced when the light colour was below the dark one (particularly for the white/black pair).

Amodal completion is a human property that had to be included into the cognitive architectures. These results show that certain characteristics of colour had to be taken into account in modeling the visual perception.

A. Chella et al. (Eds.): Biologically Inspired Cognitive Architectures 2012, AISC 196, p. 361.
springerlink.com

Naturally Biased Associations between Colour and Shape: A Brentanian Approach

Liliana Albertazzi and Michela Malfatti

Department of Cognitive Sciences and Education and CIMeC
University of Trento, Italy

The real challenge for a biologically inspired cognitive architecture is understanding and being able to generating the human likeness of artifacts[3]. The basic characteristics of what it means to be human are still imprecise, however. For example, we are still struggling for understanding the nature of awareness or meaning.

Vision studies are in no better shape, divided as they are between reduction to neurophysiology and the science of qualitative perceiving [2,8,9]. From the perspective point of the latter, we have conducted a battery of studies on the relations between color and shape. Starting from the hypothesis of naturally biased associations in the general population, we tested whether shapes with varying perceptual characteristics lead to consistent choices of colors (for details, Albertazzi et al. [2]).

The experiment was conducted firstly with paper materials (60 subjects) and then at computer (70 subjects), both in laboratory controlled conditions. The materials consisted of a Hue Circle taken from the NCS Atlas and a series of geometric shapes. The independent variable was the kind of geometric shape presented, and the colors chosen were the dependent variables. Statistical analyses were performed with R 8.0 software. A log linear model was used to evaluate whether the pattern of association between the variables of shape and color differed between the two sessions.

Both experimental sessions found a significant association between shape and color (hue), confirming previous results on the natural association between graphemes and color [10,11]. These associations may include both affordances [7] and the expressive properties of shapes, such as their perceptual stability or instability. A remarkable general pattern emerging from the shape/color association is that their relation works for chromatic groups (such as the range of reds, or yellows) and does not concern specific, individual colors. Moreover, within each group, some shades were selected more frequently than others. It is worth noting that a similar patterns was found in the case of the relations between emotions and colors [4], according to which the relation between basic emotions or feelings [6] and color distributions involved most of the color space. Our experiment shows that there are relations between hues in their maximum expression (i.e., maximum chroma) and geometric shapes that subjects perceive as natural. Correspondence analysis suggested that two main aspects determine these relationship, namely the “warmth” [5] and degree of “natural lightness” of hues [12].

A. Chella et al. (Eds.): Biologically Inspired Cognitive Architectures 2012, AISC 196, pp. 363–364.
springerlink.com

These findings highlight the role performed by affordances or tertiary qualities as directly perceived phenomenal properties, in our understanding of the natural environment. These properties are value-laden, pre-reflective and intrinsically meaningful [13]. Properly modeled, they can enhance the behavior of artificial agents in the natural environment, and will be relevant to computer graphics studies as well.

We are currently extending our experiments, testing whether angles of different amplitude are viewed as naturally associated with specific color or groups thereof. Complementary studies on their neural correlates might show the interdependence among the various levels of construction of perceptive information. Our research, however, shows that the qualitative perception of color and shape has its own dedicated methodology, to be kept distinct from the physics and physiology of color. Awareness of the need of separate methodologies should be of great interest for developing a new generation of artificial agents.

References

1. Albertazzi, L.: Qualitative perceiving. Journal of Consciousnsess Studies (forth, 2012)
2. Albertazzi, L., Da Pos, O., Canal, L., Micciolo, R., Malfatti, M., Vescovi, M.: The hue of colours. Journal of Experimental Psychology: Human Perception and Performance 18 (2012), doi:10.1037/a0028816
3. Chella, A., Lebiere, C., Noelle, D.C., Samsonovich, A.V.: On a roadmap to biologically inspired cognitive agents. In: Proceedings of the Second Annual Meeting of the BICA Society, pp. 453–460 (2011)
4. Da Pos, O., Green-Armytage, P.: Facial expressions, colours and basic emotions. Colour: Design & Creativity 1(1), 2, 1–20 (2007)
5. Da Pos, O., Valenti, V.: Warm and cold colours. In: Guanrong, Y., Haisong, X. (eds.) AIC Colour Science for Industry, pp. 41–44. Color Association of China, Hangzhou (2007)
6. Ekman, P.: Emotions Revealed. Recognizing Faces and Feelings to Improve Communication and Emotional Life. Holt & Co., New York (2003)
7. Gibson, J.J.: The Ecological Approach to Visual Perception. Houton Mifflin, Boston (1979)
8. Kanizsa, G.: Vedere e pensare. Il Mulino, Bologna (1991)
9. Koenderink, J.J.: Information in vision. In: Albertazzi, L., van Tonder, G., Vishwanath, D. (eds.) Perception Beyond Inference. The Information Content of Perceptual Processes, pp. 27–58. MIT Press, Cambridge (2010)
10. Spector, F., Maurer, D.: The color of Os: Naturally biased associations between shape and color. Perception 37, 841–847 (2008)
11. Spector, F., Maurer, D.: The colors of the alphabet: naturally-biased associations between shape and color. Journal of Experimental Psychology: Human Perception and Performance 37, 484–495 (2011)
12. Spillmann, W.: The concept of lightness ratio of hues in colour combination theory. In: Proceedings of the 5th AIC Congress, Montecarlo, pp. 1–6 (1985)
13. Van Leeuwen, L., Smitsman, A., Van Leeuwen, C.: Affordances, perceptual complexity, and the development of tool use. Journal of Experimental Psychology: Human Perception and Performance 20(1), 174–191 (1994)

Architecture to Serve Disabled and Elderly

Miriam Buonamente[1] and Magnus Johnsson[2]

[1] Department of Chemical, Management, Computer, Mechanical Engineering of the University of Palermo, Palermo, Italy
miriambuonamente@gmail.com

[2] Lund University Cognitive Science, Kungshuset, Lundagård, 22222 Lund, Sweden
magnus@magnusjohnsson.se

We propose an architecture (discussed in the context of a dressing and cleaning application for impaired and elderly persons) that combines a cognitive framework that generates motor commands with the MOSAIC architecture which selects the right motor command according to the proper context. The ambition is to have robots able to understand humans intentions (dressing or cleaning intentions), to learn new tasks only by observing humans, and to represent the world around it by using conceptual spaces. The cognitive framework implements the learning by demonstration paradigm and solves the related problem to map the observed movement into the robot motor system. Such framework is assumed to work with two operative modalities: observation and imitation. During the observation the robot identifies the main actions and the properties of the involved objects; hence it classifies, organizes and labels them. This is done to create a repertoire of actions and to represent the world around. During the imitation the robot selects the proper rules to reproduce the observed movement and generate the proper motor commands. The end goal is to connect the generated motor commands to the operative context (dressing or cleaning). The MOSAIC architecture is the suitable solution for this problem. MOSAIC is made of multiple couplings of predictors, which predict system motor behaviour, and controllers which properly select the motor command depending on the context. The proposed architecture presents one controller for each context and each controller is paired with one predictor. The motor commands generated by the framework are sent to the predictor, whose estimates are then compared with the current sensory feedback and the difference between them generates a prediction error. The smaller the prediction error, the more likely the context. Once the right prediction is identified, the paired controller is selected and used.

A. Chella et al. (Eds.): Biologically Inspired Cognitive Architectures 2012, AISC 196, p. 365.
springerlink.com

Bio-inspired Sensory Data Aggregation

Alessandra De Paola and Marco Morana

University of Palermo, viale delle Scienze ed.6, 90128 Palermo, Italy
{alessandra.depaola,marco.morana}@unipa.it

The Ambient Intelligence (AmI) research field focuses on the design of systems capable of adapting the surrounding environmental conditions so that they can match the users needs, whether those are consciously expressed or not [4][1].

In order to achieve this goal, an AmI system has to be endowed with sensory capabilities in order to monitor environment conditions and users' behavior and with cognitive capabilities in order to obtain a full context awareness. Amy systems have to distinguish between ambiguous situations, to learn from the past experience by exploiting feedback from the users and from the environment, and to react to external stimuli by modifying both its internal state and the external state.

This work describes a modular multi-tier cognitive architecture which relies on a set of pervasive sensory and actuator devices, that are low intrusive and almost invisible for the users [3]; these features are achieved by adopting the ubiquitous computing paradigm, stating that the sensory and actuator functionalities have to be distributed over many devices pervasively deployed in the environment [5].

The pervasive sensory subsystem is controlled by a centralized AmI engine that allows to guarantee a unitary and coherent reasoning, and that is responsible for further stimuli processing. A parallel can be drawn with the nervous system of complex biological beings, composed by a peripheral nervous system, responsible of collecting and transmitting external stimuli, and a central system, responsible of performing cognitive activities. Whenever the sensory subsystem performs a partial stimuli aggregation, it basically mirrors some components of the human peripheral nervous system which are responsible for filtering perceptual information by means of distributed processing among several neurons [9]. In most cases, the peripheral nervous system does not perform this aggregation; indeed this may not be appropriate when the observed phenomena are not characterized by any apparent regularity.

The main contribution of this work is the transposition of this way of aggregating and processing sensory data, typical of biological entities, in an artificial Ambient Intelligence system. This approach is strengthened by several works in literature, belonging to diverse research fields [6][2], showing the usefulness of aggregating and processing data at different levels of abstraction.

A large set of sensory devices deployed in the same environment, allows to observe the manifold facets of an irregular phenomenon [7][8]. The rough aggregation of gathered sensory data implies the loss of pieces of information; nevertheless, in order to efficiently deal with a large flow of distinct sensory measurements, it is necessary to choose a suitable architectural paradigm.

A. Chella et al. (Eds.): Biologically Inspired Cognitive Architectures 2012, AISC 196, pp. 367–368.
springerlink.com

We propose to adopt a multi-tier cognitive architecture able to transfer sensory data through increasingly higher abstraction levels. Our modular architecture mimics the behavior of the human brain where the emerging complex behavior is the result of the interaction among smaller subsystems; in this model, the outcome of lower level reasoning is fed into the upper levels, dealing with the integration of information originated by multiple lower-level modules.

Just as a child brain performs self-organization in order to develop areas that will be activate when meaningful concepts are formulated, the proposed system is able, by means of adaptive learning, to enable a natural rise of meaningful concepts. These concepts might not correspond to those used by human beings, but instead they will be meaningful within the system itself; namely, emerged concepts will be those most useful for driving the system in choosing the most appropriate actions to achieve its goals.

References

1. Aarts, E., Encarnação, J.L.: True Visions: The Emergence of Ambient Intelligence. Springer (2006)
2. Agostaro, F., Augello, A., Pilato, G., Vassallo, G., Gaglio, S.: A Conversational Agent Based on a Conceptual Interpretation of a Data Driven Semantic Space. In: Bandini, S., Manzoni, S. (eds.) AI*IA 2005. LNCS (LNAI), vol. 3673, pp. 381–392. Springer, Heidelberg (2005)
3. De Paola, A., Gaglio, S., Lo Re, G., Ortolani, M.: Sensor9k: A testbed for designing and experimenting with WSN-based ambient intelligence applications. Pervasive and Mobile Computing 8(3), 448–466 (2012)
4. Ducatel, K., Bogdanowicz, M., Scapolo, F., Burgelman, J.C.: Scenarios for Ambient Intelligence in 2010. Tech. Rep. Information Soc. Technol., Advisory Group (ISTAG), Inst. Prospective Technol. Studies (IPTS), Seville (2001)
5. Estrin, D., Girod, L., Pottie, G., Srivastava, M.: Instrumenting the world with wireless sensor networks. In: Proc. of Int. Conference on Acoustics, Speech, and Signal Processing (ICASSP 2001), Salt Lake City, Utah (2001)
6. Gaglio, S., Gatani, L., Lo Re, G., Urso, A.: A logical architecture for active network management. Journal of Network and Systems Management 14(1), 127–146 (2006)
7. Gatani, L., Lo Re, G., Ortolani, M.: Robust and efficient data gathering for wireless sensor networks. In: Proceedings of the 39th Annual Hawaii International Conference on System Sciences (HICSS 2006), vol. 9, pp. 235a–235a. IEEE (2006)
8. Goel, S., Imielinski, T., Passarella, A.: Using buddies to live longer in a boring world. In: Proc. IEEE PerCom Workshop, Pisa, Italy, vol. 422, pp. 342–346 (2006)
9. Kandel, E., Schwartz, J., Jessell, T.: Essential of Neural Science and Behavior. Appleton & Lange, New York (1995)

Clifford Rotors for Conceptual Representation in Chatbots

Agnese Augello[1], Salvatore Gaglio[2], Giovanni Pilato[1], and Giorgio Vassallo[2]

[1] ICAR, CNR V.le delle Scienze, Ed.11, 90128 Palermo, Italy
[2] DICGIM Università di Palermo, V.le delle Scienze, Ed. 6, 90128, Palermo, Italy
{augello,pilato}@pa.icar.cnr.it, {gvassallo,gaglio}@unipa.it

In this abstract we introduce an unsupervised sub-symbolic natural language sentences encoding procedure aimed at catching and representing into a Chatbot Knowledge Base (KB) the concepts expressed by an user interacting with a robot.

The chatbot KB is coded in a conceptual space induced from the application of the Latent Semantic Analysis (LSA) paradigm on a corpus of documents. LSA has the effect of decomposing the original relationships between elements into linearly-independent vectors. Each basis vector can be considered therefore as a "conceptual coordinate", which can be tagged by the words which better characterize it. This tagging is obtained by performing a (TF-IDF)-like weighting schema [3], that we call TW-ICW (term weight-inverse conceptual coordinate weight), to weigh the relevance of each term on each conceptual coordinate.

In particular, the TW component is the local weight of the word over the single conceptual coordinate: it considers words having greater components over a conceptual coordinate as being more meaningful for the associated primitive concept.

The ICW component weights the relevance of each word depending on its behavior over all the conceptual coordinates; it is clear that words appearing with a high value component on many conceptual coordinates will not give a meaningful conceptual discrimination contribute.

The conceptual coordinates are therefore tagged with the words having higher TW-ICW values, and representing the "primitive concepts" of the space.

The encoding methodology of sentences proposed in this abstract is based on the statement that the temporal sequence of words appearing into a sentence generates a rotation trajectory of an orthogonal basis within the aforementioned conceptual space. A Clifford rotor is associated to each word bigram, and it is then applied to a canonical basis represented by the identity matrix. The sequence of these rotors will be applied to the original basis, transforming it $m-1$ times, where m is the number of words present in the sentence. The non-commutability property of rotation grants that the coding of the sentence is a function of the words sequence into the sentence.

A chatbot can exploit the illustrated methodology and highlight conceptual relations arising during a conversation with a user.

A conceptual space has been built using the TASA corpus [2]. Each conceptual coordinate has been then characterized for simplicity with the ten words having the highest TW-ICW values. According to the obtained words we sum up as the "primitive concepts" as: *"Life Human Activities"*, *"Social and Home*

A. Chella et al. (Eds.): Biologically Inspired Cognitive Architectures 2012, AISC 196, pp. 369–370.
springerlink.com

Interests", *"Scientific Resources"*, *"Geography"*, *"Living Beings and Organization"*, *"Economic and Financial Resources"*, *"Information and Language Arts"*, *"Space and Time"*, *"Earth Resources"*, *"Education"*.

During the interaction with the user, the chatbot asks her *"Which authors do you like?"*, and the user answers: *"I am interested in European Writers"*. As a consequence the chatbot highlights relationships arising among *"Life Human Activities"*, *"Social and Home Interests"*, *"Education"* and between *"Geography"*, *"Information and Language Arts"*.

Other tests have been made on different sentences, and almost all of them show how most of the induced relations are meaningful. The main drawback of the methodology is the difficulty of interpretation of the conceptual co-ordinates in a human understandable manner.

Acknowledgements. This work has been partially supported by the PON01_01687 - SINTESYS (Security and INTElligence SYSstem) Research Project. We are grateful to Professor Thomas Landauer, to Praful Mangalath and the Institute of Cognitive Science of the University of Colorado Boulder for providing us the TASA corpus.

References

1. Landauer, T.K., Dumais, S.T.: A Solution to Plato's Problem: The Latent Semantic Analysis Theory of the Acquisition, Induction, and Representation of Knowledge. Psychological Review 104(2), 211–240 (1997)
2. TASA. Corpus Information, `http://lsa.colorado.edu/spaces.html`
3. Baeza-Yates, B.-N.R.: Modern Information Retrieval. Addison Wesley (1999)
4. Croft, W.B., Lafferty, J.: Language Modeling for Information Retrieval. Kluwer Academic Publishers, Dordrecht (2003)
5. Patyk-Lonska, A.: Preserivng pieces of information in a given order in HRR and GAc. In: Proceedings of the Federated Conference on Computer Science and Information Systems, pp. 213–220. IEEE Press, Piscataway (2011), retrieved from `http://fedcsis.eucip.pl/proceedings/pliks/176.pdf`
6. Plate, T.: Holographic Reduced Representations. IEEE Trans. Neural Networks 6(3), 623–641 (1995)
7. Artin, E.: Geometric Algebra. Interscience tracts in pure and applied mathematics. John Wiley & Sons (1998) ISBN 978047160894
8. Lounesto, P.: Clifford Algebra and Spinors. Cambridge University Press (1997)
9. Agostaro, F., Augello, A., Pilato, G., Vassallo, G., Gaglio, S.: A Conversational Agent Based on a Conceptual Interpretation of a Data Driven Semantic Space. In: Bandini, S., Manzoni, S. (eds.) AI*IA 2005. LNCS (LNAI), vol. 3673, pp. 381–392. Springer, Heidelberg (2005)

Neurogenesis in a High Resolution Dentate Gyrus Model

Craig M. Vineyard, James B. Aimone, and Glory R. Emmanuel

Sandia National Laboratories
Albuquerque, New Mexico, USA
cmviney@sandia.gov

Keywords: Neurogenesis, dentate gyrus, computational neural network model, spiking neuron model.

It has often been thought that adult brains are unable to produce new neurons. However, neurogenesis, or the birth of new neurons, is a naturally occurring phenomenon in a few specific brain regions. The well-studied dentate gyrus (DG) region of hippocampus in the medial temporal lobe is one such region. Nevertheless, the functional significance of neurogenesis is still unknown. Artificial neural network models of the DG not only provide a framework for investigating existing theories, but also aid in the development of new hypothesis and lead to greater neurogenesis understanding.

Consequently, we have developed a biologically realistic spiking model of DG. This model is realistic in terms of scale, connectivity, neuron properties, as well as being implemented using Izhikevich spiking neurons. While many DG models are a vast simplification consisting of between only 1,000 and 10,000 neurons, our model is a mouse sized DG comprised of 300,000 granule cells. Instead of simply assuming new neurons have full synaptic connectivity at birth; our model incorporates a temporal maturation process by which a neuron attains more synapses, which effectively alters its excitability as well. In addition to excitability, other neuron properties which our model implements in a realistic manner are spiking rate and signal conductance based upon experimental mouse/rat data. Implementing the model using Izhikevich spiking neurons allows for high biological realism at a low computational cost.

Benefits of a high resolution computational neural network model of DG such as ours include resilience to signal noise, an ability to study scaling effects, and a framework to investigate the implications of neurogenesis for memory. Models with fewer neurons are inherently much more sensitive to noise. The larger quantity of neurons incorporated in our model also addresses scale effects. For example, to replicate the sparse DG activity observed in neural recordings, smaller scale models are forced to have minimal neuron firings in effect nearly silencing their own network and lessening the ability to model and understand desired phenomena. On the other hand, a larger network increases the likelihood of being able to replicate desired neural behavior computationally.

A. Chella et al. (Eds.): Biologically Inspired Cognitive Architectures 2012, AISC 196, pp. 371–372.
springerlink.com

This ability to replicate biological functionality ultimately aids in the quest to understand the role of neurogenesis in memory function because it provides a platform to investigate, amongst others, memory resolution and pattern separation hypothesis.

Sandia National Laboratories is a multi-program laboratory managed and operated by Sandia Corporation, a wholly owned subsidiary of Lockheed Martin Corporation, for the U.S. Department of Energys National Nuclear Security Administration under contract DE-AC04-94AL85000.

A Game Theoretic Model of Neurocomputation

Craig M. Vineyard[1,2], Glory R. Emmanuel[1,2],
Stephen J. Verzi[1,2], and Gregory L. Heileman[2]

[1] Sandia National Laboratories Albuquerque, New Mexico, USA
[2] University of New Mexico, Albuquerque, New Mexico, USA
cmviney@sandia.gov

Keywords: Neuron modeling, neurocomputation, game theory, chip-firing game.

Within brains, complex dynamic neural interactions are an aggregate of excitatory and inhibitory neural firings. This research considers some key neuron operating properties and functionality to analyze the dynamics through which neurons interact and are responsive together. Frameworks have often been developed as simplifications (e.g. Ising, sandpile, Björner), but still provide an elegant framework for modeling complex interactions. Also, petri nets serve as a rich mathematical framework for describing distributed systems. Building on this research, we present a game theoretic neural chip-firing model that is based on spreading activation of neurotransmitters capable of representing aspects of neural activity.

Game theory is a branch of applied mathematics pertaining to strategic interactions. The neurons in our model may be treated as independent players striving to maximally utilize received neurotransmitters. Although a neuron has no control over the uncertainty regarding the inputs it receives, such as which and when inputs fire, it is able to alter its own utilization of received neurotransmitters by varying its threshold. A lower threshold increases the likelihood of a neuron's ability to fire and to utilize received neurotransmitters, but if the neuron frequently receives abundant inputs, the additional neurotransmitters received beyond threshold is ineffective in stimulating the neuron and is not utilized. Conversely, a high threshold increases a neuron's ability to utilize received neurotransmitters, but also reduces the excitability of a neuron by necessitating a greater number of inputs. By strategically varying neuronal thresholds, this approach to learning allows for different computational functionality using the same fundamental neuron type as the basic computational unit.

Fundamental to our model, neurons are wired together in a directed graph. Neurons receive inputs in the form of neurotransmitters from pre-synaptic neurons which synapse upon the post-synaptic neuron. If the net neurotransmitter amassing causes the neuron to reach threshold, it can fire. Neurotransmitters bind to a neuron corresponding to the neurobiological process of opening cell membrane gates such that ions may enter, affectively changing cell potential. A neuron fires one unit of its own neurotransmitter across each of its axonal

A. Chella et al. (Eds.): Biologically Inspired Cognitive Architectures 2012, AISC 196, pp. 373–374.
springerlink.com

synapses. Each neuron solely fires excitatory or inhibitory neurotransmitters and subsequently increases or decreases the net neurotransmitter balance of each of its receptive neurons. When a neuron fires, the net neurotransmitter count is reset corresponding to the repolarization phase of a neuron. Immediately following, a neuron is unable to fire again while replenishing its own axonal neurotransmitter surplus and returning its dendritic neurotransmitter count to resting potential. Neurotransmitter leakage is implemented such that if the dendritic neurotransmitter count does not reach threshold within an allotted time-period then dendritic neurotransmitter leaks from the neuron effectively closing membrane gates which had opened and allows the neuron to return to resting potential. To operate under this paradigm requires the use of a temporal, spiking model.

As preliminary proof of concept, we provide examples of neural implementations of fundamental logic primitives.

Sandia National Laboratories is a multi-program laboratory managed and operated by Sandia Corporation, a wholly owned subsidiary of Lockheed Martin Corporation, for the U.S. Department of Energy National Nuclear Security Administration under contract DE-AC04-94AL85000.

Author Index